Cloning Agricultural Plants Via In Vitro Techniques

Editor

B. V. Conger, Ph.D.
Professor
Department of Plant and Soil Science
University of Tennessee
Knoxville, Tennessee

CRC Press, Inc.
Boca Raton, Florida

Library of Congress Cataloging in Publication Data

Main entry under title:
Cloning agricultural plants
via in vitro techniques.

Bibliography: p.
Includes index.
1. Plant cell culture. 2. Plant tissue culture.
3. Cloning. 4. Plants, Cultivated. 5. Plant
propagation. I. Conger, Bob Vernon, 1938-
SB123.6.P74 631.5′3 80-23852
ISBN 0-8493-5797-7

Direct all inquiries to CRC Press, Inc., 2000 Corporate Blvd., N.W., Boca Raton, Florida, 33431.

Second Printing, 1982

International Standard Book Number 0-8493-5797-7

Library of Congress Card Number 80-23852
Printed in the United States

PREFACE

Plant cell and tissue culture generated much excitement during the 1970s concerning the potential application of the technology for improving important agricultural crop plants. This originates from the demonstration of cellular totipotency, or the ability to regenerate whole plants from single cells, and the successful creation of hybrids by somatic cell fusion in some species. There are several areas of in vitro culture which have potential practical application. However, currently the most practical application is for cloning or mass propagation of selected genotypes. This is evidenced by the large number of commercial firms engaged in propagating a variety of plants through tissue culture.

The purpose of this book is to provide a reference guide on principles and practices of cloning agricultural plants via in vitro techniques for scientists, students, commercial propagators, and other individuals who are interested in plant cell and tissue culture and especially its application for cloning. The chapters are categorized according to major classes of crop plants and each were written by a different author who has expertise for that particular crop class. Different approaches were taken by different authors and no attempt was made to achieve a uniform style or approach nor to completely avoid duplication. Basic principles as well as actual practices for specific crop species are presented. The ratio of principles to practices varies with each chapter depending on the approach taken by the author. Each chapter is followed by an extensive list of references which the reader may consult for additional information. There is, of course, a problem in such a popular and rapidly expanding field in including all current references, since many new publications appear each week. However, the inclusion of new material must cease at some point. The chapters should include most of the pertinent publications up to the end of 1979 when the manuscripts were due.

I wish to express my sincere appreciation to the other authors of this book for their excellent contributions. Appreciation is also extended to the Department of Plant and Soil Science and the Agricultural Experiment Station of the University of Tennessee for providing funds, time, and services essential for completion of this treatise. Final editing and completion of Chapters 1 and 5 were also partially supported by the Competitive Research Grants Office of the U.S. Department of Agriculture under Agreement No. 5901-0410-9-0331-0.

B. V. Conger
January 1980

THE EDITOR

Bob V. Conger, Ph.D., is a Professor of Plant and Soil Science at the University of Tennessee, Knoxville. He received his B.S. in Agronomy from Colorado State University, Fort Collins in 1963 and his Ph.D. in Genetics from Washington State University, Pullman in 1967. He was supported in his doctoral research by a National Aeronautics and Space Administration Predoctoral Traineeship.

Dr. Conger is a member of the American Association for the Advancement of Science, American Genetic Association, American Society of Agronomy, Crop Science Society of America, Tissue Culture Association, International Association for Plant Tissue Culture, and the honorary societies of Sigma Xi and Gamma Sigma Delta.

Dr. Conger has presented more than 20 papers at National and International Meetings and published more than 50 scientific papers. He has served as a member of a Food and Agriculture Organization of the United Nations/International Atomic Energy Panel on Improvement of Mutation Breeding Techniques and is currently an Associate Editor of *Environmental and Experimental Botany.* His current research interests are in the areas of plant cell, tissue culture, and breeding forage grasses.

CONTRIBUTORS

Paul J. Bottino, Ph.D.
Associate Professor
Department of Botany
University of Maryland
College Park, Maryland

Karen W. Hughes, Ph.D.
Associate Professor
Department of Botany
University of Tennessee
Knoxville, Tennessee

Ralph L. Mott, Ph.D.
Professor of Botany
Department of Botany
North Carolina State University
Raleigh, North Carolina

Robert M. Skirvin, Ph.D.
Associate Professor
Department of Horticulture
University of Illinois
Urbana, Illinois

TABLE OF CONTENTS

Chapter 1

INTRODUCTION

B. V. Conger

Not only is man absolutely dependent on plants for food, but plants are also a major source of clothing, fuel, drugs, and construction materials. Furthermore, as ornamentals they are both useful and aesthetically pleasing.[1] Cultivation and attempts to improve, maintain, and propagate useful and desirable plant genotypes predates recorded history. Also, it is not known when the first attempts were made to vegetatively propagate plants.

Plant propagation through tissue culture may have been considered as early as 1902 when Haberlandt[2] first attempted to regenerate plants from single cells. Although this pioneering work was unsuccessful, it inspired others to attempt in vitro culture of plant tissues. Very little progress was made during the 30 years following Haberlandt's paper; however, in 1934, White[3] established an actively growing clone of tomato roots. In the late 1930s, the first prolonged culture of unorganized plant tissue was reported independently in carrot by Gautheret[4] and Nobécourt,[5] and in tobacco by White.[6]

Experiments by Skoog and Miller[7] in 1957 showed that different auxin to cytokinin ratios influenced the type of growth and/or morphogenesis in tobacco. Organization and development of complete plants from a cultured mass of carrot cells was demonstrated by Steward et al.[8] in 1958. In 1962, Murashige and Skoog[9] published a defined medium for tobacco culture which has probably been cited more than any other for culture of a wide range of plant species, including both dicots and monocots. Using a defined medium, Vasil and Hildebrandt[10] obtained differentiation of completely organized tobacco plants from single cells in 1965. In 1969, Nitsch and Nitsch[11] published a method to grow hundreds of haploid tobacco plants from pollen grains.

Based on demonstrations in the early 1970s that protoplasts isolated from mesophyll cells could be induced to generate into entire plants[12] and that protoplasts could be stimulated to fuse under defined experimental conditions,[13] Carlson et al.[14] created the first parasexual hybrid by fusing protoplasts from two tobacco species in 1972. The event, perhaps more than any other, was one of the most significant from the standpoint of expanding interest, publicity, extramural funding (both public and private), and enticing researchers who had little or no previous interest to conduct research in plant cell and tissue culture.

The above represents only a few of the significant developments in plant tissue culture. For a more thorough historical background the reader is referred to the introductory chapter in the book edited by H. E. Street.[15]

There are several potential practical applications of plant cell and tissue culture in agriculture. These include:

Induction of haploid plants from anther and pollen culture — These are of great interest in plant breeding since recessive mutations induced in them can be identified immediately and doubling the chromosomes leads directly to homozygous individuals.

Protoplast culture — The primary interest here deals with the possibility of modifying the plant's genome by uptake of exogenous DNA and creating hybrids between sexually incompatible species through protoplast fusion. Although there have now been several reports of successful interspecific and intergeneric hybrids produced by the protoplast fusion (including some between sexually incompatible species)[16,17] the number of species in which plants can be regenerated from protoplasts is very small.

There have been no published reports of plant regeneration from any of the important cereal and legume crops, fruit crops, or trees.

Induction and selection of mutants — The identification of mutants in haploid cultures was mentioned above; however, this topic is also of interest in culture of somatic tissue. For example, induction and isolation of mutants from sugarcane tissue cultures which are resistant to various diseases have been especially successful.[18]

Clonal propagation or rapid multiplication of specific genotypes — The use of in vitro techniques for asexual propagation is the most advanced area of plant tissue culture. Thus, it is also the area that currently has the most practical application.[19,20,29] According to Murashige,[20] by 1978 there were at least 100 facilities engaged in commercially propagating a variety of plants through tissue culture and nearly all of them arose within the previous 5 years. Also, several new establishments continue to arise each year. The differences between tissue culture and traditional methods for cloning involve the use of smaller propagules, the provision of an aseptic and artificial environment, and substantially faster plant multiplication.[21]

Currently, cloning by either conventional or in vitro techniques is especially valuable for propagation of heterozygous, sexually incompatible, and sterile genotypes. These characteristics occur in many of our ornamental, vegetable, and fruit crops. However, to eventually exploit other novel techniques of plant tissue culture, e.g., parasexual hybridization, haploid culture, etc., in the improvement of important agronomic crops and trees the ability to rapidly clone these species in large numbers may also be necessary.

Hussey[22] and de Fossard[23] have listed a number of reasons for efficient and reliable vegetative propagation which may utilize plant tissue culture techniques. These include:

1. Rapid multiplication of new hybrid cultivars, which arise as single plants, for testing and eventual commercial production
2. Elimination of viruses from infected stocks
3. Vegetative propagation of difficult to propagate species
4. Year around propagation of clones
5. Propagation of genetically uniform parent plants in large numbers for large-scale hybrid seed production

Rapid asexual multiplication can be obtained through (1) enhanced precocious axillary shoot formation, (2) production of adventitious shoots either directly on organ explants or in callus, and (3) somatic cell embryogenesis.[21,22] Production of shoots in (1) or (2) above is followed by rooting of individual shoots to produce complete plants. The production of plantlets from callus has the potential for very rapid multiplication of large numbers. However, there are serious problems with some species in difficulty or even complete failure to regenerate plants from callus or liquid cultures and high rates of genetic and/or cytogenetic altered plants. These problems are discussed in later chapters of this book.

Asexual embryogenesis is potentially the most rapid method of cloning plants in vitro.[21] The embryos can arise directly from the explant or from an intermediary callus. However, this method, as presently known, is not a desired way to propagate plants. Its use requires synchronous development and the embryos must be protected with a coating for automated sowing in the soil, prolonged storage, and easy transport.[21] Also, there is a problem with phenotypically altered plants which may result from genetic or cytogenetic changes.

Murashige[21,24] lists a sequence of stages or steps which plant propagation through tissue culture must proceed starting with the explant through establishment of the plants in the soil. These steps have been either directly or indirectly discussed in other chapters of this book, e.g., the chapter on ornamental species discusses the various requirements for each of these stages.

In vitro techniques for cloning or mass propagation are defintely more advanced and currently have greater application in some plant species than in others. Orchids were the first plants to be propagated by tissue culture. Also, the elimination of somaic virus by shoot apex culture was described in the orchid *Cymbidium* by Morel.[25] This technique has subsequently been employed to eliminate a multitude of viruses from a variety of other plant species. Morel[25] estimated that as many as 4 million *Cymbidium* could be produced from a single explant in 1 year.

Asparagus is another crop which has been successfully propagated by in vitro methods. It has been estimated that 300,000 transplantable plants may be produced from a single shoot apex culture in a year.[26] In actual practice, however, 70,000 plants can be produced by one person working 200 days 1 year.[27] This level of production depends on an adequate supply of aseptic stock plants and the culture of 500 bud segments by one person in 1 day.

In some agricultural plants, e.g., most agronomic crops, in vitro techniques are being used very little, if at all, for propagation.[28] Furthermore, these techniques may never find application for some of these crops. The current status and potential uses, however, have been examined. Even though major advances have been made in plant tissue culture, including in vitro propagation, of many important agricultural crop plants, there are several major problems and limitations which must be overcome before these techniques can become a routine method of plant propagation in a wider range of species. Some of these include failure to regenerate plants from callus and cell cultures and, as mentioned above, maintenance of genetic and cytogenetic stability. Other problems include internal contamination by microorganisms in explants of certain plants which is difficult to eliminate and thus hinder or even preclude culture.

The purpose of this book is to review the principles and practices for cloning various classes of agricultural crop plants via in vitro techniques in terms of currently available knowledge. The requirements for tissue culture propagation are discussed in both a general sense and in some cases for specific crops in the various chapters. The associated problems as well as potentials for the future are also discussed. Although the primary emphasis is on the use of in vitro methods for cloning, other principles, practices, and potentials of plant tissue culture receive various degrees of mention in the other chapters. Also, each chapter is followed by an extensive list of references on plant cell and tissue culture for that particular class of agricultural plants.

The chapters are categorized according to various classes of economically important agricultural plants and are entitled, Ornamental Species, Vegetable Crops, Fruit Crops, Agronomic Crops, and Trees. Since each chapter was written by a different author different approaches were taken. Hopefully, the book will provide the reader with a starting point for in vitro culture of the plant or crop species in which he or she is interested and from which modifications may be made to improve or optimize culture conditions and technique. Since not all chapters may contain all information concerning general principles or procedures, e.g., stages of in vitro culture or media formulations, the reader may consult the index to find this information from another chapter(s). If more detail for a specific procedure or plant species is needed the reader may consult the original references.

REFERENCES

1. **Allard, R. W., Ed.,** *Principles of Plant Breeding,* John Wiley & Sons, New York, 1960, 1.
2. **Haberlandt, G.,** Kulturversuche mit isolierten Pflanzenzellen, *Sitzungsber. Mat. — Nat. Kl. Kais. Akad. Wiss. Wien,* 111, 69, 1902.
3. **White, P. R.,** Potentially unlimited growth of excised tomato root tips in a liquid medium, *Plant Physiol.,* 9, 585, 1934.
4. **Gautheret, R. J.,** Sur la possibilité de réalizer la culture indéfinie des tissus de tubercules de carrote, *C. R. Acad. Sci.,* 208, 118, 1939.
5. **Nobécourt, P.,** Sur la perennite de l'augmentation de volume des cultures de tissus vegetaux, *C. R. Soc. Biol.,* 130, 1270, 1939.
6. **White, P. R.,** Potentially unlimited growth of excised plant callus in an artificial nutrient, *Am. J. Bot.,* 26, 59, 1939.
7. **Skoog, F. and Miller, C. O.,** Chemical regulation of growth and organ formation in plant tissues cultured *in vitro, Symp. Soc. Exp. Biol.,* 11, 118. 1957.
8. **Steward, F. C., Mapes, M. O., and Mears, K.,** Growth and organized development of cultured cells. II. Organization in cultures grown from freely suspended cells, *Am. J. Bot.,* 45, 705, 1958.
9. **Murashige, T. and Skoog, F.,** A revised medium for rapid growth and bioassays with tobacco tissue cultures, *Physiol. Plant.,* 15, 473, 1962.
10. **Vasil, V. and Hildebrandt, A. C.,** Differentiation of tobacco plants from single isolated cells in microculture, *Science,* 150, 889, 1965.
11. **Nitsch, J. P. and Nitsch, C.,** Haploid plants from pollen grains, *Science,* 163, 85, 1969.
12. **Takebe, I., Labib, G., and Melchers, G.,** Regeneration of whole plants from isolated mesophyll protoplasts of tobacco, *Naturwissenschaften,* 58, 318, 1971.
13. **Power, J. B., Cummins, S. E., and Cocking, E. C.,** Fusion of isolated plant protoplasts, *Nature (London),* 255, 1016, 1970.
14. **Carlson, P. S., Smith, H. H., and Dearing, R. D.,** Parasexual interspecific plant hybridization, *Proc. Natl. Acad. Sci. U.S.A.,* 69, 2292, 1972.
15. **Street, H. E.,** Introduction, in *Plant Tissue and Cell Culture,* 2nd ed., Street, H. E. Ed., University of California Press, Berkeley 1977, chap. 1.
16. **Dudits, D., Hadlaczky, Gy., Bajszár, G. Y., Koncz, Cs., Lázár, G., and Horváth, G.,** Plant regeneration from intergeneric cell hybrids, *Plant Sci. Lett.,* 15, 101, 1979.
17. **Krumbiegl, G. and Schieder, O.,** Selection of somatic hybrids after fusion of protoplasts from *Datura innoxia* Mill. and *Atropa belladonna* L., *Planta,* 145, 371, 1979.
18. **Heinz, D. J., Krishnamurthi, M., Nickell, L. G., and Maretzki, A.,** Cell, tissue, and organ culture in sugarcane improvement, in *Plant Cell, Tissue, and Organ Culture,* Reinert, J. and Bajaj, Y. P. S., Eds., Springer-Verlag, Berlin, 1977, 3.
19. **Murashige, T.,** Current status of plant cell and organ cultures, *HortScience,* 12, 127, 1977.
20. **Murashige, T.,** The impact of plant tissue culture on agriculture, in, *Frontiers of Plant Tissue Culture 1978,* Thorpe, T. A., Ed., University of Calgary Printing Service, Canada, 1978, 15.
21. **Murashige, T.,** Principles of rapid propagation, in Propagation of Higher Plants through Tissue Culture, Hughes, K. W., Henke, R. R., and Constantin, M. J., Eds., Technical Information Center, U.S. Department of Energy, Springfield, Va., 1978, 14.
22. **Hussey, G.,** The application of tissue culture to the vegetative propagation of plants, *Sci. Prog.,* 65, 185, 1978.
23. **de Fossard, R. A.,** *Tissue Culture for Plant Propagators,* University of New England Printing, Armidale, 1976.
24. **Murashige, T.,** Plant propagation through tissue cultures, *Annu. Rev. Plant Physiol.,* 25, 135, 1974.
25. **Morel, G. M.,** Producing virus-free *Cymbidiums, Am. Orchid. Soc. Bull.,* 29, 495, 1960.
26. **Hasegawa, P. M., Murashige, T., and Takatori, F. H.,** Propagation of asparagus through shoot apex culture. II. Light and temperature requirements, transplantability of plants, and cytological characteristics, *J. Am. Soc. Hortic. Sci.,* 98, 143, 1973.
27. **Yang, H. J.,** Tissue culture technique developed for asparagus propagation, *HortScience,* 12, 140, 1977.
28. **Conger, B. V.,** Problems and potentials of cloning agronomic crops via in vitro techniques, in Propagation of Higher Plants through Tissue Culture, Hughes, K. W., Henke, R. R., and Constantin, M. J., Eds., Technical Information Center, U.S. Department of Energy, Springfield, Va., 1978, 62.
29. **Vasil, I. K., Ahuja, M. R., and Vasil, V.,** Plant tissue cultures in genetics and plant breeding, *Adv. Genet.,* 20, 127, 1979.

Chapter 2

ORNAMENTAL SPECIES

Karen W. Hughes

TABLE OF CONTENTS

I. INTRODUCTION

Plant cell and tissue culture techniques have, in recent years, developed into very powerful tools for propagation of ornamental species. The technology had its beginnings with Haberlandt's speculations regarding cell totipotency at the turn of the 20th century.[1] Haberlandt suggested that techniques for isolating and culturing plant tissues should be developed and postulated that if the environment and nutrition of cultured cells were manipulated, those cells would recapitulate the developmental sequences of normal plant growth. Plant tissues were first successfully cultured by White in 1934.[2] By 1939, White, Nobécourt, and Gautheret had reported the first successful callus cultures of carrot and tobacco.[3-5] In 1957, a key paper by Skoog and Miller was published in which they proposed that quantitative interactions between auxins and cytokinins determined the type of growth and/or morphogenic event that would ensue.[6] Their studies with *Nicotiana tabacum* indicated that high auxin to cytokinin ratios induced rooting while the reverse induced shoot morphogenesis. Unfortunately, this pattern of response is not universal. While manipulations of auxin to cytokinin ratios have been successful in obtaining morphogenesis in many taxa, it is now clear that many other factors affect the ability of cells in culture to differentiate into roots, shoots, or embryos.

A major stimulus for application of plant tissue culture techniques to the propagation of ornamental species may be attributed to the early work by Morel on the propagation of orchids in culture[7] and to the development and widespread use of a new medium with high concentrations of mineral salts by Murashige and Skoog.[8] Following success with rapid in vitro propagation of orchids, plant cell and tissue culture techniques were applied to other species with varying degrees of success. Herbaceous ornamentals have been relatively easy to propagate via tissue culture, but success has been elusive with some taxa including woody perennial species and with members of specific families such as the Favaceae.

II. PROPAGATION OF ORNAMENTAL SPECIES

The in vitro multiplication of ornamental plants has several advantages over conventional methods of plant propagation.

1. The number of genetically identical plants recovered from a single stock plant is greatly increased. Through tissue culture techniques, a single stock plant may produce thousands or even millions of plants depending on the capability of the culture system.[9]
2. Disease free plants may be obtained. Plants which have been propagated by tissue culture techniques are free of superficial bacteria and fungi; however, internal pathogens (viruses and viroids) and some endosymbionts may be propagated with the plant tissue. Consequently, in some families, particularly the Orchidaceae, viruses are now widespread.[10] Virus-free plants may be obtained through meristem culture (see Section IV. B).
3. Stocks may be maintained in vitro where greenhouse space for maintenance of plants is at a premium.
4. Tissue culture techniques may be used to obtain hybrids from incompatible species through either embryo or ovule culture.
5. In a few ornamental species, haploid plants have been obtained through anther culture. Haploid plants have some advantages over diploid material when mutagenesis is used in that recovery of recessive mutations is enhanced. Further, doubled haploids are homozygous and thus pure-breeding.

Species of Orchidaceae are among the most extensively studied and propagated taxa. Tissue culture techniques have been used for rapid multiplication, for the production of virus-free orchids through meristem culture, and for germinating seeds. Several excellent reviews concerning the application of tissue culture technology to orchids are available and details will not be repeated here.[9,11-13] For reference purposes, genera of orchids which have been propagated through tissue culture are listed in Table 1.

Ornamental species from more than forty different families, excluding the Orchidaceae, have been propagated using tissue culture techniques. In some taxa, propagation has been applied on a commercial scale. In other taxa, propagation has been shown to be possible although not necessarily feasible from a commercial standpoint. Recent reviews by Murashige[9,154] on the propagation of higher plants and by Holdgate[155] on the commercial aspects of propagation are available. Fern species which show demonstrated potential for propagation through tissue culture are listed in Table 2. Flowering plant species, other than orchids, are listed in Table 3.*

III. PROPAGATION STAGES

Murashige has separated the steps in clonal multiplication into three stages: Stage I — establishment of an aseptic culture; Stage II — multiplication; and Stage III — rooting and preparation of the propagule for transfer to soil.[9]

A. Stage I

The first step in any plant cell or tissue culture system is to obtain a suitable explant. Almost any plant tissue or organ may be used for an explant depending on the purpose of the culture system and the species involved (Tables 2, 3, and Section IV. B). Surface contaminants, e.g., bacteria and fungi, must be removed prior to culturing. Disinfection may involve combinations of the following procedures:

1. The explant is washed in running tap water for 1 to 2 hr. This procedure sharply reduces the level of surface contaminants and, at least in the Gesneriaceae, does not seem to affect the viability of the tissues.[385] If the explant is especially pubescent, a prewash with detergent may help to wet the surface. Extended washing may break up colonies of surface bacteria and fungi making them more accessible to sterilizing agents. It may also reduce the population size of the surface contaminants.
2. Plant tissues are immersed in an antiseptic solution to kill remaining surface contaminants. A dilution of commercial laundry bleach (5 to 6% sodium hypochlorite) in sterile water is an effective antiseptic although the dilution level and time of exposure must be determined for each type of explant. A few drops of a detergent such as Tween 20 (polyoxyethylene sorbitan monolaurate) may be added to break the surface tension of the tissues. Tween 20 may be autoclaved in water. A 70% solution of ethanol may be used for sterilization; however, there is some evidence that ethanol may seriously affect plant growth response.
3. Following sterilization procedures, the explant is rinsed several times in sterile distilled water and all damaged tissue removed. The explant is subdivided where necessary and transferred to nutrient medium. Tissues that brown rapidly may be dipped in a solution of critic and abscoric acid (100 to 150 mg ℓ^{-1} each) to retard oxidation.[154]

* Table 3 will be found at the end of the text.

Table 1
ORCHID GENERA IN WHICH TISSUE CULTURE TECHNIQUES HAVE BEEN SUCCESSFUL

Genus	Ref.
Acianthus	14
Aerides	15, 16
Aranda	17—22
Aranthera	22
Arundina	23
Ascofinetia	24
Bletilla	25
Brassavola	15, 16
Brassocattleya	26
Broughtonia	15, 16
Caladenia	14
Calanthe	27, 28
Calochilus	14
Calopogon	29
Cattleya	15, 16, 26, 28, 30—59
Cymbidium	7, 26, 43, 44, 49, 60—85
Cypripedium	29, 74, 86, 87
Dendrobium	15, 16, 28, 44, 50, 88—99
Dipodium	14
Diuria	14
Doritis	15, 16
Encyclia [*Epidendrum*]	100, 101
Epidendrum	15, 16, 102—107
Eriochilus	14
Eulophidium	26
Glossodia	14
Goodyera	29
Habenaria	29
Haemaria	108
Laeliocattleya	26, 102, 103, 109, 110
Liparis	29
Listera	29
Lycaste	44
Microtis	14
Miltonia	28, 43, 44
Neostylis	18
Neottia	111
Odontoglossum	44
Odontonia	27
Oncidium	15, 16, 27
Ophrys	27, 112
Paphiopedilum [*Cypripedium*]	26, 28, 106, 113—117
Phaius	44
Phalaenopsis	15, 26, 28, 32, 50, 113, 118—136
Pogonia	29
Potinara	26
Pterostylis	14
Renanthera	15
Rhynchostylis	137, 138
Schomburgkia	48
Sophrolaeliocattleya	26, 113
Spathoglottis	139, 140
Spiranthes	29
Thelymitra	14
Vanda	15, 17, 55, 138, 141—151

Table 1 (continued)
ORCHID GENERA IN WHICH TISSUE CULTURE TECHNIQUES HAVE BEEN SUCCESSFUL

Genus	Ref.
Vascostylis	24
Vanilla	86, 152
Vuylstekeara	27
Zeuxine	153
Zygopetalum	27

Table 2
FERN GENERA IN WHICH TISSUE CULTURE TECHNIQUES HAVE BEEN SUCCESSFUL[a]

Genus and species	Explant source	Morphogenic response	Ref.
Adiantaceae			
Adiantum cuneatum Langsd.	Rhizome tip	Adventitious shoots	9
Adiantum tenerum Swartz	Homogenized gametophytic tissues	Sporophytic plants	157
Blechnaceae			
Woodwardia fimbriata Grant c.f.	Rhizome tip	Adventitious shoots	9, 158
Cyatheaceae			
Alsophila australis R. Br.	Rhizome tip	Adventitious shoots	9, 158
Davalliaceae			
Davallia bullata Wall.	Homogenized plants from culture	Sporophytic plants	159
Nephrolepsis cordifolia (L.) Presl.	Runner tips	Sporophytic plants	160
Nephrolepis exaltata (L.) Schott var. *bostoniensis*	Runner tips	Adventitious shoots	9, 158, 161, 162
Nephrolepis falcata f. *furcans* Proctor	Runner tip	Adventitious shoots	163
Dennstaedtiaceae			
Microlepia strigosa (Thunb.) K. Presl.	Rhizome tip	Adventitious shoots	9, 158
Dyopteridoidea			
Cyrtomium falcatum Presl.	Homogenized gametophyte tissues	Sporophyte plants	157
Platycerioideae			
Platycerium bifurcatum C. Chr.	Homogenized gametophytic tissues	Sporophytic plants	157, 159
Platycerium stemaria Beauvois	Shoot tips	Adventitious shoots	164
Pteridoideae			
Pteris argyraea T. Moore	Rhizome tip	Adventitious shoots	9
Pteris cretica L.	Leaf stem	Gametophytes, sporophytes	165
Pteris ensiformis Burm.	Homogenized gametophyte tissues	Sporophytic plants	157
Pteris vittata L.	Rhizome callus	Sporophytes, gametophytes	166
Thelypteridaceae			
Cyclosorus dentatus (Forsk) Ching	Root apex explants	Gametophytes, sporophytes	167

[a] Assignment of genera to families follows that of Crabbe, Jermy, and Mickel.[156]

Some plant tissues have internal contaminants that can not be removed by standard techniques.[386,387] The addition of benomyl or benolate at 10 mg ℓ^{-1} to the medium may prove effective in reducing fungal contamination; however, benomyl can affect subsequent plan development adversely.[388,389] Antibiotics may be added to inhibit bacterial growth, but antibiotics are often harmful to plant cells, particularly those which affect the ribosomes. Since antibiotics are heat labile, they must be filter sterilized.

B. Stage II

Several alternate procedures have been used for the clonal multiplication of plants including the so called "meristem culture", development of axillary and terminal buds, induction of adventitious shoots, and somatic embryogenesis.

1. Development of Axillary and Terminal Buds

Axillary and terminal buds may be induced to develop in vitro. A single bud may produce only a single shoot, or depending on the species and the medium, may produce multiple shoots. Occasionally, callus formation and the development of shoots from meristematic areas in the callus occur. If shoots formed from the buds, in turn, develop buds along their axis, the procedure may be continued indefinitely. Clonal multiplication by this procedure is limited to the number of buds available; however, the procedure may be applied to some woody taxa where adventitious bud development or somatic embryogenesis has not been successful.

Propagation through "meristem culture" is a special class of the above, but with much wider application. The true apical meristem is restricted to a very small area of actively dividing cells, devoid of leaf primordia, at the very tip of the shoot. Culture of the true apical meristem is rarely successful (see Section IV.B) and in practice, a larger shoot is usually excised and subcultured. The term "meristem culture" has been erroneously used to describe shoot tip culture.[9,10]

2. Adventitious Shoot Development

In many species, plant organs, e.g., roots, shoots, or bulbs may be induced to form on tissues which normally do not produce these organs. Such adventitious organogenesis has more potential than the induction of axillary buds for mass clonal propagation of plants. A single leaf, for example, may produce thousands of buds or shoots, each genetically identical to the explant. Numerous taxa are multiplied by this procedure (Tables 2 and 3).

3. Somatic Embryogenesis

The greatest potential for clonal multiplication is through somatic embryogenesis where, technically, a single isolated cell can produce first an embryo, then a complete plant. Somatic embryogenesis has been demonstrated in several highly morphogenic species to date including carrot, *Nigella, Antirrhinum,* and *Petunia.* Somatic embryogenesis may occur in suspension cultures or occasionally in callus. In the examples of somatic embryogenesis known to date, induction of embryogenesis requires exposure to an auxin, often 2,4-D, followed by a reduction in auxin levels. Embryo induction also required a source of reduced nitrogen.[390] Somatic embryogenesis has recently been reviewed by Kohlenbach,[390] Raghavan,[391] and Narayanaswamy.[392]

C. Stage III

Shoots induced in culture may spontaneously produce roots, but more often it is necessary to transfer shoots to another medium to induce rooting. In some taxa such as *Saintpaulia* and *Episcia,* shoots will root when transferred directly to the soil.[316,321]

Plants removed from culture generally lose excessive amounts of water from their leaves when transferred to soil. Consequently, the survival of unprotected plants is poor.[393] Anatomical investigations with cauliflower and carnation plants grown in vitro indicated that vascular development was incomplete[393] and that epicuticular wax was reduced.[393,394] Both of these factors may contribute to the wilting that occurs after transplanting. Wilting problems may be circumvented in several ways. Murashige[9,154] and Hasegawa et al.[395] reported that high light intensities (3000 to 10,000 lx) improved survival of plantlets. Survival is also improved by covering the plants with a plastic tent or glass and misting for the first 2 weeks in soil. The cover should be gradually removed to allow the plants to acclimatize to greenhouse conditions.

IV. FACTORS AFFECTING TISSUE CUTURE SUCCESS

Numerous factors affect the successful induction of morphogenesis in plant cell and tissue cutures. Those in which there are a sufficient number of studies for discussion are given below. Where possible, examples from ornamental species have been cited; however, other examples have been included where necessary.

A. Media

One of the most important factors governing the growth and morphogenesis of plant tissues in culture is the composition of the medium. Several plant tissue and cell culture media are in common use including formulations devised by White,[396] Murashige and Skoog,[8] and Gamborg, et al. (B5 medium).[397] The Murashige and Skoog formulation is often used when morphogenesis is required. B5 medium, particularly in combination with 2,4-D is used when rapid cell proliferation is desirable. Two excellent reviews by Huang and Murashige[398] and by Gamborg et al.[399] summarize the various media and their components.

In general, plant tissue culture media consist of mineral salts, a carbon and energy source, vitamins, and growth factors. Other organic compounds may be included.

1. Mineral Salts

The salt composition of several media has been reviewed by Huang and Murashige.[398] The Murashige and Skoog formulation, probably the most popular medium for horticultural studies, is high in nitrates, potassium, and ammonia. Dilutions of the salt formulation have been used for some taxa. The B5 medium developed for soybean cell culture is also widely used and is characterized by a high potassium nitrate level. The Schenk and Hildebrandt medium is similar to B5, but with slightly higher levels of mineral salts. Whites medium is often used in modified form since levels of potassium and nitrogen in the original formulation are too low to maintain maximum callus and suspension culture growth for many species.[399]

Iron is an essential medium components for growth and morphogenesis of many species. Nitsch[400] demonstrated that plantlets would not form from microspores of *Nicotiana tabacum* unless iron was added to the medium. Iron forms such as citrate and tartrate are difficult to dissolve and may precipitate shortly after preparation. Murashige and Skoog used an EDTA chelated form of iron to bypass this problem[8] and Nitsch[400] demonstrated that the chelated form was more effective in embryo induction than ferric citrate. EDTA chelates are not entirely stable in liquid cultures particularly after autoclaving and may combine with other compounds to form a precipitate after a few days.[401] Steiner and Van Winden have developed techniques for preparation of an iron chelate that does not precipitate and is therefore available to the plant tissue.[402]

2. Carbon and Energy Source

The preferred carbon source for most plant tissue culture is sucrose. Glucose and fructose may be substituted in some cases, but most other sugars are poor carbohydrate sources for the plant. Tran Thanh Van has investigated the effects of various sugars on flower formation from thin layer epidermal sections of tobacco.[403] At 1/12 *M* the best response was obtained with either saccharose plus glucose or saccharose plus fructose. At 1/6 *M*, glucose alone and saccharose plus glucose were equal in ability to induce flower production. Thorpe has shown that in tobacco, at least part of the sucrose requirement is osmotic. When sucrose is reduced and mannitol substituted to give an equivalent weight to volume osmoticum, there is no reduction in shoot formation; however, when mannitol is not added, shoot production is reduced. Combinations giving osmotic values lower than 2% or higher than 4% caused a significant reduction in shoot formation.[404] In contrast, Barg and Umiel reported that tobacco callus grown on medium with very low sucrose levels (1%) developed a dark green color and produced a higher number of shoots in a shorter period of time. Their results also implicate osmoticum levels as a factor in morphogenesis.[405] Differences between their results and those of Thorpe may be due to differences in genotypes used in the two studies or to differences in the types and levels of growth regulators used. Organogenesis in the Liliaceae is inhibited at high sucrose levels.[223] At 30 gℓ^{-1}, 100% of the surviving shoot-tip explants produced bulblets, but at 90 g ℓ^{-1}, only three of twenty-eight explants produced bulblets. The remaining explants produced callus.

Fern differentiation may be regulated by sucrose levels. Sulkylan and Mehra have shown that low sucrose levels induced gametophytes and high sucrose levels induced sporophyte roots in the fern *Nephrolepis cordifolia*.[160] Kshirsagar and Mehta have also demonstrated that sporophyte differentiation in the fern *Pteris vittata* was dependent upon sucrose concentration.[166]

Rhizogenesis, in contrast to shoot induction, may be stimulated by high sucrose levels. In *Lilium* sucrose at 90 g ℓ^{-1} increased root dry weight.[223] In *Episcia*, root formation was greatly enhanced by adding sucrose at 30 g ℓ^{-1} to the Stage III medium.[316]

3. Vitamins

Linsmaier and Skoog demonstrated that, for *Nicotiana tabacum*, most vitamins are not essential for callus growth.[406] Pyridoxine, nicotinic acid, and biotin could be deleted from the medium without loss of growth. Riboflavin inhibited growth. Folic acid and p-amino-benzoid acid (PABA) increased growth, but were not essential. Thiamine was essential for growth of tobacco callus and is usually included in plant tissue culture media at levels between 0.1 mg ℓ^{-1} and 1.0 mg ℓ^{-1}. Absorbic acid increased growth only when levels of thiamine were suboptimal. Where the nutritional requirements for a given taxa have not been established, vitamins are often added to the medium as a precautionary measure.

4. Auxins and Cytokinins

Early studies by Skoog and Miller[6] which indicated that auxin to cytokinin ratios determined the type and extent of organogenesis have greatly influenced subsequent investigations. While no universal ratio for root or shoot induction exists, both an auxin and a cytokinin are usually added to the medium in order to obtain morphogenesis. There is considerable variability between taxa and between genotypes in the auxin and cytokinin levels required for morphogenesis (Section IV.H). Appropriate auxin and cytokinin levels must be determined for each species or variety under study. Even different sources within the same plant may vary in their auxin and cytokinin requirements (Section IV.B.2).

The commonly used auxins are 3-indoleacetic acid (IAA), 3-indolebutyric acid (IBA), 2,4-dichlorophenoxyacetic acid (2,4-D), 2,4,5-trichlorophenoxyacetic acid (2,4,5-T), 4-chlorophenoxyacetic acid (CPA), and 4-amino-3,5,6-trichloropicolinic acid (Pichloram or TCP). Based on stem curvature assays 2,4-D has eight to twelve times the activity of IAA, 2,4,5-T has four times the activity, and CPA has two times the activity.[407] 2,4-D, 2,4,5-T, CPA, and Pichloram are often used to induce rapid callus proliferation; however CPA is less effective in maintaining a high growth rate of cells than 2,4-D and 2,4,5-T.[408] High levels of auxin and in particular 2,4-D tend to suppress morphogenesis.[289,409-412] NAA is less inhibitory than IAA. Recovery of morphogenic potential after a period of induced callusing may be be better if Pichloram rather than 2,4-D is used to induce cell proliferation.[413]

In a few instances, low levels of 2,4-D apparently promote morphogenesis. Rao et al. showed that stem and leaf explants of *Petunia inflata* produced embryos at concentrations ranging from 100 to 2000 $\mu M\ \ell^{-1}$.[372] The best embryo formation occurred at 200 μg and 500 μg ℓ^{-1}. If 2,4-D was omitted from the medium, no embryos were produced. 2,4-D was more effective than NAA in inducing shoot formation in *Narcissus;* however, the reverse was true for ten other monocotyledonous species studied by Hussey.[197] In *Haworthia,* roots were induced by 2,4-D, but shoot development was suppressed.[209] Differences in response to 2,4-D between monocots and dicots may be due to the metabolism of 2,4-D into a physiologically inactive derivative in monocots, while in dicots physiologically active amino acid conjugates are formed.[414,415] 2,4-D has been shown to impair synthesis of the middle lamella,[295] to increase cell wall monomer synthesis,[416] and to affect microtubules.[417,418] These effects may be responsible for the friable nature of 2,4-D induced callus.

Cytokinens are 6-benzyladenine (also 6-benzylamino purine, BA, BAP), 6-γ-γ-dimethylallyaminopurine (also isopentenyladenine, 2iP, IPA), 6-furfuryl-aminopurine (kinetin), and 6-(4-hydroxy-3-methyl-trans-2-butenylamino) purine (zeatin). Many plant tissues have an absolute requirement for cytokinin. Other are cytokinin independent. Mok's studies with *Phaseolus lunatus* suggest that genotype-specific cytokinin dependence is under genetic control and that the number of genes is small.[419] Meins and Binns, on the other hand, have shown that cytokinin independence which develops in cell cultures of *Nicotiana tabacum* may be under developmental control.[420] Thus, cytokinin requirements of plant cells and tissues in culture may be both genetic and epigenetic.

While both auxin and cytokinin autotrophy are possible, it is rare that plant tissues will grow and proliferate in the absence of both. Genetic tumors and *Agrobacterium* induced tumors are an exception.

5. Other Organic Compounds

Given the essential raw materials, green plants should be able to synthesize all amino acids used in protein synthesis. In cases where nutritional requirements have not been established, mixtures of amino acids may be added as casein hydrolysate between 0.05 and 0.1%.[398] L-glutamine and L-asparagine are often added as an additional ammonia source at 100 mg ℓ^{-1}. Adenine sulfate, when added to the medium, often can enhance growth and shoot formation.[421] Adenine effects have been reviewed by Murashige.[9]

The discovery by Pollard et al. that *myo*-inositol was present in coconut milk and had growth promoting activity led to the inclusion of inositol in plant tissue culture media.[422] *Myo*-inositol is involved in the synthesis of phospholipids, cell wall pectins,[423,424] and in the cytoplasmic membrane systems.[425] It promotes growth in a variety of species including *Pelargonium hortorum* at levels between 50 and 100 mg ℓ^{-1},[409] and is essential for callus growth in *Fraxinus pennsylvanica*,[426] and for differentiation in *Haworthia*.[427]

The liquid endosperm of the coconut (coconut milk) promotes growth and differentiation in a wide variety of excised plant tissues including *Datura* embryos,[428] tobacco pith,[429] and *Pharbitis nil* apical meristem.[290] A number of cell division factors are present in coconut milk including diphenylurea,[430] 9-β-D-ribofuranosylzeatin,[431] and a compound which co-chromatographs with zeatinriboside.[432] Coconut milk contains a large number of free amino acids including phenylalanine which has cell division activity in soybean assays.[432] Coconut milk is often added at 10% v/v of the medium.

The addition of activated charcoal (AC) to plant tissue culture media may have either beneficial or harmful effects. Growth, organogenesis, and embryogenesis are stimulated in a variety of species and tissue including *Funaria* protonema,[433] ivy,[239] tobacco microspores,[434] orchids,[435,436] palm embryos,[436] ginger shoot tips,[436] onion,[437,438] and wild carrot.[437,438] AC inhibits growth of soybean,[437] *Haplopappus*,[437] and tobacco callus.[439]

AC effects may be attributed to three factors: (1) darkening of the medium,[440] (2) adsorption of inhibitory compounds,[433,435,436,438,441] and (3) adsorption of plant growth hormones from the medium.[439,442] Weatherhead et al. demonstrated that AC which had been used in tissue culture medium contained 5-hydroxymethyl-furfural (HMF), a compound produced by sucrose degradation during autoclaving.[442] HMF when it is not bound to AC, is mildly inhibitory to the growth of *Nicotiana tabacum* anther cultures. Fridborg et al. hypothesized that AC absorbs phenylacetic acid and p-OH-benzoic acid from the medium since wild carrot suspension cultures without AC contained high levels of these plant produced compounds, but the medium containing AC did not.[443] p-OH-benzoic acid is inhibitory to carrot embryogenesis. Clearly, at least part of the AC activity is due to removal of inhibitory compounds. Further, there is evidence that removal of essential growth hormones from the medium is probably responsible for the inhibition of cultures which require exogenous growth hormones for continued growth. IAA and 2iP are bound to AC very quickly.[439] NAA, kinetin, and BA are also bound to AC.[442] Both removal of inhibitory compounds and removal of auxins and cytokinins may be responsible for the observed induction of embryogenesis as high auxin levels inhibit embryogenesis in many species. AC is usually added between 0.5 and 3%. Since the quality and source of AC may affect results, AC should be acid washed and neutralized prior to use.[444]

B. The Explant

An explant is a piece of tissue or organ which is removed from the plant for purposes of culture. Success in culturing the explant is influenced by a number of factors inherent to the explant including the genotype of the explant (Section IV.H), the size of the explant, its physiological age, and the tissue or organ source of the plant.

1. Explant Size

As a general rule, very small explants have low surival rates in culture. This is true for both tissues and callus. In studies with carnation, shoot apices measuring 0.09 mm were incapable of morphogenesis. Apices up to 0.2 mm without any underlying periapical tissue are capable of slight growth. Isolates of 0.35 mm produced the largest number of normal shoots; however, shoot production was reduced when the explant included periapical tissue rather than leaf primordia. When the explant was as large as 0.5 mm, shoot production was again reduced, possibly due to the presence of subapical tissue.[284] The number of shoots produced from a shoot-tip explant may also be influenced by the size of the explant. *Chrysanthemum* shoot tips between 0.2 and 0.5 mm and shoot meristems between 0.1 and 0.2 mm produced only a single shoot. Larger explants, 0.5 to 1.55 mm in length produced multiple shoots.[246]

Explant size is rarely a problem unless the purpose of the culture system is to obtain virus-free plants. At one time the apical meristem area was believed to be virus-free, but more recent studies have demonstrated that viruses may be found even in the meristematic area.[445] The virus population increases in tissues subjacent to the apical meristem. Virus-free plants may be obtained by excising very small stem apices and culturing these in combination with heat treatments to eliminate residual viruses, but survival of these explants is low.

2. Source of the Explant

Early work with tobacco cell cultures led to the concept that, unlike animal cells, plant cells were totipotent, i.e., they retained embryogenic competence. In practice, plant tissues which retain embryogenic competence are limited and the ability to produce embryos will vary from tissue to tissue and from one species to another. Plant cells may loose embryogenic capability but retain the ability to form plant organs, a process known as determination to investigators working with animal systems. The degree of determination will vary from cell type to cell type. Less determined cells may retain a degree of morphogenic ability, e.g., callus derived from *Nigella sativa* roots produced only roots, while callus derived from stem and leaf tissues produced only shoots.[357] It is unlikely, however, that highly differentiated cells and tissues can revert to an embryonic state.

Within any plant, tissues differ in their degree of determination and thus in their ability to undergo morphogenesis. Takayama and Misawa examined the ability of differing explants of *Lilium auratum* and *Lilium speciosum* to produce bulbs in vitro.[223] Explants were taken from leaves, penduncles, bulb scales, petals, anthers, and from bulb scales of bulbs grown in culture. Explants from stamens and anthers produced neither bulbs nor roots. (Leaf explants did not survive the culture conditions.) Fifty-three percent of the penduncle explants, 75% of the petal explants, and 95% of the bulb scale explants produced bulbs. When the explant was taken from bulbs grown in culture, the success rate was 100%. This conditioning effect, i.e., an improved morphogenic response with explants from plants grown in culture has been noted in other species including several species of Gesneriaceae.[385]

Tissues that exhibit morphogenic competence in one taxa may not be equally competent in other taxa. Bulbs or corms, leaves, inflorescence stems, and ovaries of twelve monocotyledonous species were used as explant sources by Hussey.[197] While plantlets were obtained from all four explant sources in species of the Liliaceae, only bulbs and inflorescence stems were productive in the Iridaceae and Amaryllidaceae. Variation in explant response between species of the same family was also observed, e.g., explants from *Hyacinthus, Muscari, Ornithogalum,* and *Scilla,* all members of the Liliaceae, produced plantlets and callus when cultured, but *Tulipa* explants produced neither callus nor plants.[197]

Studies on morphogenesis in higher plants is further complicated by differing nutritional and growth-regulator requirements for morphogenesis with different types of explants. Carnation hypocotyl and shoot-tip explants were shown by Petru and Landa[282] to differ in media requirements for growth and organogenesis. Buys et al.[446] also suggested that different explants required different nutrients.

3. Physiological Age

The physiological age of the explant influences the type and extent of morphogenesis. The youngest and less differentiated tissues are found in plant meristems and the cuture of this plant tissue has been successful in a wide range of species. In general, for tissues other than meristems, young tissues have a higher degree of morphogenic competence than older tissues. Leaf disks from young leaves of *Passiflora suberosa* produce flowers at higher frequencies than leaf disks from older leaves.[345] Shoots could be produced from stem segments of *Iris* but only from the youngest tissue near the bud.[205] Similar patterns of differentiation were observed in *Gladiolus*.[201] Pierik and co-workers found that callus could be induced from young tissues and embryos of *Anthurium* but not from the mature tissues.[186,188] In woody species, embryo and seedling tissues have a higher capacity for regeneration.[447-449] Pierik and Steegmans observed that the ability of *Rhododendron* stem segments to produce roots decreased with increasing age of the stem.[450] Juvenile phase ivy tissues, embryos and embryo organs, but not mature phase tissues had high regenerative capability; however, the potential for regeneration was reduced when the tissues callused.[240] This observation supports observations in other species that morphogenic potential may be lost or reduced when callusing occurs.[9]

C. Light

The culture of plant cells on medium containing sucrose should bypass the need for photosynthesis, and, in fact, cells of numerous species are routinely cultured in the dark; however, several studies have demonstrated that light plays an important role in inducing organogenesis. In the Iridaceae, callus cultured in the dark will produce shoots when transferred to the light.[196,199,202] Similar results were observed in studies with the Liliaceae, *Heloniopsis orientalis*.[213,214] Constant darkness reduced the number of spears produced in cultures of *Asparagus*.[395] Experiments with *Nicotiana tabacum* indicated that light was a critical factor in the initiation of shoots and that it also influenced root growth;[451] however, in many species, light inhibits root growth.[450,452,453]

Light-triggered induction of organogenesis may be related, either directly or indirectly, to the increased accumulation of starch in specific cells through the photosynthetic processes. Thorpe and Murashige have shown that there is a significant accumulation of starch in the tobacco callus cells which ultimately give rise to bud primordia.[454] Cells which did not give rise to bud primordia did not accumulate starch. This accumulation was observed in both light and dark grown shoot-forming cultures, but was generally higher in light grown tissues.[455] Thorpe and Meier[455] and Thorpe[456] have suggested that starch may serve as a readily available source of energy for shoot production. This was supported by the observation that shoot-forming tissues showed increased respiratory activity. Alternately, starch may act as an osmotic agent.[455] (see Section IV.A.2). Starch accumulation was increased in the presence of various cytokinins, particularly kinetin.[455,456] Kato has shown that bud formation in *Heloniopsis orientalis* is controlled both by a photosynthetic system and by plant hormone levels.[214] Treatment with 3-(3,4-dichlorophenyl)-1,1-dimethyl urea (DCMU), and inhibitor of the Hill reaction, and another photosynthesis inhibitor, 3-amino-1,2,4-triazole (AT), greatly inhibited light-induced bud formation in *Heloniopsis* on medium without sugars. Morphactin induced bud formation in the dark which Kato suggests, may be due to a stimulating effect on carbohydrate synthesis.[214] These studies, further implicate photosynthesis and starch accumulation as factors in morphogenesis; however, further studies are necessary to establish if a direct link between these two factors is responsible for morphogenesis.

Light effects may be subdivided into photoperiod, light intensity, and wavelength. Morphogenetic requirements for light in an in vitro system may be satisfied by one or more of these factors; however, few studies are available where these factors are considered separately. Light effects have been reviewed recently by Murashige.[9,154]

1. Photoperiod

Early studies with *Bryophyllum tubiflorum* and *B, daigremontianum* (Kalanchoë) indicated that photoperiod affected bud formation.[457,458] Similar results were obtained with *Begonia* × *cheimantha,*[459] *Streptocarpus,*[333] and *Begonia* × *hiemalis.*[264]

The effective photoperiod for morphogenesis varies between taxa. A 12-hr photoperiod was optimal for shoot production from *Helianthus* tuber sections,[460] and 16 hr of light was most favorable for induction of flowers from thin-layer epidermal sections of tobacco.[403] Maximum bud formation was induced in geranium callus with a photoperiod of 15 to 16 hr.[306] Shorter and longer periods of light reduced the frequency of budding. With continuous illumination, the callus did not turn green and buds were not produced. These results were consistent with results reported for parsley that a dark period was required to induce root and embryoid formation.[461] *Pharbitis nil* apical meristems, on the other hand, will grow and produce plantlets with photoperiods ranging from 16 to 24 hr of light, and morphogenesis is not inhibited by total light.[290] Photoperiod response in culture may be influenced by changes in endogenous levels of auxin and cytokinin. Heide[459] and Heide and Skoog[462] demonstrated that day-length influenced the endogenous auxin and cytokinin levels in *Begonia* and *Bryophyllum.* Further, plants with photoperiod requirements may manifest this requirement in culture.[9,154]

2. Wavelength

The induction of developmental events in plants by specific wavelengths of light and reversal or inhibition of these events by other wavelengths has been of interest to plant scientists since the early 1900s. Many such photomorphogenic events are controlled by phytochrome, a ubiquitous plant pigment which has an absorption maximum of 660 nm (P_r). With exposure to red light, P_r is converted to its far-red form (P_{fr}) which has an absorption maximum of 730 nm. Reversion of P_{fr} to P_r may occur slowly in the dark or rapidly with exposure to far-red light. Phytochrome may control developmental response by differential gene activation, differential gene repression, or both.

A second category of response is blue-light mediated. Action spectra for some blue light responses have a maximum at 365 nm which would implicate a flavin-type receptor. Other spectra have only a small peak at 365 nm which might argue for a carotenoid receptor. The evidence at present favors a flavin-type receptor.

The major photosynthetic pigments, chlorophyll a and chlorophyll b, have absorption spectra in both the red and blue light wavelengths. Chlorophyll a has absorption peaks at 440 and 680 nm. Chlorophyll b has absorption peaks at 470 and 650 nm. Both red and blue light are necessary for photosynthesis. The other major photosynthetically active pigments in green plants, carotenoids, and the xanthophyll lutein, absorb broadly in the 400 to 500 nm range.

Since wavelength-specific morphogenic responses are common in the plant kingdom, it is not surprising that they should be observed in vitro. The role of light of different wavelengths on bud formation in *Heloniopsis orientalis* was investigated by Kato.[214] White (a mixture of wavelengths), red, and blue lights were effective in inducing bud formation but not green light. Red light stimulated flower formation in tobacco thin-layer epidermal sections, while darkness and far-red light stimulated root production. Bud and cellus formation were not influenced by wavelength of light.[403] Red light increased bud formation in the moss *Funaria* and short periods of far-red light partially reversed that effect.[463] Bud induction of pine cotyledons was maximized at 660 nm, while purple and near-UV light had no effect.[464] In contrast, experiments by Weis and Jaffe indicated that the critical portion of the light spectrum for shoot initiation from tobacco pith was the blue region and red light was without effect.[465] Seibert also reported that blue light in the region of 467 nm was effective in inducing shoots in tobacco callus.[466] Purple light was also effective. These experiments indicate that photomorphogenesis does occur in vitro, possibly by interacting with phytochrome chlorophylls and other photo-absorbing pigments.

3. Light Intensity

Murashige has reported that optimum light intensity for plant tissues in culture may differ from the requirements of the plants themselves.[9] Optimum light for cultured tissues in Stage I or II was around 1000 lx.[9,395] For Stage III, light intensity between 3000 and 10,000 lx was required. The higher light intensities improved survival of plants transferred to soil.

Light intensity has been shown to affect the type of growth in culture. Increased light intensity (from 3000 to 6000 lx) caused an increase in shoot formation in one cultivar of *Begonia* × *hiemalis,* but no increase in dry weight. Two other cultivars responded to the increased light with increased biomass but no increase in shoot production.[263] An increase in light intensity from 1000 to 3000 lx caused differentiation of cladophylls on *Asparagus* spears.[395]

D. Temperature

Early studies on the growth of plant cells in culture indicated that the optimum range was between 26 and 28°C, but that species requirements differed considerably.[467,468] Studies on the growth of suspension cultures of *Ipomoea* at temperatures from 15 to 34°C indicated that maximum growth of the cultures occurred between 25 and 32°C. Growth of *Narcissus*[185] and induction of roots on *Rhododendron* explants[450] was best at 25°C. Bulb formation in *Lilium auratum,* on the other hand, was best at 20°C and decreased progressively up to 30°C. Lower temperatures were not tested.[223] In practice, plant cultures are grown around 25°C and few studies investigating temperature requirements are available. Temperature effects have been reviewed by Murashige.[9,154]

It seems likely that temperature requirements of the intact plant may be reflected in the temperature response of cultured tissues from that plant. In some species, a cold period is required to overcome dormancy. Pine cultures grown for long periods of time will gradually slow their growth. In some cases, growth can be reestablished by 2 to 3 months of cold treatment mimicking winter conditions.[385,469] *Eschscholzia californica* seeds require a cold period to overcome dormancy of the seed. This cold requirement is also expressed in the development of plants from embryos in vitro. A 2-week treatment at 6°C was necessary to induce development of embryoids into plantlets. Thus, embryos of *Eschscholzia* exhibit a phenomenon of dormancy that is very similar to the seed.[343]

A low-temperature treatment may trigger specific morphogenetic events. A cold treatment of 17°C was necessary for the in vitro induction of gametangia in *Physcomitrella patens* and 7°C in *Funaria hygrometrica* and *Physcomitrium pyriforme.* No gametangia were formed at 24°C.[470] A low-temperature pretreatment was found to increase shoot production from petiole segments of *Begonia* × *cheimantha.* At constant temperature, best shoot induction was obtained between 18 and 21°C, but a pretreatment at 15 to 18°C for 2 weeks improved the number and size of shoots developing. A heat pretreatment severely inhibited shoot formation. Fonnesbech concluded that the formation of shoots takes place in the first 2 weeks of culture and that high temperatures are detrimental to processes which induce shooting.[256,257]

Plant tissue cultures may be affected by alternating day/night temperature cycles. Gautheret observed that root morphogenesis from *Helianthus tuberosus* tuber explants was best with alternating temperatures.[460]

Temperature may influence the type of morphogenic event. *Phalaenopsis* flower stalk buds cultured at 25°C exhibited position effects, i.e., reproductive shoots developed from the upper nodes and vegetative shoots developed from the lower buds. At 28°C, all buds grew vegetatively regardless of position. These results are consistent with studies done on intact plants. Tanaka and Sakanishi concluded that the *Phalaenopsis* flower stalk buds, with the exception of the uppermost bud, are not predetermined to grow vegetatively or reproductively and that temperature was a controlling factor.[471] Temperature control of morphogenesis in lettuce has also been observed.[472] At 17°C, lettuce explants produced shoots only at high kinetin levels. At 28°C, shoots were produced only on medium without kinetin, provided that adequate levels of auxins were included in the media. These experiments implicate kinetin and temperature interactions as important in regulation of morphogenesis.

E. Gas Phase

Variability in tissue culture response may be due, in part, to the influence of the gas phase of the culture. Beasley and Eaks have found that the common practice of flaming flasks for sterilization purposes induces variable and often high levels of ethylene into the flask.[473] This occurs with both gas and alcohol flames. Most of the gas diffused out of the flask within 2 hr, but if flask closures are tight, high levels of ethylene could persist for a long time. Beasley and Eaks suggested that high levels in their flasks caused excessive callus formation on cotton ovules cultured with gibberellic acid (GA) and reduced fiber formation when cultured with IAA. Since ethylene in flasks may diffuse into the incubator space, ethylene effects may be seen in sealed incubators. Significant differences in response of cotton ovules were noted when ovules were grown in sealed incubators and incubators that were opened regularly.[473]

Ethylene has a wide range of effects, some stimulatory and some inhibitory. There are a large number of instances where the action of ethylene is antagonistic to morphogenesis and promotes unorganized prliferation of cells.[409] It is now clear that growing cell cultures produce ethylene[474] and that the production in tissues is stimulated by auxins, particularly 2,4-D and IAA. Many so called "IAA effects" may in reality be ethylene effects, including the reduction of greening observed in many cultures with the addition of auxin.[475] Carbon dioxide, also produced by plant cultures, may be a competitive inhibitor of ethylene in vivo, possibly reacting at the same cell binding site.

Dalton and Street have examined the effect of di[illegible] closures on accumulation of oxygen, carbon dioxide, and ethylene in the gas phase of spinach cultures. They determined that a closure of polyethylene film provided greater diffusion resistance to these gases than a foil closure.[475] Reduction of growth was observed in the polyethylene closed flasks and was apparently due to both carbon dioxide and ethylene increases in the gas phase. Thomas and Murashige have also examined the accumulation of volatile metabolities in cultures of *Nicotiana tabacum* and *Phoenix dactylifera*.[476,477] Cell cultures in tubos sealed with vaccine caps to prevent gas exchange exhibited marked increases in carbon dioxide, ethylene, ethanol, and acetaldehyde. Significantly, cultures with well-developed shoots did not produce ethanol and only rarely acetaldehyde. Addition of 2,4-D to the culture medium suppressed organogenesis and increased ethanol production. Tisserat and Murashige demonstrated that ethanol reversibly inhibited somatic embryogenesis in carrot.[478] These studies indicate that ethanol production and differentiation may be inversely related.

F. Polarity

Polarity effects in plants and plant tissues are well documented and wide ranging.[479,480] In view of the ubiquitous nature of polarity, it is not surprising that in vitro organogenesis may exhibit polarity effects, i.e., that the orientation of the explant on the medium with respect to its orientation in the plant may influence organogenesis. Polarity effects may be caused by several factors including a chemical gradient in the plant such as that observed for auxin transport in *Phaseolus*[81] or anatomical differences between the explant tissues.[292]

Polarity effects have been observed in the culture of inflorescence stalk sections from *Gladiolus*,[201] *Asparagus*,[482] and *Amaryllis* hybrids[483] and on stem segments of *Rhododendron*.[450] Shoot induction was greater when explants were placed on medium with their morphological base away from the surface (reversed polarity). *Gladiolus* explants cultured with their morphological base facing the medium also produced shoots but to a lesser degree and shoot induction was delayed. The requirement for reversed polarity is absolute in *Narcissus*. Organogenesis will only occur when scape sections are inverted on the medium.[184]

Polarity effects have been shown to affect organogenesis from bud-scale explants in many of the Liliaceae. The greatest organogenic potential occurred in *Lilium longiflorum* explants from the basal portion of the bulb scale and along the abaxial (outer) edge of the explant.[227] A similar pattern was observed in *Hyacinth*;[484] however, in *Lilium speciosum*, organogenesis always arose from the basal adaxial areas of the bulb scales regardless of the orientation of the explant in relation to the medium.[233] Polarity was observed in shoot and root induction from detached leaves of *Echeveria elegans* in that shoots were invariably produced on the adaxial side of the leaf.[292]

G. Subculture

1. Loss of Morphogenic Potential

There is increasing evidence that extended subculturing of callus or cell suspension cultures can lead to a reduction in morphogenic potential.[289,303,316,485,486] Experiments with *Convolvulus* callus showed that after three or more subcultures, some lines lost morphogenic ability.[289] Reinert and Backs observed a gradual loss of morphogenic potential in long-term carrot cultures.[486] However, restoration of regenerative ability in carrot cultures with the addition of kinetin was reported by Wochok and Wetherell.[487] A similar pattern was observed in root tips of *Isatis tinctoria* that exhibited reduced ability to form shoots with repeated subculturing. Shooting ability was restored with addition of kinetin to the medium.[269] In contrast, some tissue and cell cultures retain morphogenic potential over extended periods of subculturing. *Chrysanthemum* propagated by callus containing meristematic areas retained morphogenic potential over a 3½-year period.[244] Similar stability was observed in *Lilium longiflorum* cultures.[488]

Loss of morphogenic potential could occur in several ways:

1. Shoot production from callus may be initiated from organized centers of cell division (meristemoids) which are derived from the initial explant. These organized centers may gradually be lost through repeated subculturing and rapid unorganized cell division. This may be true of subcultured *Pelargonium* callus which does not form vascular bundles.[303]
2. Loss of morphogenic potential might be the result of reduced endogenous levels of growth regulators. This view is supported by studies with *Isatis tinctoria* and carrot cultures.[269,487]
3. Accumulation of chromosome abnormalities may inhibit developmental processes. There is a considerable body of evidence that chromosome aberrations occur in culture including changes in number (aneuploidy and polyploidy) and rearrangements (delections, translocations, isochromosome formation, etc.)[448-492]

Torrey[493] and Murashige and Nakano[494,495] have shown that a progressive loss of morphogenic ability was correlated with increasing aneuploidy; however, the positive correlation does not necessarily imply a direct cause and effect relationship. Other factors may be affecting chromosome instability and loss or morphogenic potential. There is some evidence that chromosomal variability in callus may be reduced in regenerating tissues and in plants derived from the callus.[490,491] Attentuation of the variability may result from either selective morphogenesis of the cells in culture (i.e., only a few cells retain morphogenic potential and will regenerate) or reversion of the cells in culture towards their original karyotype.

2. Variability from Culture

Continuous subculture often increases the frequency of aberrant plants arising from the cultures.[193,243,302,496,497] Often the variability is associated with callus formation.[284,302] The range of variability is wide. Phenotypic changes observed in plants derived from callus of the scented geranium included changes in leaf morphology, loss of anthocyanin color, loss of pubescence, dwarfing, changes in essential oil constituents, and flower morphology. The variation observed from culture was much higher than that from cutting. Variation may result from segregation of chimeras,[498] changes in chromosome number and structure,[499-501] genetic mutations (either nuclear or cytoplasmic), reversion to a juvenile stage, nonheritable physiological adaptations, and segregation of extra-nuclear genomes. While variability may be undesirable from a propagator's point of view, it serves as a pool from which desirable traits may be selected.[502]

H. Genotype

The genotype of the plant chosen for propagation may influence response in culture. Within a species, some genotypes appear to propagate easily while others fail to respond. Genotype-specific effects have been reported for *Anthurium andraeanum*[186] and *Anthurum scherzerianum*.[188] Only one third of the *A. andraeanum* genotypes were capable of forming callus. Three fourths of the *A. scherzerianum* genotypes formed callus and subsequently produced plants. Differences were observed in the induction of callus from anthers in several cultivars of geranium.[308] The percentage of anthers producing callus varied from 10 to 62%. Hussey observed that the morphological response of *Gladiolus* stem segments varied with the cultivar.[201] Clonal variation in response was observed in *Narcissus*[185] and in callus induction from scape sections of day lilies.[217] Genotypic differences were observed between three cultivars of *Begonia* × *hiemalis*. Explants from the three cultivars differed in survival in culture, in the proportion of shoots which survived transfer to soil, and in morphogenic response to increased light intensity.[263]

In some cases, the differences in response by different taxa may reflect differences in nutritional or hormonal requirements for differentiation. Cultivar differences in the hormonal levels for shoot initiation were observed in *Begonia* × *hiemalis*,[259] *Episcia cupreata* varieties,[316] and geranium callus.[304]

While the nature of differences in response between different genotypes is unknown, estimates of the number of genes involved can be made. Izhar and Power identified variations in growth response on various media of protoplasts derived from highly inbred *Petunia* lines.[503] Plants from two taxa with different hormone requirements were crossed to obtain the F_1 and backcrossed to both parents. Genetic analysis indicated that only a few genes controlled differences in hormonal requirements. Their data also suggested that different stages in protoplast development were controlled by different genes. The genes determining cytokinin response in *Petunia* were studied by

Hanson et al.[504] Two inbred lines with different optima for shoot development were crossed. The F_1 hybrid exhibited the low cytokinin requirement of one parent. F_2 and backcross analysis indicated that at least two different genes were involved in determining the level of cytokinin response.

I. Season

Seasonal variation in morphogenic response of the explant has been observed in *Lilium speciosum.* Bulb-scale explants will proliferate and differentiate freely in culture without added auxin, but only if the explant is obtained in the spring or the summer.[233] Morphogenic ability seems to be limited to periods of vegetative growth. Seasonal variation has also been observed in explants from artichoke tubers. Spontaneous in vitro proliferation occurred only during periods when the plant was in a vegetative condition. At other times the explant could be induced to proliferate only when IAA was added to the medium. Seasonal response has been observed in the induction of haploid plants from anthers in tobacco.[505]

V. SUMMARY AND FUTURE CONSIDERATIONS

Clearly, there are a large number of factors which influence morphogenic response in ornamental species. This manuscript has attempted to identify major factors and to summarize the pertinent literature in each area; however, much more research is needed to establish the genetic and physiological basis for regulation of morphogenesis. In some respects, at least one genus of the ornamental plants, *Petunia,* has considerable potential for this type of research.

An ideal in vitro system for genetic and physiological studies should have the following characteristics: (1) both callus and suspension cultures may be easily established, (2) regeneration of plants through organogenesis should occur with high frequency, (3) somatic embryogenesis should be inducible, (4) procedures for plating single cells (usually protoplasts) and recovering plants of single cell origin should be available, and (5) haploids should be available. No current in vitro system satisfies all of the above criteria. Tobacco, which is often used as a model system, meets most of the criteria, but cell cultures do not undergo somatic embryogenesis. Somatic embryogenesis can be induced in suspension cultures of wild carrot, another species often used as a model system; however, haploids are very difficult to obtain. Most of the above criteria are met by *Petunia. Petunia* is a member of the Solanaceae, a family with very high regenerative ability. Haploid plants have been obtained from both anther culture[506] and the culture of individual microspores.[507] Protoplasts have been isolated, successfully cultured and plants have been obtained from protoplast-derived calli.[508-511] Additionally, the technology for protoplast fusion and recovery of fusion products (somatic hybrids) has been developed.[512-514] Thus, petunia compares favorably with tobacco and carrot as a model system for molecular studies.

The regulation of morphogenesis may occur at several levels: (1) at the level of transcription, i.e., at the level of the gene itself, (2) post-transcriptionally, but prior to translation as recently suggested by Davidson and Britten,[515] (3) at the translational level, i.e., at the level of protein synthesis on the ribosomes, or (4) post-translationally, i.e., by regulation of metabolic pathways. Several approaches to the problem of morphogenesis and differentiation may be taken. One approach is to compare cells undergoing morphogenesis with cells in which morphogenesis is not occurring. This approach has recently been taken by Varma and Dougall, who examined DNA, RNA, and protein levels in carrot cultures.[516] Another approach is to examine determination processes in a single gene or a set of genes. This approach has been taken by Meins and Binns with cytokinin habituation in tobacco.[420] Regardless of the approach used, a clear understanding of the processes leading to differentiation and morphogenesis in higher plants is essential, both in terms of basic knowledge and for the practical aspects of crop improvement.

ACKNOWLEDGMENTS

I would like to thank Susan Wilson and Susan Terry for their help in compiling the references. I would also like to thank my student, Bob Barni, and my colleagues, Drs. Paula Nakosteen, Beth Mullin, and Jim Caponetti, for their critical review of the manuscript.

Table 3
FLOWERING PLANT GENERA IN WHICH TISSUE CULTURE TECHNIQUES HAVE BEEN SUCCESSFUL

Genus and species	Explant source	Morphogenic response	Ref.
	Monocotyledonae		
Agavaceae			
Agave sp.	Seed fragments	Shoots	168
Cordyline terminalis L. Kunth	Stem tip	Multiple shoots	169, 170
	Stem segments from shoot tip grown in vitro	Adventitious shoots	169, 171
	Callus from young leaves	Plantlets	172
C. banksii Hook. f.	Shoot tips	Multiple shoots	173
Dracaena deremensis Engler	Dormant lateral bud	Shoot	174
	Stem segments and subcultured callus	Adventitious shoots	174
D. godseffiana Sander	Shoot tips	Multiple shoots	170
Dracaena sp.	Shoot tip	Adventitious shoots	9
Furcraea gigantea Venten.	Callus from leaf base	Adventitious shoots and roots	175
Sanseveria trifasciata Prain	Leaves	Adventitious shoots	176
Yucca sp.	Lateral buds	Multiple shoots	177
	Shoot tip	Adventitious shoots, embryos	9
Amaryllidaceae			
Hippeastrum hybridum	Bulb scales	Buds and roots	178
	Bulb scale	Adventitious bulbs	9
Hippeastrum sp. (*Amaryllis* sp.)	Tissue from dormant bulbs	Bulblets	179
	Meristematic areas of scale, leaf, or stem	Adventitious shoots	173
	Scape	Bulbs	180
	Callus from bulb and floral peduncle	Adventitious shoots	181
Hippeastrum spp. (*Amaryllis* spp.)	Leaf base, scape, peduncle, bulb scale, ovary	Rooted shoots	179
Narcissus biflorus Curt.	Anthers	Embryoids	182
N. triandrus L.	Bulb scale segments	Callus, bulbs	183
Narcissus sp.	Leaf base, inverted scape segments, ovary	Adventitious shoots and roots	184
	Leaf base, inverted scape	Adventitious shoots	184,185
	Meristematic areas of scale, leaf, or stem	Adventitious shoots	173
Nerine sp.	Meristematic areas of scale, leaf, or stem	Adventitious shoots	173

Table 3 (continued)
FLOWERING PLANT GENERA IN WHICH TISSUE CULTURE TECHNIQUES HAVE BEEN SUCCESSFUL

Genus and species	Explant source	Morphogenic response	Ref.
Araceae			
Anthurium andraeanum Lind.	Callus from embryo, leaf, and petiole	Adventitious shoots	186
	Node explants from aseptically grown plants	Adventitious shoots	171
	Young leaves	Shoots	187
A. scherzerianum Schott.	Callus from leaf	Adventitious shoots	188
	Young leaves	Shoots	187
Dieffenbachia picta Schott.	Shoot tip	Shoots	189
	Lateral buds	Shoots	177
Philodendron lacerum (Jacq.) Schott. Sc.	Node	[No details reported]	171
P. oxycardium Koch. & Sello.	Node	[No details reported]	171
Philodendron sp.	Shoot tip	Adventitious shoots	9
Scindapsis aureus Engler	Node	[No details reported]	171
Spathiphyllum aureus	Node and leaves from aseptically grown plants	[No details reported]	171
S. clevelandii	Shoot tips and lateral buds	Rooted shoots	190
	Inflorescence, buds, stem segments	Adventitious shoots	191
Spathiphyllum sp.	Node	[no details reported]	171
Syngonium podophylum Schott.	Lateral buds	Multiple shoots	170
	Shoot tips and axillary buds	Multiple shoots	192
Bromeliaceae			
Aechmea distichantha Lem.	Leaves and defoliated stems from axillary bud derived plants grown in vitro	Adventitious shoots	193
A. fasciata Baker	Buds from base and terminal shoot	Multiple shoots	194
	Leaves and defoliated stems from axillary bud derived plants grown in vitro	Adventitious shoots	193
Aechmea sp. hybrid	Leaves and defoliated stems from axillary bud derived plants grown in vitro	Adventitious shoots	193
Ananas comosus Merill	Buds from base and terminal shoot	Multiple shoots	194
Cryptanthus bivitattus Regel	Buds from base and terminal shoots	Multiple shoots	194
Cryptanthus sp.	Leaves and defoliated stems from axillary bud derived plants grown in vitro	Adventitious shoots	193
Cryptanthus spp.	Stem buds	Multiple shoots	195
Cryptbergia meadii Hort.	Buds from base and terminal shoots	Multiple shoots	194

Table 3 (continued)
FLOWERING PLANT GENERA IN WHICH TISSUE CULTURE TECHNIQUES HAVE BEEN SUCCESSFUL

Genus and species	Explant source	Morphogenic response	Ref.
Dyckia sulphurea C. Koch	Buds from base and terminal shoots	Multiple shoots	194
Guzmania sp.	Buds from base and terminal shoots	Multiple shoots	194
Iridaceae			
Freesia sp.	Callus from corms, stems, leaves, pedicles, flower bud, and anther	Adventitious shoots	196
	Corms sprouted on medium	Adventitious shoots	197
	Shoot tips, adventitious shoots from stem segments	Axillary shoots	198
Freesia sp. hybrids	Bulb, callus from stem	Adventitious shoots	197
Gladiolus hortulans Bailey	Cormel stem tips	Plantlets, stems (plants, subcultured for axillary stems)	199
	Buds from corms	Plantlets, stems (plants, subcultured for axillary stems)	200
Gladiolus sp.	Inflorescence stalk — callus	Shoots	201
	Corms sprouted on medium	Adventitious shoots	197
	Shoot tips, adventitious shoots from stem segments	Axillary shoots	198
	Inflorescence sections, bulb	Adventitious shoots	197
Iris hollandica	Young inflorescence stems from freshly sprouted bulbs	Adventitious shoots	202
Iris sp.	Excised embryos	Plantlet	203
	Callus from inflorescence segments	Adventitious shoots	204
	Callus from stem segments near flower bud	Adventitious shoots	205
	Shoot tips, adventitious shoots from stem segments	Axillary shoots	198
	Callus from shoot apex	Embryos	206
Neomarica coerulea (Ker-Gawl.) Sprague	Node explants	Adventitious shoots	207
Schizostylus coccinea Backh. & Harvey	Inflorescence sections	Adventitious shoots	197
Sparaxis bicolor hybrids	Inflorescence sections, bulb	Adventitious shoots	197
Sparaxis sp.	Corms sprouted on medium	Adventitious shoots at point of removal of axillary and main shoots	197
Liliaceae			
Aloe pretoriensis Pole-Evans	Calli from seed segments	Adventitious shoots	208
Asparagus myriocladus Thunb.	Node explants	Adventitious shoots	171

Table 3 (continued)
FLOWERING PLANT GENERA IN WHICH TISSUE CULTURE TECHNIQUES HAVE BEEN SUCCESSFUL

Genus and species	Explant source	Morphogenic response	Ref.
Haworthia angustifolia Haw.	Callus from gynoecia and inflorescence segments	Adventitious shoots	209
H. atrofusca	Callus from gynoecia and inflorescence segments	Adventitious shoots	209
H. chloracantha	Callus from gynoecia and inflorescence segments	Adventitious shoots	209
H. maughanii Poelln.	Callus from gynoecia and inflorescence segments	Adventitious shoots	209
H. mirabilis Haw.	Callus	Buds	210
H. retusa (L.) Haw.	Callus from gynoecia and inflorescence segments	Adventitious shoots	209
Haworthia truncata × *H. setata*	Callus from gynoecia and inflorescence segments	Adventitious shoots	209
H. turgida Haw.	Callus from gynoecia and inflorescence segments	Adventitious shoots	209
	Ovary wall	Adventitious shoots	211
H. variegata L. Bolus	Callus from gynoecia and inflorescence segments	Adventitious shoots	209
Haworthia spp.	Callus from gynoecia and inflorescence segments	Adventitious shoots	209
Haworthia sp.	Inflorescence axis bearing flowers, callus from inflorescence axis	Adventitious shoots	212
Heloniopsis orientalis (Thunb.) T. Tanaka	Leaves	Adventitious shoots	213, 214
Hemerocallis sp.	Callus (origin not indicated)	Adventitious shoots	215
	Petals	Adventitious shoots	216
	Callus from inflorescence slices	Adventitious shoots	217
	Shoot tips	Multiple shoots	218
	Scape tissues	Adventitious shoots	179
Hosta decorara Bailey	Shoot tip	Multiple shoots	219
Hosta sp.	Flower sections	Adventitious shoots	220
Hyacinthus orientalis L.	Tissue culture derived bulblets	Plantlets	221
	Bud scale sections	Bulblets	222
H. hybrids	Bulb, leaf, inflorescence stem, ovary	Adventitious shoots, callus	197
Lilium auratum Lindl.	Bulb scale segments	Bulbs	223
L. candidum L.	Floral segments	Bulblets	224
L. leichtlinii Hook.	Leaves	Plantlets	225
L. longiflorum Thunb.	Callus from stem apices	Plantlets	226
	Bulb scale segments	Bulblets	227—231
	Seed derived callus	Plants (chromosomal mosaics)	232
L. regale Wils.	Dormant bulb explants	Bulblets	179
L. speciosum Thunb.	Bulb scale segments	Bulbs	225
	Bulb scale cores	Adventitious shoots and roots	233
L. hybrids	Callus from bulb scales	Adventitious shoots	234
	Hybrid embryos	Hybrid plants	235

Table 3 (continued)
FLOWERING PLANT GENERA IN WHICH TISSUE CULTURE TECHNIQUES HAVE BEEN SUCCESSFUL

Genus and species	Explant source	Morphogenic response	Ref.
Lilium sp.	Meristematic areas of scale, leaf, and stem	Adventitious shoots	198
	Bulb cross sections	Bulbs, plants	236
Muscari botryoides Mill.	Bulb, leaf, inflorescence stem, ovary	Adventitious shoots, callus	237
Ornithogalum thyrsoides Jacq.	Bulb, leaf, inflorescence stem, ovary	Adventitious shoots, callus	197
	Ovaries, sepals, bulb scales, leaves, infloresc-ence stem	Adventitious shoots	197, 237
Scilla sibirica Haw.	Bulb, leaf, inflorescence stem, ovary	Adventitious shoots, callus	197
	Dicotyledoneae		
Araliaceae			
Hedera helix L.	Stem from mature phase	Embryos	238, 239
	Callus from juvenile phase stem	Adventitious shoots and roots	238, 239
	Embryo cotyledons	Shoots	240
Asteraceae			
Chrysanthemum cinerariaefolium Vis.	Shoot tips	Multiple shoots	241
	Capitulum	Shoots	242
C. morifolium Ramat.	Callus from shoot tips	Adventitious shoots	243, 244
	Petioles and pedicles	Adventitious buds	245
	Shoot tips, young inflorescences	Multiple and adventi-tious shoots	246
Chrysanthemum sp.	Callus from stem explants	Adventitious shoots	247
	Shoot tips	Rooted shoots	248
C. boreale × *japonense*	Maternal ovary following cross	Hybrid plants	249
Dahlia pinnata Gillian	Shoot tip	Virus-free plants	250
Gerbera jamesonii Bolus.	Capitulum explants	Adventitious shoots	251
	Shoot tips, lateral shoots	Lateral shoots	252
Gerbera sp.	Young florets and meris-tem	Plantlets	253
Gynura sarmentosa	Buds	Plants	254
Balsaminaceae			
Impatiens "Pinkinini"	Apical meristem	Single shoots	255
Begoniaceae			
Begonia × *cheimantha* Hort.	Petiole	Adventitious shoots, roots	256, 257
	Leaf disks, petioles	Buds	258
Begonia evansiana Andr.	Leaf disks, petioles, floral stalks	Buds	258
Begonia × *hiemalis* Hort. (Rieger Begonia)	Leaf, petiole	Adventitious shoots	259, 260
	Flower peduncle	Adventitious shoots	261
	Petiole	Adventitious shoots — roots, shoots and roots	262—264
B. multiflora Benth.	Leaf disk, petioles, floral stalks	Buds	258

Table 3 (continued)
FLOWERING PLANT GENERA IN WHICH TISSUE CULTURE TECHNIQUES HAVE BEEN SUCCESSFUL

Genus and species	Explant source	Morphogenic response	Ref.
	Dicotyledoneae		
	Shoot tips	Multiple shoots	282
	Frozen shoot tips	Shoots	283
	Shoot tips	Shoots	284, 285
Gypsophilia paniculata L.	Callus from leaves, stems and shoot tips	Shoots	286
Convolvulaceae			
Convolvulus arvensis L.	Root segments	Roots and shoots	287
	Plated cell suspension cultures	Buds from callus, roots from callus, roots could be induced to form shoots	288
	Stem explants	Shoots	289
Pharbitis nil (Choisy)	Apical meristem	Roots and shoots	290
	Stem explant, shoot tip	Floral primordia	291
Crassulaceae			
Crassula sp.	Leaf	Adventitious shoots	9
Echeveria elegans Bgr.,	Excised leaves	Roots and shoots	292
Kalanchöe bossfeldiana Poelln.	Leaf, shoot tip, stem	Plantlets	293
Kalanchöe sp.	Leaf	Adventitious shoots	9
	Leaf disks, petiole sections, shoot tips	Adventitious shoots, multiple shoots	294
Sedum telephium L.	Callus from sections of immature leaves	Adventitious shoots	295
Ericaceae			
Rhododendron sp. (azaleas)	Shoot tips	Axillary shoots	296
Rhododendron sp.	Shoot tips	Axillary shoots	297
Euphorbiaceae			
Acalphya wilkesiana Müll. Arg.	Buds	Multiple shoots	298
Euphorbia pulcherrima Wild. ex. Klotzsch	Petiole and internode, callus derived from internode	Adventitious shoots	299
	Embryos germinated in vitro	Adventitious shoots on seedling hypocotyl	300
Fabaceae			
Clianthus formosus (G. Don.) Ford and Vickery	Callus from seedling	Adventitious roots and shoots	301
Geraniaceae			
Pelargonium adcifolium	Callus from stem, root, and petiole	Shoots	302
P. capitatum (L.) L'Her. ex. Ait.	Callus from stem, root, and petiole	Shoots	302
P. crispum (L.) L'Her. ex. Ait.	Callus from stem, root, and petiole	Shoots	302
P. domesticum L. H. Bailey	Callus from stem, root, and petiole	Shoots	302
P. domesticum × *P. denticulatum*	Callus from stem, root, and petiole	Shoots	302
P. graveolens L'Her. ex. Ait.	Callus from stem, root, and petiole	Shoots	302
	Callus from stem sections	A selected clone gave rise to a high polyploid plant	302

Table 3 (continued)
FLOWERING PLANT GENERA IN WHICH TISSUE CULTURE TECHNIQUES HAVE BEEN SUCCESSFUL

Genus and species	Explant source	Morphogenic response	Ref.
	Dicotyledoneae		
B. rex Putz.	Leaf	Adventitious shoots	265
	Superficial and deep internode tissues	Buds from superficial tissues	266
B. socotrana Hook.	Leaf disks, petioles, flora stalks	Buds	258
B. sutherlandii Hook. f.	Leaf disks, petioles flora stalks	Buds	258
B. teuscheri × *coccinea* Hort.	Leaf disks, petioles, floral stalks	Buds	258
Begonia sp.	Shoot tip	Adventitious shoots	9
Berberidaceae			
Nandina sp.	Lateral buds	Axillary shoots	267
Brassicaceae			
Cheiranthus cheiri L.	Seedling derived callus and suspension cultures	Roots and embryoids and plantlets	268
Isatis tinctoria L.	Root tips	Shoots	269
Lobularia maritima (Desv.)	Stem segments	Buds, roots	270
Cactaceae			
Chamaecereus sylvestrii Britt and Rose	Areole	Plantlets	271
Epiphyllum grandiflora	No data	Shoots	272
E. hybrid	Areole	Plantlets	271
Hatiora salicorniodes	Areole	Plantlets	271
Hylocereus calearatus	No data	Shoots	272
Lobivia binghamiana Backeb.	Areole	Plantlets	271
Mammillaria elongata D.C.	Tubercle	Shoots	272—274
	Tubercle	Plantlets	271
M. woodsii Craig	Stem pith	Plantlets	275
Myrtillocactus geometrizans (Mart.) Cons. (T.)	Seedling stem fragments	Buds, roots	276
Opuntia basilaris Engelm. & Bigel.	Areole	Plantlets	271
Opuntia polyacantha Haw.	Areole	Shoots	272, 277, 278
	Areole	Plantlets	271
Pachycereus pringlei Britt & Rose	Areole	Plantlets	271
Pereskia aculeata Mill.	Areole	Plantlets	271
Rhipsalis terres (Vell.) Steud	No data	Shoots	272
Schlumbergera bridgesii Lőfgren	Shoot tips	Adventitious shoots	279
S. gaertneri Brit.	Shoot tips	Adventitious shoots	279
Selenicereus grandiflorus Britt. & Rose	Areole	Plantlets	271
Weberocereus biolleyi (A. Webb.) Britt & Rose	No data	Shoots	272
Caryophyllaceae			
Dianthus caryophyllus L.	Whole shoot tips and callus from subdivided shoot tips	Single and multiple shoots	280
	Multiple shoots from shoot tip culture	Axillary shoots	281
	Hypocotyl	Adventitious rooted shoots	282

Table 3 (continued)
FLOWERING PLANT GENERA IN WHICH TISSUE CULTURE TECHNIQUES HAVE BEEN SUCCESSFUL

Genus and species	Explant source	Morphogenic response	Ref.
	Dicotyledoneae		
Pelargonium × *hortorum* L. H. Bailey	Pith callus	Shoots and roots	303
	Stem tips	Shoots, no multiplication attempted, no virus symptoms	304
	Callus derived from stem tips	Shoots, apparently virus free	305
	Calli from stem tips, internodal portions, pith with vascular tissue, and petiole	Shoots	306
	Anthers	Plantlets without virus symptoms	307
	Haploid callus from anthers	Shoots, haploid plants	308, 309
Pelargonium peltatum (L.) L'Hér. Eex. Ait.,	Callus from stem tips	Shoots, apparently disease free	310
	Shoot tips	Virus-free shoots	311
P. peltatum × *P. zonale* hybrids	Callus from stem tip	Shoots, apparently virus free	311
	Shoot tips	Virus-free shoots	311
P. zonale (L.) L'Hér. ex. Ait.	Callus from stem tips	Shoots, apparently disease free	310
	Shoot tips	Virus-free shoots	311
	Meristem	Virus-free shoots	312
Gesneriaceae			
Chirita sinensis Lindl.	Leaf	Not given	313
Episcia cupreata (Hook) Hanst.	Leaf	Adventitious shoots	314
	Stem tips, leaf sections	Adventitious shoots	315
	Callus from leaf and stolon	Adventitious shoots and roots	316, 317
Episcia dianthaflora H. E. Moore & R. G. Wills	Callus from leaf and stolon	Adventitious shoots and roots	316
Episcia × *wilsonii*	Stem tips, leaf sections	Adventitious shoots	315
Kohleria amabilis (Planch & Lindon) Fritsch	Leaf and petiole	Adventitious shoots	318
Nautilocalyx sp.	Leaf midrib sections	Adventitious shoots and roots	319
Saintpaulia ionantha H. Wendl.	Leaf sections	Adventitious shoots	320
	Anthers	Haploid plantlets	321
	Petiole sections	Adventitious shoots	322—324
	Petiole sections	Roots	325
	Leaf sections	Adventitious shoots	326, 327
	Thin superficial layers, petiole segments, and leaf disks	Adventitious shoots and roots	328
	Shoot tips	Plants	328
	Floral parts	Plantlets	329

Table 3 (continued)
FLOWERING PLANT GENERA IN WHICH TISSUE CULTURE TECHNIQUES HAVE BEEN SUCCESSFUL

Genus and species	Explant source	Morphogenic response	Ref.
	Dicotyledoneae		
Sinningia speciosa (Lodd.) Hiern (Gloxinia)	Apex of vegetative shoots	Adventitious shoots	330
	Shoot tip	Axillary shoots	9
	Leaf	Adventitious shoots	331
	Pedicle segments	Adventitious shoots	331
Streptocarpus nobilis C. B. Clarke	Leaf disks	Flower buds	332
Streptocarpus × hybridus Voss	Leaf disks	Adventitious shoots and roots	333
Streptocarpus (mixed hybrids)	Pedicels	Adventitious shoots	334
Haemodoraceae			
Anigozanthos flavidus D.C.	Apical meristem	Multiple shoots	301
	Lateral bud explants	Shoots	335
A. manglesii D. Don.	Lateral bud explants	Shoots	335
Macropidia fulginosa (Hook.) Druce	Lateral bud explants	Shoots	335
Labiatae			
Coleus blumei Benth	Leaf disks	Adventitious shoots	336
Mentha viridis L.	Terminal bud	Shoot	337
Lentibulariaceae			
Pinguicula moranesis	Leaves	Plantlets	338
Loganiaceae			
Buddleia davidii Franch	Meristem	Virus-free plants	339
Moraceae			
Ficus lyrata Warb.	Callus from leaf explants	Shoots	340
Nyctaginaceae			
Bougainvillea glabra Choisy	Shoot tip	Multiple shoots	341
Oleaceae			
Forsythia sp.	Meristem	Virus-free stem	342
Papaveraceae			
Eschscholzia californica Cham.	Embryoids from placental tissue	Plantlets	343, 344
Passifloraceae			
Passiflora caerulea L.	Leaf disks	Adventitious shoots	345
P. edulis Sims	Leaf disks	Adventitious shoots	345, 346
P. foetida L.	Leaf disks	Adventitious shoots	345
P. mollissima Bailey	Axillary buds	Shoots	346
P. suberosa	Leaf disks	Adventitious shoots	345
Piperaceae			
Peperomia caperata Yunck.	Leaf and stem segments	Adventitious shoots	347
Peperomia "red ripple"	Leaf disks	Adventitious shoots	348
P. longifolia	Stem	Ebryoids	349
Piper nugrum	Stem	Ebryoids	349
Plumbaginaceae			
Plumbago indica L.	Internode segments	Callus, roots, shoots, inflorescenses, and flowers	350
Polemoniaceae			
Phlox drummondii Hook.	Callus from flower buds	Shoots with terminal flower	351
Phlox paniculata L. (*P. decussata* Lyon ex. Pursh)	Meristem	Rooted shoot	352
	Shoots	Axillary shoots	353
Phlox subulata L.	Shoots	Axillary shoots	353

Table 3 (continued)
FLOWERING PLANT GENERA IN WHICH TISSUE CULTURE TECHNIQUES HAVE BEEN SUCCESSFUL

Genus and species	Explant source	Morphogenic response	Ref.
	Dicotyledoneae		
Portulacaceae			
Portulaca grandiflora Hook	Callus from excised stem	Shoots	354
Primulaceae			
Cyclamen persicum	Tuber explants	Adventitious shoots	355
Ranunculaceae			
Anemone coronaria L.	Shoot apical meristem	Shoots	356
	Peduncle sections	Shoots	356
Clemantis montana Buch.-Ham. ex D.C.	Shoot tip	Multiple shoots	173
Nigella sativa L.	Calli from stem and leaf	Adventitious shoots	357
	Calli from roots	Roots	357
	Callus (source not given)	Embryoids	358
Rosaceae			
Rosa hybrida L.	Shoot tips, lateral buds	Multiple shoots	359
	Shoots	Rooted shoots	360
Rosa sp.	Pith from water shoots	Buds	361
Saxifragaceae sp.	Shoot tip	Axillary shoots	9
Scrophulariaceae			
Antirrhinum majus L.	Protoplasts from leaves	Embryoids	362
	Callused stem segments	Embryoids	362, 363
	Stem	Embryos and plantlets from embryos	364
Limnophilia chinensis	Callus from internode segments	Shoots	365
Linaria vulgaris Mill.	Roots	Buds	366
Mazus pumilus (Burm. f.) van Steenis	Internode segments from inflorescence, calli from inflorescence	Roots or shoots	367
Torenia fournieri Linden ex. E. Fourn.	Stem and leaf sections	Roots or buds	368, 369
Verbascum thapsus L.	Internode segments	Buds	370
Solanaceae			
Petunia × *hybrida* Hort.	Leaf sections and stem sections	Callus, roots, buds, embryos	371
	Stem internodes and leaf disks	Embryos, adventitious shoots	372
	Pollen grains	Haploid, diploid, triploid shoots	373
	Leaf sections	Adventitious shoots	374
	Internode stem segments	Embryos (plants), roots, shoot primordia	375
	Callus from shoot apical meristem	Shoot buds	376
	Callus from terminal shoots, leaves	Shoots	376
	Green leaf sections from mosaic-infected plants	Virus-free plants	377
Petunia inflata R. E. Fries	Leaf sections and stem sections	Callus, roots, buds, embryos	371
	Stem internodes and leaf disks	Embryos, shoots	372
	Internode and stem segments	Embryos (plants), roots, shoot primordia	375

Table 3 (continued)
FLOWERING PLANT GENERA IN WHICH TISSUE CULTURE TECHNIQUES HAVE BEEN SUCCESSFUL

Genus and species	Explant source	Morphogenic Response	Ref.
	Dicotyledoneae		
P. parodii	Calli from leaf protoplasts	Adventitious shoots	378
P. parviflora	Calli from leaf protoplasts	Adventitious shoots	379
Physalis peruviana L.	Leaf	Buds and plantlets	380
Salpiglossis sinuata Ruiz & Pav.	Anthers	Plants from sporophytic tissue	381
	Leaf disks	Adventitious shoots	382, 383
Solanum dulcamara L.	Leaf	Buds and plantlets	380
S. nigrum L.	Leaf	Buds and plantlets	380
	Protoplasts	Plantlets	384

REFERENCES

1. **Haberlandt, G.**, Kulturversuche mit isolierten Pflanzenzellen, *Sitzungsber Math-Naturwiss. Kl. Kais. Akad. Wiss. Wien.*, 11, 69, 1902.
2. **White, P. R.**, Potentially unlimited growth of excised tomato root tips in a liquid medium, *Plant Physiol.*, 9, 585, 1934.
3. **White, P. R.**, Potentially unlimited growth of excised plant callus in an artificial nutrient, *Am. J. Bot.*, 26, 59, 1939.
4. **Nobécourt, P.**, Sur la perennite de l'augmentation de volume des cultures de tissus vegetaux, *C. R. Soc. Biol.*, 130, 1270, 1939.
5. **Gautheret, R. J.**, Sur la possibilite de realiser la culture indefinie des tissus de tubercules de carotte, *C. R. Acad. Sci.*, 208, 118, 1939.
6. **Skoog, F. and Miller, C. O.**, Chemical regulation of growth and organ formation in plant tissues grown, *in vitro, Symp. Soc. Exp. Biol.*, 11, 118, 1957.
7. **Morel, G. M.**, Producing virus free *Cymbidium, Am. Orch. Soc. Bull.*, 29, 495, 1960.
8. **Murashige, T. and Skoog, F.**, A revised medium for rapid growth and bioassays with tobacco cultures, *Physiol. Plant.*, 15, 473, 1962.
9. **Murashige, T.**, Plant propagation through tissue cultures, *Annu. Rev. Plant Physiol.*, 25, 135, 1974.
10. **Langhans, R. W., Horst, R. K., and Earle, E. D.**, Disease-free plants via tissue culture propagation, *HortScience*, 12, 25, 1977.
11. **Rao, A. N.**, Tissue culture in the orchid industry, in *Applied and Fundamental Aspects of Plant Cell and Tissue Culture*, Reinert, J. and Bajaj, Y. P. S., Eds., Springer-Verlag, New York, 1977, 44.
12. **Sato, D. M., Barnes, D., and Murashige, T.**, Bibliography on clonal multiplication of orchids through tissue culture, *Tissue Culture Assoc. Man.*, 4, 783, 1978.
13. **Arditti, J.**, Clonal propagation of orchids by means of tissue culture — a manual, in *Orchid Biology, Reviews and Perspectives*, Arditti, J., Ed., Comstack Publishing Associates, New York, 1977, 202.
14. **McIntyre, D. K., Veitch, G. J., and Wrigley, J. W.**, Australian terrestrial orchids from seed. II. Improvement in techniques and further successes, *Am. Orchid Soc. Bull.*, 43, 52, 1974.
15. **Sagawa, Y. and Valmayor, H. L.**, Embryo culture of orchids, in Proc. 5th World Orchid Conf., DeGarmo, L. R., Ed., Long Beach, 1966, 99.
16. **Valmayer, H. L. and Sagawa, Y.**, Ovule culture in some orchids, *Am. Orchid Soc. Bull.*, 36, 766, 1967.
17. **Chong-Jin, G. and Lian-Hua, L.**, Effects of growth regulators on morphogenesis of orchid callus tissues, *Plant Physiol.*, 61, 47, 1978.

18. **Loh, C.-S., Rao, A. N., and Goh, C.-J.**, Clonal propagation from leaves in the orchid, *Aranda, J. Singapore Natl. Acad. Sci.*, 4, 97, 1975.
19. **Goh, C.-J.**, Meristem culture of *Aranda* Deborah, *Malays. Orchid Rev.*, 12, 9, 1973.
20. **Goh, C.-J. and Lie, L.-H.**, Effects of growth regulators on morphogenesis of orchid callus tissues, *Plant Physiol. Suppl.*, 61, 47, 1978.
21. **Teo, C. K. H. and Teo, P. S.**, Clonal propagation of orchids by tissue culture, *Planter*, 50, 213, 1974.
22. **Cheah, K. T. and Sagawa, Y.**, *In vitro* propagation of *Aranda* Wendy Scott and *Aranthera* James Storei, *HortScience*, 13, 661, 1978.
23. **Mitra, G. C.**, Studies on seeds, shoot tips, and stem discs of an orchid grown in aseptic culture, *Indian. J. Exp. Biol.*, 9, 79, 1971.
24. **Intuwong, O. and Sagawa, Y.**, Clonal propagation of Sarcanthine orchids by aseptic culture of inflorescences, *Am. Orchid Soc. Bull.*, 42, 209, 1973.
25. **Ichihashi, S.**, Studies on the media for orchid seed germination. III. The effects of total ionic concentration, cation/anion ratio, NH_4/NO_3 ratio, and minor elements on the growth of *Bletilla striata, J. Jpn. Soc. Hortic. Sci.*, 47, 524, 1979.
26. **Rosa, D. M. and Laneri, U.**, Modification of nutrient solutions for germination and growth *in vitro* of some cultivated orchids and the vegetative propagation of some *Cymbidium* cultivars, *Am. Orchid Soc. Bull.*, 46, 813, 1977.
27. **Bertsch, W.**, A new frontier: Orchid propagation by meristem tissue culture, *Am. Orchid Soc. Bull.*, 36, 32, 1967.
28. **Karasawa, K.**, On the media with banana and honey added for seed germination and subsequent growth of orchids, *Orchid Rev.*, 74, 313, 1966.
29. **Stoutamire, W. P.**, Seeds and seedlings of native orchids, *Mich. Bot.*, 3, 107, 1964.
30. **Kusumoto, M.**, Effects of combinations of growth regulators and of organic supplements on the growth of *Cattleya* plantlets cultured *in vitro, J. Jpn. Soc. Hortic. Sci.*, 47, 492, 1979.
31. **Kusumoto, M.**, Effects of combinations of growth regulators and of organic supplements on the proliferation and organogenesis of *Cattleya* protocorm like bodies cultured *in vitro, J. Jpn. Soc. Hortic. Sci.*, 47, 502, 1979.
32. **Champagnat, M., Morel, G., and Mounetous, B.**, La multiplication végétative des *Cattleya*, A partir de jeunes feuilles cultivées aseptiquement *in vitro, Ann. Sci. Nat. Bot. Biol. Veg.*, 11, 97, 1970.
33. **Lindemann, E. G. P., Gunckel, J. E., and Davidson, O. W.**, Meristem culture of *Cattelya, Am. Orchid Soc. Bull.*, 39, 1002, 1970.
34. **Arditti, J.**, The use of tomato juice media for orchid seed germination, *Malays. Orchid Rev.*, 8, 67, 1966.
35. **Arditti, J.**, The effects of niacin, adenine, ribose, an niacinamide coenzymes on germinating orchid seeds and young seedlings, *Am. Orchid Soc. Bull.*, 35, 892, 1966.
36. **Arditti, J.**, The effect of tomato juice and its fractions on the germination of orchid seeds and on seedling growth, *Am. Orchid Soc. Bull.*, 35, 175, 1966.
37. **Champagnat, M. and Morel, G.**, Multiplication végétative des *Cattleya* a partir de bourgeons cultivés, *in vitro, Bull. Soc. Bot. France Mem.* 111, 1969.
38. **Farrar, M. D.**, Growing orchids seedlings, *Am. Orchid Soc. Bull.*, 32, 888, 1963.
39. **Hirsh, D. H.**, Gibberellates: Stimulant for *Cattleya* seedlings in flasks, *Am. Orchid Soc. Bull.*, 28, 342, 1959.
40. **Ito, I.**, Ultra low temperature stage of orchid pollinia and seeds, *Jpn. Orchid Soc. Bull.*, 11, 4, 1965.
41. **Knudson, L.**, Nutrient solution for orchids, *Bot. Gaz. (Chicago)*, 112, 528, 1951.
42. **Lawrence, D. and Arditti, J.**, A new medium for the germination of orchid seed, *Am. Orchid Soc. Bull.*, 33, 766, 1964.
43. **Morel, G.**, Tissue culture — A new means of clonal propagation in orchids, *Am. Orchid Soc. Bull.*, 33, 473, 1964.
44. **Morel, G. M.**, Clonal propagation of orchids by meristem culture, *Cymb. Soc. News*, 20, 3, 1965.
45. **Lindemann, E. G., Gunckel, J. E., and Davidson, O. W.**, Meristem culture of *Cattleya, Am. Orchid Soc. Bull.*, 39, 1002, 1970.
46. **Noggle, G. R. and Wynd, F. L.**, Effects of vitamins on germination and growth of orchids, *Bot. Gaz. (Chicago)*, 104, 455, 1943.
47. **Rossifer, E. S.**, Growth stimulants for growing orchid seeds, *Am. Orchid Soc. Bull.*, 29, 904, 1960.
48. **Scully, R. M., Jr.**, Aspects of meristem culture in the *Cattleya* Alliance, *Am. Orchid Soc. Bull.*, 36, 103, 1967.
49. **Tsukamoto, Y., Kano, K., and Katsuura, T.**, Instant media for orchid seed germination, *Am. Orchid Soc. Bull.*, 32, 354, 1963.
50. **Valmayor, H. L.**, Further investigations into nutrient media, in Proc. 7th World Orchid Conf., Ospina, H. M., Ed., Medellin, Colombia, 1974, 211.

51. **Zeigler, A. R., Sheehan, T. S., and Poole, R. T.**, Influence of various media and photoperiod on growth and amino acid content of orchid seedlings, *Am. Orchid Soc. Bull.*, 36, 195, 1967.
52. **Arditti, J.**, Germination and growth of orchids on banana fruit tissue and some of its extracts, *Am. Orchid Soc. Bull.*, 37, 112, 1968.
53. **Pierik, R. L. M. and Steegmans, H. H. M.**, The effect of 6-benzylamino-purine on growth and development of *Cattleya* seedlings grown from unripe seeds, *Z. Pflanzenphysiol.*, 68, 228, 1972.
54. **Raghavan, V.**, Effects of certain organic nitrogen compounds on growth *in vitro* of seedlings of *Cattleya, Bot. Gaz. (Chicago)*, 125, 260, 1964.
55. **Raghavan, V. and Torrey, J. G.**, Inorganic nitrogen nutrition of the seedlings of the orchid *Cattleya, Am. J. Bot.*, 51, 264, 1964.
56. **Reinert, R. A. and Mohr, H. C.**, Propagation of *Cattleya* by tissue culture of lateral bud meristems, *Proc. Am. Soc. Hortic. Sci.*, 91, 664, 1967.
57. **Ichihashi, S. and Kako, S.**, Studies on clonal propagation of *Cattleya* through tissue culture method. I. Factors affecting survival and growth of shoot meristem of *Cattleya in vitro, J. Jpn. Soc. Hortic. Sci.*, 42, 264, 1973.
58. **Ichihashi, S. and Kako, S.**, Studies on clonal propagation of *Cattleya* through tissue culture method. II. Browning of *Cattleya, J. Jpn. Soc. Hortic. Sci.*, 46, 325, 1977.
59. **Vanseveren, N.**, Quelques phénomènes d'organogénèse dans des plantules de *Cattleya* Londl. (Orchidaceae) obtenues par propagation meristematique *in vitro, Bull. Soc. R. Bot. Belt.*, 102, 211, 1969.
60. **Champagnat, M.**, Particularities du developement des apex de *Cymbidium* cultivés *in vitro, Bull. Soc. Fr. Physiol. Veg.*, 11, 225, 1965.
61. **Champagnat, M., Morel, G., Chabut, P., and Cognet, A. M.**, Recherches morphologiques et histologiques sur la multiplication vegetale de quelques orchidess du genre *Cymbidium, Rev. Gen. Bot.*, 73, 706, 1966.
62. **Steward, F. C. and Mapes, M. O.**, Morphogenesis in aseptic cell cultures of *Cymbidium, Bot. Gaz. (Chicago)*, 132, 65, 1971.
63. **Homes, J.**, Shoot differentiation in orchid protocorms cultured *in vitro, Abstr. Int. Congr. Plant Tissue and Cell Culture*, Thorpe, T. A., Ed., University of Calgary, Canada, 1978, 157.
64. **Wilfret, G. J.**, Formation of protocorm-like bodies on excised *Cymbidium* shoot tips, *Am. Orchid Soc. Bull.*, 35, 823, 1966.
65. **Wimber, D. E.**, Clonal multiplication of *Cymbidiums* through tissue culture of shoot meristem, *Am. Orchid Soc. Bull.*, 32, 105, 1963.
66. **Wimber, D. E.**, Additional observations on clonal multiplication of *Cymbidiums* through culture of shoot meristems, *Cymb. Soc. News*, 20, 7, 1965.
67. **Ueda, H. and Torikata, H.**, Organogenesis in meristem cultures of *Cymbidium*. I. Studies on the effects of growth substances added to culture media under continuous illumination, *J. Jpn. Soc. Hortic. Sci.*, 37, 240, 1968.
68. **Ueda, H. and Torikata, H.**, Organogenesis in the meristem tissue cultures of *Cymbidums*. II. Effects of growth substances on the organogenesis in dark culture, *J. Jpn. Soc. Hortic. Sci.*, 38, 188, 1969.
69. **Ueda, H. and Torikata, H.**, Organogenesis in the meristem cultures of *Cymbidiums*. III. Histological studies on the shoot formation at the rhizome tips of *Cymbidium goeringii* Reichb. F. cultured, *in vitro, J. Jpn. Soc. Hortic. Sci.*, 38, 262, 1969.
70. **Ueda, H. and Torikata, H.**, Organogenesis in the meristem cultures of *Cymbidiums*. IV. Study on the cytokinin activity in the extracts from the protocorms, *J. Jpn. Soc. Hortic. Sci.*, 39, 202, 1970.
71. **Ueda, H. and Torikata, H.**, Organogenesis in the meristem cultures of *Cymbidiums*. V. Anatomical and histochemical studies on phagocytosis in the micorrhizome of *Cymbidium goeringii, J. Jpn. Soc. Hortic. Sci.*, 39, 50, 1970.
72. **Ueda, H. and Torikata, H.**, Organogenesis in the meristem cultures of *Cymbidiums*. VI. Effects of light and culture medium on adventitious root formation, *J. Jpn. Soc. Hortic. Sci.*, 39, 73, 1970.
73. **Ueda, H. and Torikata, H.**, Effects of light and culture medium on adventitious root formation by *Cymbidiums* in aseptic culture, *Am. Orchid Soc. Bull.*, 41, 322, 1972.
74. **Kaho, K.**, Acceleration of the germination of so-called "hard to germinate orchid" seeds, *Am. Orchid Soc. Bull.*, 37, 690, 1968.
75. **Champagnat, M., Morel, G., and Gambade, G.**, Particularites morphologiques et pouvoir de regeneration de *Cymbidium virescens* cultive *in vitro, Bull. Soc. Bot. Fr.*, 115, 236, 1968.
76. **Fonnesbech, M.**, Growth hormones and propagation of *Cymbidium in vitro, Physiol. Plant.*, 27, 310, 1972.
77. **Fonnesbech, M.**, Organic nutrients in the media for propagation of *Cymbidium in vitro, Physiol. Plant.*, 27, 360, 1972.
78. **Morel, G. and Champagnat, M.**, Wachstum and entwicklung von *Cymbidium viresuns, Die Orchidee*, 17, 250, 1966.
79. **Sagawa, Y., Shoji, T., and Shoji, T.**, Clonal propagation of *Cymbidiums* through shoot meristem culture, *Am. Orchid Soc. Bull.*, 35, 118, 1966.

80. **de Fossard, R. A.**, *Tissue Culture, For Plant Propagators,* University of New England Printery, Armidale, 1976, 352.
81. **Werckmeister, P.**, Uber die lichtinduktion der geotropen orientiergung von luft-und bodenwurzeln in gewebekulturen von *Cymbidium, Ber, Dtsch. Bot. Ges.,* 83, 19, 1970.
82. **Werckmeister, P.**, Die steureung von vermehrung (proliferation) und wachstum in der meristem kultur von *Cymbidium* und die verwendung eines kohlenährmediums, *Die Orchidee,* 21, 126, 1970.
83. **Werckmeister, P.**, Light induction of geotropism and the control of proliferation and growth of *Cymbidium* in tissue culture, *Bot. Gaz. (Chicago),* 132, 346, 1971.
84. **Rucker, W.**, Wirkung von hydroxyharnstoff auf entwicklung und differenzierung *in vitro* kultivierter protokorme von *Cymbidium, Z. Pflanzenphysiol.,* 76, 229, 1975.
85. **Kusumoto, M.**, Effects of combinations of growth regulating substances and of organic matter on the propagation and organogenesis of *Cymbidium* protocorms, cultured *in vitro* (Japanese), *J. Jpn. Soc. Hortic. Sci.,* 47, 391, 1978.
86. **Hegarty, C. P.**, Observations on the germination of orchid seed, *Am. Orchid Soc. Bull.,* 24, 457, 1955.
87. **Liddell, R. W.**, Notes on germinating *Cypripedium* seed. I, *Am. Orchid Soc. Bull.,* 22, 195, 1953.
88. **Gilliland, H. B.**, On the symmetry of the orchid embryo, in *Proc. Centennary and Bicentennary Conf. Biology,* Singapore, 1958, 276.
89. **Nimoto, D. H. and Sagawa, Y.**, Ovule development in *Dendrobium, Am. Orchid Soc. Bull.,* 30, 813, 1961.
90. **Marston, M. E.**, Vegetative propagation of plants using tissue culture techniques, Nottingham University School of Agricultural Report, England, 1966, 77.
91. **Ito, I.**, Germination of seeds from immature pod and subsequent growth of seedlings in *Dendobium nobile* Lindl., *Sci. Rep. Saikyo Univ.,* 7, 35, 1955.
92. **Sagawa, Y. and Shoji, T.**, Clonal propagation of *Dendrobiums* through shoot meristem culture, *Am. Orchid Soc. Bull.,* 36, 856, 1967.
93. **Israel, H. W.**, Production of *Dendrobium* seedlings by aseptic culture of excised ovularies, *Am. Orchid Soc. Bull.,* 32, 441, 1963.
94. **Singh, H. and Sagawa, Y.**, Vegetative propagation of *Dendrobium* by flower stalk cuttings, *Hawaii Orchid J.,* 1, 19, 1972.
95. **Vajrabhaya, M. and Vajrabhaya, T.**, Variations of *Dendrobium* airising in meristem propagations, in Proc. 7th World Orchid Conf., Ospina, H. M., Ed., Medellin, Colombia, 1972, 231.
96. **Kim, K. K., Kunisaki, J. T., and Sagawa, Y.**, Shoot tip culture of *Dendrobiums, Am. Orchid Soc. Bull.,* 38, 1077, 1970.
97. **Sanguthai, O., Sanguthai, S., and Kamemoto, H.**, Chromosome doubling of a *Dendrobium* hybrid with colchicine in meristem culture, *Hawaii Orchid J.,* 2, 12, 1973.
98. **Intuwong, O. and Sagawa, Y.**, Clonal propagation of *Dendrobium* Golden Wave and other nobile types, *Am. Orchid Soc. Bull.,* 44, 319, 1975.
99. **Mosich, S. K., Ball, E. A., and Arditti, J.**, Clonal propagation of *Dendrobium* by means of node cultures, *Am. Orchid Soc. Bull.,* 43, 1055, 1974.
100. **Frei, J. K.**, Effect of bark substrate on germination and early growth of *Encyclia tampensis* seeds, *Am. Orchid Soc. Bull.,* 42, 701, 1973.
101. **Frei, J. K,. Ramona, O. P., Fodor, C., and Haynick, J. L.**, The suitability of certain barks as growth media for orchids, *Am. Orchid Soc. Bull.,* 44, 51, 1975.
102. **Churchill, M. E., Ball, E. A., and Arditti, J.**, Production of orchid plants from seedling leaf tips, *Orchid Dig.,* 34, 271, 1970.
103. **Churchill, M. E., Arditti, J., and Ball, E. A.**, Clonal propagation of orchids from leaf tips, *Am. Orchid Soc. Bull.,* 40, 109, 1971.
104. **Churchill, M. E., Ball, E. A., and Arditti, J.**, Tissue culture of orchids. II. Methods for root tips, *Am. Orchid Soc. Bull.,* 41, 726, 1972.
105. **Churchill, M. E., Ball, E. A., and Arditti, J.**, Tissue culture of orchids. I. Methods for leaf tips, *New Phytol.,* 72, 161, 1973.
106. **Stewart, J. and Button, J.**, Rapid vegetative multiplication of *Epidendrum* O'brienianum *in vitro* and the greenhouse, *Am. Orhid Soc. Bull.,* 45, 922, 1976.
107. **Stewart, J. and Button, J.**, Development of callus from *Epidendrum* root tips culture, *in vitro, Am. Orchid Soc. Bull.,* 47, 607, 1978.
108. **Teo, C. K. H.**, Clonal propagation of *Haemaria discolor* by tissue culture, *Am. Orchid Soc. Bull.,* 47, 1028, 1978.
109. **Knudson, L.**, Flower production by orchid grown non-symbiotically, *Bot. Gaz. (Chicago),* 89, 192, 1930.
110. **Arditti, J.**, Niacin biosynthesis in germinating *Laeliocattleya* orchid embryos and young seedlings, *Am. J. Bot.,* 54, 291, 1967.

111. **Champagnat, M.**, Recherches sur la multiplication vegetative de *Neottia nidus-avis, Rich. Ann. Sci. Natl. Bot.*, 12, 209, 1971.
112. **Champagnat, M. and Morel, G.**, La culture in vitro des tissues de tubercules d'Ophrys, *C. R. Acad. Sci.*, 274, 3379, 1972.
113. **Ernst, R., Arditti, J., and Healey, P. L.**, The nutrition of orchid seedlings, *Am. Orchid Soc. Bull.*, 39, 599, 1970.
114. **Ernst, R.**, The use of activated charcoal in asymbiotic seedling culture of *Paphiopedilum, Am. Orchid Soc. Bull.*, 43, 35, 1974.
115. **Bubeck, S. K.**, A study of *Paphiopedilum* meristem culture, Ph.D. dissertation, Rutgers University, New Brunswick, N. J., 1973.
116. **Stewart, J. and Button, J.**, Tissue culture studies in *Paphiopedilum, Am. Orchid Soc. Bull.*, 44, 591, 1975.
117. **Flamee, M.**, Influence of selected media and supplements on the germination and growth of *Paphiopedilum* seedlings, *Am. Orchid Soc. Bull.*, 47, 419, 1978.
118. **Ernst, R.**, Effect of select organic nutrient additives on growth *in vitro* of *Phalaenopsis* seedlings, *Am. Orchid Soc. Bull.*, 36, 694, 1967.
119. **Ernst, R.**, Effect of carbohydrate selection on the growth rate of freshly germinated *Phaelaenopsis* and *Dendrobium* seeds, *Am. Orchid Soc. Bull.*, 36, 1068, 1967.
120. **Sagawa, Y.**, Vegetative propagation of *Phalaenopsis* by stem cuttings, *Am. Orchid Soc. Bull.*, 30, 808, 1961.
121. **Ernst, R.**, Studies in asymbiotic culture of orchids, *Am. Orchid Soc. Bull.*, 44, 12, 1975.
122. **Intuwong, O., Kumisaki, J. T., and Sagawa, Y.**, Vegetative propagation of *Phalaenopsis* of flower stalk cuttings, *Hawaii Orchid J.*, 1, 13, 1972.
123. **Kotomori, S. and Murashige, T.**, Some aspects of aseptic propagation of orchids, *Am. Orchid Soc. Bull.*, 34, 484, 1965.
124. **Scully, R. M.**, Stem propagation of *Phalaenopsis, Bull. Pacific Orchid Soc.*, 23, 13, 1965.
125. **Scully, R. M.**, Stem propagation of *Phalaenopsis, Am. Orchid Soc. Bull.*, 35, 40, 1966.
126. **Tse, A. T., Smith, R. J., and Hackett, W. P.**, Adventitious shoot formation on *Phalaenopsis* nodes, *Am. Orchid Soc. Bull.*, 40, 807, 1971.
127. **Rotor, G.**, A method of vegetative propagation of *Phalaenopsis* species and hybrids, *Am. Orchid Soc. Bull.*, 18, 738, 1949.
128. **Arditti, J., Healey, P. L., and Ernst, R.**, The role of mycorrhiza in nutrient uptake of orchids. II. Extracellular hydrolysis of oligosaccharides by asymbiotic seedlings, *Am. Orchid Soc. Bull.*, 41, 503, 1972.
129. **Intuwong, O. and Sagawa, Y.**, Clonal propagation of *Phalaenopsis* by shoot tip culture, *Am. Orchid Soc. Bull.*, 43, 893, 1974.
130. **Tanaka, M., Hasegawa, A., and Goi, M.**, Studies on the clonal propagation of monopodial orchids by tissue culture. I. Formation of protocorm-like bodies from leaf tissue in *Phalaenopsis* and *Vanda, J. Jpn. Soc. Hortic. Sci.*, 44, 47, 1975.
131. **Tanaka, M. and Sakanishi, Y.**, Factors affecting the growth of *in vitro* cultured lateral buds from *Phalaenopsis* flower stalks, *Sci. Hortic.*, 8, 169, 1978.
132. **Flamee, M. and Boesman, G.**, Clonal multiplication of *Phalaenopsis*-hybrids by means of sections of the flower stalk, *Med Ed. Fac. Landbouww et. Rijksuniv. Gent.*, 42, 1865, 1977.
133. **Tin-Yau Tse, A., Smith, R. J., and Hackett, W. P.**, Adventitious shoot formation on *Phalaenopsis* nodes, *Am. Orchid Soc. Bull.*, 40, 807, 1971.
134. **Zimmer, K. and Pieper, W.**, Zur vegetativen vermehrung von *Phalaenopsis in vitro, Die Orchidee*, 28, 118, 1977.
135. **Reisinger, D. M., Ball, E. A., and Arditti, J.**, Clonal propagation of *Phalaenopsis* by means of flower stalk node cultures, *Orchid Rev.*, 84, 45, 1976.
136. **Arditti, J., Ball, E. A., and Reisinger, D. M.**, Culture of flower-stalk buds: A method for vegetative propagation of *Phalaenopsis, Am. Orchid Soc. Bull.*, 46, 236, 1977.
137. **Vajrabhaya, M. and Vajrabhaya, T.**, Tissue culture of *Rhynchostylis gigantea,* a monopoidal orchid, *Am. Orchid Soc. Bull.*, 39, 907, 1970.
138. **Teo, C. K. H., Kunisaki, J. T., and Sagawa, Y.**, Clonal Propagation of strap-leaf *Vanda* by shoot tip culture, *Am. Orchid. Soc. Bull.*, 42, 402, 1973.
139. **Beechey, N.**, *Spathoglottis plicata,* a Malayan orchid, *Am. Orchid Soc. Bull.*, 39, 900, 1970.
140. **Beechey, N.**, Propagation of orchids from aerial roots, *Am. Orchid Soc. Bull.*, 39, 1085, 1970.
141. **Goh, C. J.**, Some effects of auxin on orchid seed germination, *Malays. Orchid Rev.*, 9, 115, 1970.
142. **Goh, C. J.**, Tissue culture of *Vanda* Miss Joaquim, *J. Singapore Natl. Acad. Sci.*, 2, 31, 1970.
143. **Goh, C. J.**, The influence of pH on orchid culture, *Malays. Orchid Rev.*, 10, 32, 1971.
144. **Ong, P. Y.**, Physiological studies on pod development and seed germination of *Vanda* Miss Joaquim, Honors Research Exercise, University of Singapore, Malaysia, 1969.

145. **Rao, A. N.,** Organogenesis in callus cultures of orchid seeds, in *Plant Tissue and Organ Culture, Symposium,* Maheshwari, P. and Rangaswamy, N. S., Eds., International Society of Plant Morphologists, Delhi, India, 1963, 332.
146. **Rao, A. N.,** Occurrence of polyembryony in *Vanda* during *in vivo* and *in vitro* conditions, *Experientia,* 20, 388, 1964.
147. **Rao, A. N., and Avadhani, P. N.,** Effect of Chlorox on the germination of *Vanda* seeds, *Curr. Sci.,* 32, 467, 1963.
148. **Rao, A. N. and Avadhani, P. N.,** Some aspects of in vitro culture of *Vanda* seeds, in Proc. 4th World Orchid Conf., Singapore, 1964, 194.
149. **Sagawa, Y. and Sehgal, O. P.,** Aseptic stem propagation of *Vanda* Miss Joaquim, *Pacific Orchid Soc. Bull.,* 25, 17, 1967.
150. **Kunisaki, J. T., Kim, K., and Sagawa, Y.,** Shoot tip culture of *Vanda, Am. Orchid Soc. Bull.,* 41, 435, 1972.
151. **Sanguthai, S. and Sagawa, Y.,** Induction of polyploidy in *Vanda* by colchicine treatment, *Hawaii Orchid J.,* 2, 17, 1973.
152. **Withner, C. L.,** Ovule culture and growth of *Vanilla* seedling, *Am. Orchid Soc. Bull.,* 24, 380, 1955.
153. **Arekal, G. D. and Karanth, K. A.,** *In vitro* seed germination of *Zeuxine strateumatica* Schlr. (= *Z. sulcata* Lindley), Orchidaceae, *Curr. Sci.,* 47, 552, 1978.
154. **Murashige, T.,** Manipulation of organ initiation in plant tissue cultures, *Bot. Bull. Acad. Sin.,* 18, 1, 1977.
155. **Holdgate, D. P.,** Propagation of ornamentals by tissue culture, in *Applied and Fundamental Aspects of Plant Cell, Tissue and Organ Culture,* Reinert, J. and Bajaj, Y. P. S., Eds., Springer-Verlag, New York, 1977, 18.
156. **Crabbe, J. A., Jermy, A. C., and Mickel, J. M.,** A new arrangement for the pteridophyte herbarium, *Fern Gaz.,* 11, 141, 1975.
157. **Knauss, J.,** A partial tissue culture method for pathogen-free propagation of selected ferns from spores, *Proc. Fla. State Hortic. Soc.,* 89, 363, 1976.
158. **Harper, K. L.,** Asexual multiplication of *Leptosporangiate* ferns through tissue culture, M.S. thesis, University of California, Riverside, 1976.
159. **Cooke, R. C.,** Homogenization as an aide in tissue culture propagation of *Platycerium* and *Davalia, HortScience,* 14, 21, 1979.
160. **Sulklyan, D. S. and Mehra, P. N.,** *In vitro* morphogenetic studies in *Nephrolepis cordifolia, Phytomorphology,* 27, 396, 1977.
161. **Burr, R. W.,** Mass propagation of Boston fern through tissue culture, *Proc. Int. Plant Prop. Soc.,* 25, 122, 1975.
162. **Burr, R. W.,** Mass propagation of ferns through tissue culture, *In Vitro,* 12, 309, 1976.
163. **Beck, M. J.,** The effects of kinetin and naphthalene acetic acid ratios on *in vitro* shoot multiplication and rooting in the fish tail fern, M. S. thesis, The University of Tennessee, Knoxville, 1980.
164. **Hennen, G. R. and Sheehan, T. J.,** In vitro propagation of *Platycerium stemaria* (Beauvois) Desv., *HortScience,* 13, 245, 1978.
165. **Bristow, J. M.,** The controlled *in vitro* differentiation of callus derived from a fern *Pteris cretica* L. into gametophytic and sporophytic tissues, *Dev. Biol.,* 4, 361, 1962.
166. **Kshirsagar, M. K. and Mehta, A. R.,** *In vitro* studies in ferns: growth and differentiation in rhizome callus of *Peteris vittata, Phytomorphology,* 28, 50, 1979.
167. **Mehra, R. N. and Plata, H. K.,** *In vitro* controlled differentiation of the root callus of *Cyclosorus dentatus, Phytomorphology,* 21, 367, 1971.
168. **Groenwald, E. G., Wessels, D. C. J., and Koeleman, A.,** Callus formation and subsequent plant regeneration from seed tissue of an *Agave* species, *Z. Pflanzenphysiol.,* 81, 369, 1977.
169. **Kunisaki, J. T.,** *In vitro* propagation of *Cordyline terminalis* L. Kunth, *HortScience,* 10, 601, 1975.
170. **Miller, L. R. and Murashige, T.,** Tissue culture propagation of tropical foliage plants, *In Vitro,* 12, 797, 1976.
171. **Kunisaki, J. T.,** Tissue culture propagation of tropical ornamental plants, *HortScience,* 12, 17, 1977.
172. **Mee, G. W. P.,** Propagation of *Cardyline terminalis* from callus culture, *HortScience,* 13, 660, 1978.
173. **Hussey, G.,** 1978, The application of tissue culture to the vegetative propagation of plants, *Sci. Prog. (London),* 65, 185, 1978.
174. **Deberg, P.,** Intensified vegetative multiplication of *Dracaena deremensis, Acta Hortic.,* 54, 83, 1975.
175. **Lakshmanan, K. K. and Janardhanan, K.,** Morphogenesis in leaf culture of *Furcraea gigantea, Phytomorphology,* 27, 85, 1977.
176. **Berry-Lowe, S. and Pertuit, A.,** Tissue culture of *Sansevieria, HortScience,* 12, 239, 1977.
177. **Litz, R. E. and Conover, R. A.,** Tissue culture propagation of some foliage plants, *Proc. Fla. State Hortic. Soc.* 90, 301, 1977.
178. **Mii, M., Mori, T., and Iwase, N.,** Organ formation from the excised bulb scales of *Hippeastrum hybridum in vitro, J. Hortic. Sci.,* 49, 241, 1974.

179. **Graves, R. H., Reid, R. H., and Mathes, M. C.**, Tissue culture of the monocots *Lilium, Hemerocallis,* and *Hippeastrum,* in Propagation of Higher Plants through tissue Culture: A Bridge Between Research and Application, Conf. No. 7804111, Hughes, K. W., Henke, R., and Constantin, M., Eds., National Technical Information Service, U.S. Department of Commerce, Springfield, Va., 1978, 248.
180. **Pajerski, L. and Ascher, P. D.**, Propagation of *Amaryllis (Hippaestrum)* from cultured scape tissue, *HortScience,* 12, 393, 1977.
181. **Bapat, V. A. and Narayanaswamy, S.**, Growth and organogenesis in explanted tissues of *Amaryllis* in culture, *Bull. Torrey Bot. Club,* 103, 53, 1976.
182. **Koul, A. K. and Karihaloo, J. L.**, *In vitro* embryoids from anthers of *Narcissus biflorus* Cur., *Euphytica,* 26, 97, 1977.
183. **McChesney, J. M.**, *In vitro* culture of the monocot *Narcissus triandrus* L. cv. Thalia, *Trans. Kansas Acad. Sci.,* 74, 44, 1971.
184. **Seabrook, J. E. A., Cumming, B. G., and Dionne, L. A.** The *in vitro* induction of adventitious shoot and root apices on *Narcissus* (daffodil and narcissus) cultivar tissue, *Can. J. Bot.,* 54, 814, 1976.
185. **Seabrook, J. E. A. and Cumming, B. G.**, Propagation of *Narcissus* (daffodil) through tissue cuture, *In Vitro,* 14, 356, 1978.
186. **Pierik, R. L. M., Steegmans, H. H. M., and Van Der Meys, J. A. J.**, Plantlet formation in callus tissue of *Anthurium andraenum* Lind., *Sci. Hortic.,* 2, 193, 1974.
187. **Fersing, G. and Lutz, A.**, 1977, Etude comparative de la multiplication vegetative *in vitro* de doux especes horticoles d'*Anthurium: A. andreanum* et *A. scherzerianum, C. R. Acad. Sci.,* 284, 2231, 1977.
188. **Pierik, R. L. M. and Steegmans, H. H. M.**, Vegetative propagation of *Anthurium scherzerianum* shoot through callus cultures, *Sci. Hortic.,* 4, 291, 1976.
189. **Knauss, J. F.**, A tissue culture method for producing *Dieffenbachia picta* cv. "perfection" free of fungi and bacteria, *Proc. Fla. State Hortic. Soc.,* 89, 293, 1976.
190. **Strode, R. E. and Oglesby, R. P.**, Parameters in commercial propagation of *Spathiphyllum clevelandii* through tissue culture, *In Vitro,* 14, 356, 1978.
191. **Fonnesbech, M. and Fonnesbech, A.**, In vitro propagation of *Spathiphyllum, Sci. Hortic.,* 10, 2, 1979.
192. **Makino, R. K. and Makino, P. J.**, Propagation of *Syngonium podophyllum* cultivars through tissue culture, *In Vitro,* 14, 357, 1978.
193. **Zimmer, K. and Pieper, W.**, Methods and problems of clonal propagation of bromeliads, *in vitro, Acta Hortic.,* 64, 25, 1976.
194. **Jones, J. B. and Murashige, T.**, Tissue culture propagation of *Aechmea fasciata* Baker and other bromeliads, *Proc. Int. Plant Prop. Soc.,* 24, 117, 1974.
195. **Davidson, S. E. and Donnan, A.**, *In vitro* propagation of *Cryptanthus* spp., *Proc. Fla. State Hortic. Soc.,* 90, 303, 1977.
196. **Bajaj, Y. P. S. and Pierik, R. L. M.**, Vegetative propagation of *Fressia* through callus cultures, *Neth. J. Agric. Sci.,* 22, 153, 1974.
197. **Hussey, G.**, Totipotency in tissue explants of some members of the Liliaceae, Iridaceae, and Amaryllidaceae, *J. Exp. Bot.,* 26, 253, 1975.
198. **Hussey, G.**, *In vitro* propagation of some monocotyledons of horticultural importance, in *Abstr., 4th Int. Congr. Plant Tissue and Cell Culture,* Thorpe, T. A., Ed., University of Calgary, Canada, 1978, 156.
199. **Simonsen, J. and Hildebrandt, A. C.**, *In vitro* growth and differentiation of *Gladiolus* plants from callus cultures, *Can. J. Bot.,* 49, 1817, 1971.
200. **Hussey, G.**, *In vitro* propagation of *Gladiolus* by precocious axillary shoot formation, *Sci. Hortic.,* 6, 287, 1977.
201. **Ziv, M., Halevy, A. H., and Shilo, R.**, Organs and Plantlets regeneration of *Gladiolus* through tissue culture, *Ann. Bot. (London),* 34, 671, 1970.
202. **Hussey, G.**, Propagation of Dutch iris by tissue culture, *Sci. Hortic.,* 4, 163, 1976.
203. **Stoltz, L. P.**, Growth regulator effects on growth and development of excise mature iris embryos, *in vitro, HortScience,* 12, 495, 1977.
204. **Meyer, M. M., Jr., Fuchigami, L. H., and Roberts, A. N.**, Propagation of tall bearded irises by tissue culture, *HortScience,* 10, 479, 1975.
205. **Weiler, J. and Emershad, R.**, Tissue culture for iris hybridizers, *Bull. Am. Iris Soc.,* 58, 36, 1977.
206. **Reuther, G.**, Embryoid differentiation patterns in callus of the genus *Iris* and *Asparagus,* in *Abstr., 4th Int. Congr. Plant Tissue and Cell Culture,* Thorpe, T. A., Ed., University of Calgary, Canada, 1978, 43.
207. **Hosoki, T.**, Propagation of tropical plants by tissue culture, Ph.D. dissertation, University of Hawaii, Honolulu, 1975.

208. **Groenewald, E. G., Koeleman, A., and Wessels, D. C. J.,** Callus formation and plant regeneration from seed tissue of *Aloe pretoriensis* pole Evans, *Z. Pflanzenphysiol.*, 75, 270, 1974.
209. **Kaul, K. and Sabharwal, P. S.,** Morphogenetic studies on *Haworthia:* Establishment of tissue culture and control of differentiation, *Am. J. Bot.*, 59, 377, 1972.
210. **Pandy, K. N., Sabharwal, P. S., and Calkins, J.,** Effects of ionizing radiation (^{60}Co gamma rays) on growth and morphogenesis of *Haworthia mirabilis* Haw. callus tissues, *In Vitro*, 15, 246, 1979.
211. **Majumdar, S. K.,** Production of plantlets from the ovary wall of *Haworthia turgida* var. *pallidifolia, Planta*, 90, 212, 1970.
212. **Majumdar, S. K. and Sabharwal, P. S.,** Induction of vegetative buds on inflorescence of *Haworthia in vitro, Am. J. Bot.*, 55, 705, 1968.
213. **Kato, Y.,** Induction of adventitious buds on undetached leaves, excised leaves and leaf fragments of *Heloniopsis orientalis, Physiol. Plant.*, 42, 39, 1978.
214. **Kato, Y.,** The involvement of photosynthesis in inducing bud formation on excised leaf segments of *Heloniopsis orientalis* (Liliaceae), *Plant Cell Physiol.*, 19, 791, 1978.
215. **Heuser, C. W.,** Tissue culture propagation of *Hemerocallis, HortScience*, 11, 321, 1976.
216. **Heuser, C. W. and Apps, D. A.,** *In vitro* plantlet formation from flower petal explants of *Hemerocallis* cv. Chipper Cherry, *Can. J. Bot.*, 54, 616, 1976.
217. **Meyer, M. M.,** Propagation of day lilies culture, *HortScience*, 11, 485, 1976.
218. **Oglesby, R. P.,** Tissue cultures of ornamentals and flowers: problems and perspectives, in Propagation of Higher Plants Through Tissue Culture: A Bridge Between Research and Application, Conf. No. 7804111, Hughes, K., Henke, R., and Constantin, M., Eds., National Technical Information Service, U.S. Department of Commerce, Springfield, Va., 1978, 59.
219. **Hammer, P. A.,** Tissue culture propagation of *Hosta decorata* Bailey, *HortScience*, 11, 309, 1976.
220. **Meyer, M. M.,** Propagation of *Hosta* by *in vitro* techniques, *HortScience*, 11, 309, 1976.
221. **Pierik, R. L. M. and Ruibing, M. A.,** Regeneration of bublets on bulb scale segments of hyacinth *in vitro, Neth. J. Agric. Sci.*, 21, 129, 1973.
222. **Pierik, R. L. M. and Post, A. L. M.,** Rapid vegetative propagation of *Hyacinthus orientalis* L., *in vitro, Sci. Hortic.*, 3, 293, 1975.
223. **Takayama, S. and Misawa, M.,** Differentiation in *Lilium* bulbscales grown *in vitro* — effect of various cultural conditions, *Physiol. Plant.*, 46, 184, 1979.
224. **Coquen, P. C. and Astie, M.,** Ontogenese de bulbilles vegerarives obtenued par culture de pieces florales isolees chez le *Lilium candidum* L., *Bull. Bot. Soc. Fr.*, 124, 51, 1977.
225. **Kato, Y. and Yasutake, Y.,** Plantlet formation and differentiation of epidermal tissues in green callus cultures from excised leaves of *Lilium, Phytomorphology*, 27, 390, 1977.
226. **Sheridan, W. F.,** Tissue culture of the monocot Lilium, *Planta*, 82, 189, 1968.
227. **Hackett, W. P.,** Control of bulblet formation on bud scales of *Lilium longiflorum, HortScience*, 4, 171, 1969.
228. **Gupta, P. P., Sharma, A. K., and Chaturvedi, H. C.,** Multiplication of *Lillium longiflorum* Thunb. by aseptic culture of bulb scales and their segments, *Indian J. Exp. Biol.*, 16, 940, 1978.
229. **Stimart, D. P. and Ascher, P. D.,** Tissue culture of bulb scale sections for asexual propagation of *Lilium longiflorum* Thunb., *J. Am. Soc. Hortic. Sci.*, 103, 182, 1978.
230. **Hackett, W. P.,** Aseptic multiplication of lily bulblets from bud scales, *Proc. Int. Plant Prop. Soc.*, 105, 105, 1969.
231. **Hackett, W. P.,** Control of bulblet formation on bud scales of *Lilium longiflorum, HortScience*, 4, 171, 1969.
232. **Bennici, A.,** Cytological chimera in plants regenerated from *Lilium longiflorum* tissues grown, *in vitro, Z. Pflanzenzuecht.*, 82, 349, 1979.
233. **Robb, S. M.,** The culture of excised tissue from bulb scales of *Lilium speciosum* Thunb., *J. Exp. Bot.*, 8, 348, 1957.
234. **Simmonds, J. A. and Cummings, B. G.,** Propagation of *Lilium* hybrids. II. Production of plantlets from bulb-scale callus cultures for increased propagation rates, *Sci. Hortic.*, 5, 161, 1976.
235. **Asano, Y.,** Studies on crosses between distantly related species of lilies. III. New hybrids obtained through embryo culture, *J. Jpn. Soc. Hortic. Sci.*, 47, 401, 1978.
236. **Anderson, W. C.,** Rapid propagation of *Lilium* cv. Red Carpet, *In Vitro*, 13, 145, 1977.
237. **Hussey, G.,** Plantlet regeneration from callus and parent tissue in *Ornithogalum thyrosoides, J. Exp. Bot.*, 27, 375, 1976.
238. **Banks, M. S.,** Phase change in *Hedera helix*, in *Abstr., 4th Int. Congr. Plant Tissue and Cell Culture*, Thorpe, T. A., Ed., University of Calgary, Canada, 1978, 45.
239. **Banks, M. S. and Hackett, W. P.,** Differentiation from *Hedera helix, Plant Physiol. Suppl.*, 61, 245, 1978.
240. **Banks, M. S., Christensen, M. R., and Hackett, W. P.,** Callus and shoot formation in organ and tissue cultures of *Hedera helix* L., English Ivy, *Planta*, 145, 205, 1979.

241. **Grewal, S. and Sharma, K.**, Clonal multiplication of medicinal plants by tissue growth, 4, Pyrethrum plant *(Chrysanthemum cinerariaefolium* Vis) regeneration from shoot tip culture, *Indian J. Exp. Biol.*, 16, 1119, 1978.
242. **Roest, S. and Bokelmann, G. S.**, Vegetative propagation of *Chrysanthemum cinerariaefolium, Sci. Hortic.*, 1, 120, 1973.
243. **Ben-Jaacov, J. and Langhans, R. W.**, Rapid multiplication of *Chrysanthemum* plants by stem tip proliferation, *HortScience*, 7, 289, 1972.
244. **Earle, E. D.**, Propagation of *Chrysanthemum* in vitro. II. Production, growth and flowering of plantlets from tissue culture, *J. Am. Soc. Hortic. Sci.*, 99, 352, 1974.
245. **Custers, J. B. M. and Franken, J.**, Some observations on adventitious bud formation of *Chrysanthemum morifolium* Ram. cv. "Super White" *in vitro, Abstr., 4th Int. Congr. Plant Tissue and Cell Culture,* Thorpe, T. A., Ed., University of Calgary, Canada, 1978, 32.
246. **Wang, S.-O. and Ma, S.-S.**, Clonal multiplication of *Chrysanthemum in vitro, J. Agric. Assoc. China*, 101, 64, 1978.
247. **Hill, G. P.**, Shoot formation in tissue cultures of *Chrysanthemum* "Bronze Pride", *Physiol. Plant.*, 21, 386, 1968.
248. **Evert, D. R. and Holt, M. A.**, Aseptic culture of chrysanthemums in the plant propagation class, *Proc. Int. Plant Prop. Soc.*, 24, 444, 1974.
249. **Watanabe, K.**, Successful ovary culture and production of F_1 hybrids and androgenic haploids in Japanese chrysanthemum species, *J. Hered.*, 68, 317, 1977.
250. **Mullin, R. H. and Schlegel, D. E.**, Meristem-tip culture of *Dahlia* infected with dahlia mosaic virus, *Plant Dis. Rep.*, 62, 565, 1978.
251. **Pierik, R. L. M., Steegmans, H. H. M., and Marelis, J. J.**, *Gerbera* plantlets from *in vitro* cultivated capitulum explants, *Sci. Hortic.*, 1, 117, 1973.
252. **Murashige, T., Serpa, M., and Jones, J.**, Clonal multiplication of *Gerbera* through tissue culture, *HortScience*, 9, 175, 1974.
253. **Beauchesne, G.**, *In vitro Gerbera* propagation from a flower bud, in *Abstr., 4th Int. Congr. Plant Tissue and Cell Culture,* Thorpe, T. A., Ed., University of Calgary, Canada, 1978, 153.
254. **Cailloux, M.**, Simultaneous use of three cytokinins gives better yields with *Gynura sarmentosa,* in Propagation of Higher Plants Through Tissue Culture: A Bridge Between Research and Application, Conf. No. 7804111, Hughes, K., Henke, R., and Constantin, C. Eds., National Technical Information Service, U.S. Department of Commerce, Springfield, Va., 1978, 248.
255. **Koenigsberg, S. and Langhans, R.**, Tissue culture studies with the New Guinea-Java hybrid *Impatiens, HortScience,* 11, 321, 1976.
256. **Fonnesbech, M.**, The influence of NAA, BA and temperature on shoot and root development from *Begonia × cheimantha* petiole segments grown *in vitro, Physiol. Plant.*, 32, 49, 1974.
257. **Fonnesbech, M.**, Temperature effects on shoot and root development from *Begonia × cheimantha* petiole segments grown, *in vitro, Physiol. Plant.*, 32, 282, 1974.
258. **Ringe, F. and Nitsch, J. P.**, Conditions leading to flower formation on excised *Begonia* fragments cultured *in vitro, Plant Cell Physiol.*, 9, 639, 1968.
259. **Mikkelsen, E. P. and Sink, K. C,**. Tissue culture of Rieger Begonias, *HortScience,* 11, 321, 1976.
260. **Mikkelsen, E. P. and Sink, K. C.**, *In vitro* propagation of Rieger Elatior Begonias, *HortScience,* 13, 252, 1978.
261. **Applegren, M.**, Regeneration of *Begonia hiemalis in vitro, Acta Hortic.*, 64, 31, 1976.
262. **Welander, T.**, *In vitro* organogenesis in explants from different cultivars of *Begonia × Hiemalis, Physiol. Plant.*, 41, 142, 1977.
263. **Welander, T.**, *In vitro* propagation of clones from different cultivars of *Begonia × hiemalis, Swed. J. Agric. Res.*, 8, 181, 1978.
264. **Hilding, A. and Welander, T.**, Effect of some factors on propagation of *Begonia × hiemalis in vitro, Swed. J. Agric. Res.*, 6, 191, 1976.
265. **Shigematsu, K. and Matsubara, H.**, The isolation and propagation of the mutant plant from sectorial chimera induced by irradiation in *Begonia rex., J. Jpn. Soc. Hortic. Sci.*, 41, 196, 1972.
266. **Bigot, M. C.**, Comparaison des aptitudes pour le bourgeonnement de tissus superficiels et de tissus profonds cultives *in vitro,* Cas de la tige d'un lis hybride, cultivar Enchantment, *C. R. Acad. Sci. Ser. D,* 278, 1027, 1974.
267. **Matsuyama, J.**, Tissue culture propagation of *Nandina, In Vitro,* 14, 357, 1978.
268. **Khanna, P. and Staba, E. J.**, *In vitro* physiology and morphogenesis of *Cheiranthus cheiri* var. Clott of Gold and *C. cheiri* var. Goliath, *Bot. Gaz. (Chicago),* 131, 1, 1970.
269. **Danckwardt-Lilliestrom, C.**, Kinetin induced shoot formation from isolated roots of *Isatis tinctoria, Physiol. Plant.*, 10, 794, 1957.
270. **Khanna, R. and Chopra, R. N.**, Regulation of shoot-bud and root formation from explants of *Lobularia maritima, Phytomorphology,* 27, 267, 1977.

271. **Mauseth, J. D.,** A new method for the propagation of cacti: sterile culture of axillary buds, *Cact. Succulent J.,* 51, 186, 1979.
272. **Johnson, J. L. and Emino, E. R.,** Propagation of selected cactus species by tissue culture, *HortScience,* 12, 404, 1977.
273. **Johnson, J. L. and Emino, E. R.,** Tissue Culture propagation of *Mammilaria elongata* as influenced by plant growth regulators, *HortScience,* 12, 394, 1977.
274. **Johnson, J. L. and Emino, E. R.,** *In vitro* propagation of *Mammillaria elongata, HortScience,* 14, 605, 1979.
275. **Kolar, Z., Bartek, J., and Vyskot, B.,** Vegetative propagation of the cactus *Mamillaria woodsii* Craig through tissue cultures, *Experimentia,* 32, 668, 1976.
276. **Colomas, J. and Bulard, C.,** Comportement en culture *in vitro* des tissues de tige de *Myrtillocactus geometrizans* (Mart.) Cons. (T.) et biosynthese de betalaines, *Bull. Bot. Soc. Fr.,* 124, 385, 1977.
277. **Johnson, J. L. and Emino, E. R.,** Tissue culture propagation of *Opuntia polyacantha* as influenced by plant growth regulators, *HortScience,* 12, 239, 1977.
278. **Johnson, J. L. and Emino, E. R.,** Tissue culture propagation in the cactaceae, *Cact. Succulent J.,* 51, 275, 1979.
279. **Johnson, R. T., Koenigsberg, S. S., and Langhans, R. W.,** Tissue culture propagation of Christmas and Easter Cactus, *HortScience,* 11, 303, 1976.
280. **Hackett, W. P. and Anderson, J. M.,** Aseptic multiplication and maintenance of differentiated carnation shoots derived from shoot apices, *Proc. Am. Soc. Hortic. Sci.,* 90, 365, 1967.
281. **Earle, E. D. and Langhans, R. W.,** Carnation propagation from shoot tips cultured in liquid medium, *HortScience,* 10, 608, 1975.
282. **Petru, E. and Landa, Z.,** Organogenesis in isolated carnation plant callus tissue cultivated, *In vitro, Biol. Plant.,* 16, 450, 1974.
283. **Seibert, M.,** Shoot initiation from carnation shoot apices frozen to −196°C., *Science,* 191, 1178, 1976.
284. **Gukasyan, I. A., Butenko, R. G., Petoyan, S. A., and Sevost'yanova, T. A.,** Morphogenesis of isolated apices remontant carnation on an artificial medium, *Sov. Plant Physiol.,* 24, 130, 1977.
285. **Narayan, S. M.,** Morphogenesis in excised shoot apical meristem of *Dianthus caryophyllus* L., *Diss. Abstr. B,* 37, (6), 2645B, 1976.
286. **Kusey, W. E. and Hammer, P. A.,** Tissue culture propagation of *Gypsophila paniculata* "Bristol Fairy", *HortScience,* 12, 151, 1977.
287. **Bonnet, H. T. and Torrey, J. G.,** Chemical control of organ formation in root segments of *Convolvulus* cultured *in vitro, Plant Physiol.,* 40, 1228, 1965.
288. **Earle, E. D. and Torrey, J. G.,** Morphogenesis in cell colonies grown from *Convolvulus* cell suspensions plated on synthetic media, *Am. J. Bot.,* 52, 891, 1965.
289. **Hill, G. P.,** Morphogenesis in stem-callus cultures of *Convolvulus arvensis* L., *Ann. Bot. (London),* 31, 437, 1967.
290. **Bapat, V. A. and Rao, P. S.,** Shoot apical meristem culture of *Pharbitis nil, Plant Sci. Lett.,* 10, 327, 1977.
291. **Templeton, K. M. and Cline, M. G.,** Flowering of shoot apices of *Pharbitis nil* cultured *in vitro, In Vitro,* 13, 178, 1979.
292. **Raju, M. V. S. and Mann, E. H.,** Regenerative studies on the detached leaves of *Echeveria elegans,* Anatomy and regeneration of leaves in sterile culture, *Can. J. Bot.,* 48, 1887, 1970.
293. **Smith, R. H. and Nightingale, A. E.,** *In vitro* propagation of *Kalanchoe, HortScience,* 14, 20, 1979.
294. **Karp. R. and Sink, K. C.,** Induction of adventitious shoots on *Kalanchoe* explants, *in vitro, HortScience,* 11, 321, 1976.
295. **Brandao, I. and Salema, R.,** Callus and plantlets development from cultured leaf explants of *Sedum telephium* L., *Z. Pflanzenphysiol.,* 85, 1, 1977.
296. **Ma, S. S. and Wang, S. O.,** Clonal multiplication of azaleas through tissue culture, *Acta Hortic.,* 78, 209, 1977.
297. **Anderson, W. C.,** Tissue culture propagation of rhododendrons, *In Vitro,* 14, 334, 1978.
298. **Stoltz, L. P.,** In vitro propagation of *Acalphya wilkesiana, HortScien ce,* 14, 702, 1979.
299. **Langhe, E. D., Debergh, P., and Van Rijk, R.,** *In vitro* culture as a method for vegetative propagation of *Euphorbia pulcherrima, Z. Pflanzenphysiol.,* 71, 271, 1974.
300. **Nataraja, K., Chennaveerajah, M. S., and Girigowda, P.,** *In vitro* production of shoot buds in *Euphorbia pulcherrima, Curr. Sci.,* 42, 577, 1973.
301. **McIntyre, D. K. and Whitehorne, G. J.,** Tissue culture in the propagation of Austrialian plants, *Proc. Int. Plant Prop. Soc.,* 24, 262, 1974.
302. **Skirvin, R. M. and Janick, J.,** Tissue culture-induced variation in scented *Pelargonium* spp., *J. Am. Soc. Hortic. Sci.,* 101, 281, 1976.
303. **Chen, H. R. and Galston, A. W.,** Growth and development of *Pelargonium* pith cells *in vitro.* II. Initiation of organized development, *Physiol. Plant.,* 20, 533, 1967.

304. **Pillai, S. K. and Hildebrandt, A. C.,** *In vitro* differentiation of geranium *(Pelargonium hortorum,* Bailey) plants from apical meristems, *Phyton (Buenos Aires),* 25, 81, 1968.
305. **Pillai, S. K. and Hildebrandt, A. C.,** Geranium plants differentiated *in vitro* from stem-tip and callus cultures, *Plant Dis. Rep.,* 52, 600, 1968.
306. **Pillai, S. K. and Hildebrandt, A. C.,** Induced differentiation of geranium plants from undifferentiated callus, *in vitro, Am. J. Bot.,* 56, 52, 1969.
307. **Abo El-Nil, M. M. and Hildebrandt, A. C.,** Differentiation of virus-symptomless geranium plants from anther callus, *Plant Dis. Rep.,* 55, 1017, 1971.
308. **Abo El-Nil, M. M., Hildebrandt, A. C., and Evert, R. F.,** Effect of auxin cytokinin interaction on organogenesis in haploid callus of *Pelargonium hortorum, In Vitro,* 12, 602, 1976.
309. **Abo Il-Nil, M. M. and Hildebrandt, A. C.,** Origin of androgenetic callus and haploid geranium plants, *Can. J. Bot.,* 51, 2107, 1973.
310. **Reuther, G.,** Untersuchungen zur vermehrung von pelargonienvarietaten durch gewebekultur, *Gartenbauwissenschaft,* 4, 181, 1975.
311. **Theiler, R.,** Tissue culture of *Pelargonium* shoot tips (German), *Gartenbau,* 98, 887, 1977.
312. **Hakkaart, F. A. and Hartel, G.,** Virus eradication from some *Pelargonium zonale* cultivars by meristem-tip culture, *Neth. J. Plant Pathol.,* 85, 39, 1979.
313. **Allen, H. M. and Mott, R. L.,** Rapid propagation of *Chirita sinensis* (Gesneriaceae) through tissue culture, *HortScience,* 10, 150, 1975.
314. **Johnson, B. B.,** *In vitro* propagation of *Episcia cupreata, HortScience,* 13, 596, 1978.
315. **Bilkey, P. C. and McCown, B. H.,** In vitro culture and propagation of *Episcia* sp. (Flame violets), *J. Am. Soc. Hortic. Sci.,* 104, 109, 1979.
316. **Pearson, A. E.,** Effects of auxin and cytokinins on morphogenesis of three varieties of *Episcia,* M. S. thesis, The University of Tennessee, Knoxville, 1979.
317. **Pearson, A. E. and Hughes, K. W.,** Propagation of *Episcia* by tissue culture, in *Abstr. 4th Int. Congr. Plant Tissue and Cell Culture,* Thorpe, T. A., Ed., University of Calgary, Canada, 1978, 152.
318. **Hughes, K. W. and Barni, B.,** Regeneration of *Kohleria amabilis* (Gesneriaceae) from leaf and petiole explants, in Propagation of Higher Plants Through Tissue Culture, Conf. No. 7804111, Hughes K., Henke, R., and Constantin, M., Eds., National Technical Information Service, U. S. Department of Commerce, Springfield, Va., 1978, 248.
319. **Venverloo, C. J.,** Regulation of root and shoot formation on small *Nautilocalyx* explants, in *Abstr., 4th Int. Congr. Plant Tissue and Cell Culture,* Thorpe, T. A., Ed., University of Calgary, Canada, 1978, 28.
320. **Kakulczanka, K. and Suszynska, G.,** Regenerative properties of *Saintpaulia ionantha* Wendl. leaves cultured, *in vitro, Acta Soc. Bot. Pol.,* 41, 503, 1972.
321. **Hughes, K., Bell, S. L., and Caponetti, J. D.,** Anther derived haploids of the African violet, *Can. J. Bot.,* 53, 1442, 1975.
322. **Grunewaldt, J.,** Die *in vitro* regeneration aus blattstielquerschnitten von *Saintpaulia ionantha* H. Wendl., *Gartenbauwissenschaft,* 41, 145, 1976.
323. **Boccon-Gibod, J., Daux, P., and Pottier, J. C.,** Clonal propagation of *Saintpaulia ionantha* Wendl. *in vitro:* determination of an index propagation's speed, in *Abstr. 4th Int. Congr. Plant Tissue and Cell Culture,* Thorpe, T. A., Ed., University of Calgary, Canada, 1978, 151.
324. **Bilkey, P. C., McCown, B. H., and Hildebrandt, A. C.,** Micropropagation of African violet from petiole cross-sections, *HortScience,* 13, 37, 1978.
325. **Jungnickel, F.,** Regenerative properties of isolated petioles of *Saintpaulia ionantha* influenced by 6-benzylaminopurine or -napthalene acetic acid, *Biochem. Physiol. Pflanz.,* 170, 457, 1976.
326. **Start, N. D. and Cumming, B. G.,** *In vitro* propagation of *Saintpaulia ionantha* Wendl., *HortScience,* 11, 204, 1976.
327. **Cooke, R. C.,** Tissue culture propagation of African violets, *HortScience,* 12, 549, 1977.
328. **Flores, H. F., Fierro, C. A., and Koo, F. K. S.,** *In vitro* culture and radiation studies of *Saintpaulia ionantha* Wendl., in *Abstr. 4th Int. Congr. Plant Tissue and Cell Culture,* Thorpe, T. A., Ed., University of Calgary, Canada, 1978, 151.
329. **Vazquez, A. M. and Short, K. C.,** Morphogenesis in cultured floral parts of African violets, *J. Exp. Bot.,* 29, 1265, 1978.
330. **Haramaki, C.,** Tissue culture of gloxinia, *Proc. Int. Plant Prop. Soc.,* 21, 442, 1971.
331. **Johnson, B. B.,** In vitro propagation of *Gloxinia* from leaf explants, *HortScience,* 13, 149, 1978.
332. **Handro, W.,** Structural aspects of neo-formation of floral buds on leaf disks of *Streptocarpus nobilis* cultured *in vitro, Ann. Bot. (London),* 41, 303, 1977.
333. **Applegren, M. and Heide, O.,** Regeneration in *Streptocarpus* leaf disks and its regulation by temperature and growth substances, *Physiol. Plant.,* 27, 417, 1972.
334. **Raman, K.,** Rapid multiplication of *Streptocarpus* and *Gloxinia* from *in vitro* cultured pedicel segments, *Z. Pflanzenphysiol.,* 83, 411, 1977.

333. **Applegren, M. and Heide, O.,** Regeneration in *Streptocarpus* leaf disks and its regulation by temperature and growth substances, *Physiol. Plant.,* 27, 417, 1972.
334. **Raman, K.,** Rapid multiplication of *Streptocarpus* and *Gloxinia* from *in vitro* cultured pedicel segments, *Z. Pflanzenphysiol.,* 83, 411, 1977.
335. **Ellyard, R. K.,** In vitro propagation of *Anigozanthos manglesii, Anigozanthos flavidus* and *Macropidia fulginosa, HortScience,* 13, 662, 1978.
336. **Hervey, A. and Robbins, W. J.,** Development of plants from leaf disks of variegated *Coleus* and its relation to patterns of leaf chlorosis, *In Vitro,* 14, 294, 1978.
337. **Codaccioni, M. and Laisne, G.,** Influence de la composition du milieu de culture sur la morphologie des plants du *Mentha viridis* L. cultive *in vitro, C. R. Acad. Sci. Ser. D,* 286, 29, 1978.
338. **Adams, R. M.,** Koenigsberg, S. S., and Langhans, R. W., *In vitro* propagation of the butterwort *Pinguicula moranesis, HortScience,* 14, 701, 1979.
339. **Duron, M. and Moran, J. C.,** Amelioration de l'etat sanitaire de *Buddleia davidii* "Opera" par culture de meristemes, *Ann. Phytopathol.,* 10, 371, 1978.
340. **Deberg, P. C., DeWael, G., and Zielinkska, E.,** Mass propagation of some glass house ornamentals, in *Abstr. 4th Int. Congr. Plant Tissue and Cell Culture,* Thorpe, T. A., Ed., University of Calgary Canada, 1978, 153.
341. **Chaturvedi, H. C., Sharma, A. K., and Prasad, R. N.,** Shoot apex culture of *Bougainvillea glabra* "Magnifica", *HortScience,* 13, 36, 1978.
342. **Duron, M.,** Utilization of *in vitro* culture for improvement of the healthy state of *Forsythia* (Vahl.) cultivars, *C. R. Acad. Sci. Ser. D,* 284, 183, 1977.
343. **Kavathekar, A. K., Ganapathy, P. S., and Johri, B. M.,** *In vitro* responses of embryoids of *Eschscholzia californica, Biol. Plant.,* 20, 98, 1978.
344. **Kavathekar, A. K., Ganapathy, P. S., and Johri, B. M.,** Chilling induces development of embryoids into plantlets in *Eschscholzia, Z. Pflangenphysiol.,* 81, 358, 1977.
345. **Scorza, R. and Janick, J.,** Flowering from *Passiflora suberosa* leaf disks *in vitro,* in Propagation of Higher Plants Through Tissue Culture: A Bridge Between Research and Application, Conf. No. 7804111, Hughes, K., Henke, R., and Constantin, M., Eds., Technical Information Service, U. S. Department of Commerce, Springfield, Va., 1979, 249.
346. **Moran Robles, M. J.,** *In vitro* vegetative multiplication of axillary buds of *Passiflora edulis* var flavicarpa Degener and *P. mollissima* Bailey (French), *Fruits,* 33, 693, 1978.
347. **Berry, S.,** The effect of auxins, vitamins and light on *Peperomia caperata* tissue cultures, in Propagatio of Higher Plants Through Tissue Culture: A Bridge Between Research and Application, Conf. No. 7804111, Hughes, K., Henke, R., and Constantin, C., Eds., National Technical Information Service, U.S. Department of Commerce, Springfield, Va., 1978, 247.
348. **Henry, R. J.,** *In vitro* propagation of *Peperomia* "Red Ripple" from leaf disks, *HortScience,* 13, 150, 1978.
349. **Scaramella, P.,** Embryogeny in *Piper nugrum* and *Peperomia longifolia* explants. Connections wtih mono and dicotyledons, in *Abstr., 4th Int. Congr. Plant Tissue and Cell Culture,* Thorpe, T. A., Eds., University of Calgary, Canada, 1978, 43.
350. **Nitsch, C. and Nitsch, J. P.,** The induction of flowering *in vitro* in stem segments of *Plumbago indica* L. I. The production of vegetative buds, *Planta,* 72, 355, 1967.
351. **Konar, R. N. and Konar, A.,** Plantlet and flower formation in callus cultures from *Phlox drummondii, Phytomorphology,* 16, 379, 1966.
352. **Oleson, M. N. and Fonnesbech, M.,** Phlox plants from shoot tips, *Acta Hortic.,* 54, 95, 1975.
535. **Schnabelrauch, L. S. and Sink, K. C.,** *In vitro* propagation of *Phlox subulata* and *Phlox paniculata, HortScience,* 14, 607, 1979.
354. **Konar, R. N.,** *In vitro* studies on *Portulaca grandiflora* Hook., *Z. Pflanzenphysiol.,* 86, 443, 1978.
355. **Geier, T.,** *In vitro* growth and morphogenesis of explants taken from various organs of *Cyclamen persicum,* in *Abstr., 4th Int. Congr. Plant Tissue and Cell Culture,* Thorpe, T. A., Ed., University of Calgary, Canada, 1978, 31.
356. **Sutter, L. E. and Langhans, R. W.,** Micropropagation of *Anemone coronaria,* in Propagation of Higher Plants Through Tissue Culture: A Bridge Between Research and Application, Conf. No. 7804111, Hughes, K., Henke, R., and Constantin, M., Eds., National Technical Information Service, U.S. Department of Commerce, Springfield, Va., 1978, 249.
357. **Banergee, S. and Gupta, S.,** Embryogenesis and differentiation in *Nigella sativa* leaf callus *in vitro, Physiol. Plant.,* 38, 115, 1976.
358. **Gupta, S. and Banerjee, S.,** Source of nitrogen supporting growth and differentiation in *Nigella sativa,* in *Abstr. 4th Int. Congr. Plant Tissue and Cell Culture,* Thorpe, T. A., Ed., University of Calgary, Canada, 1978, 44.
359. **Hasegawa, P. M.,** *In vitro* propagation of Rose, *HortScience,* 14, 610, 1979.
360. **Skirvin, R. M. and Chu, M. C.,** Propagation of roses with tissue culture, *Ill. Res.,* 21, 3, 1979.

361. **Jacobs, G., Bornman, C. H., and Allen, P.**, Tissue culture studies on rose, use of pith explants, *S. Afr. J. Agric. Sci.*, 11, 673, 1968.

362. **Poirier-Hamon, S., Rao, P. S., and Harada, H.**, Culture of mesophyll protoplasts and stem segments of *Antirrhinum majus* (Snapdragon): growth and organization of embryoids, *J. Exp. Bot.*, 25, 752, 1974.

363. **Rao, P. S., Bapat, V. A., and Harada, H.**, Gamma radiation and hormonal factors controlling morphogenesis in organ cultures of *Antirrhinum majus* L. cv. Red Majestic., *Z. Pflanzenphysiol.*, 80, 144, 1976.

364. **Sangwan, R. S. and Harada, H.**, Chemical regulation of callus growth, organogenesis, plant regeneration, and somatic embryogenesis in *Antirrhinum majus* tissue and cell cultures, *J. Exp. Bot.*, 26, 868, 1975.

365. **Sangwan, R. S., Norreel, B., and Harada, H.**, Effects of kinetin and gibberellin A_3 on callus growth and orange formation in *Limnophila chinensis* tissue culture, *Biol. Plant*, 18, 126, 1976.

366. **Charlton, W. A.**, Bud initiation in excised roots of *Linaria vulgaris*, *Nature (London)*, 207, 781, 1965.

367. **Raste, A. P. and Ganapathy, P. S.**, In vitro behaviour of inflorescence segments of *Mazus pumilus*, *Phytomorphology*, 20, 367, 1971.

368. **Chlyah, H.**, Néoformation dirigée a partir de fragments d'organes de *Torenia fournieri* (Lind) cultivés, *in vitro*, *Biol. Plant.* 15, 80, 1973.

369. **Kamada, H. and Harada, H.**, Influence of several growth regulators and amino acids on *in vitro* organogenesis of *Torenia fournieri* Lind., *J. Exp. Bot.*, 30, 27, 1979.

370. **Caruso, J. L.**, Bud formation in excised stem segments of *Verbascum thapsus*, *Am. J. Bot.*, 58, 429, 1971.

371. **Handro, W., Rao, P. S., and Harada, H.**, Controlé hormonal de la bourgeons racines et embryons sur des explantats de feuilles et de tiges de *Pentunia* cultivés *in vitro*, *C. R. Acad. Sci. Ser. D*, 275, 2861, 1972.

372. **Rao, P. S., Handro, W., and Harada, H.**, Hormonal control and differentiation of shoots, roots and embryos in leaf and stem cultures of *Petunia inflata* and *Petunia hybrida*, *Physiol. Plant.*, 28, 458, 1973.

373. **Sangwan, R. W. and Norreel, B.**, Induction of plants from pollen grains of *Petunia* cultured *in vitro*, *Nature (London)*, 257, 222, 1975.

374. **Daykin, M., Langhans, R. W., and Earle, E. D.**, Tissue culture propagation of the double petunia, *HortScience*, 11, 35, 1976.

375. **Sangwan, R. S. and Harada, H.**, Chemical factors controlling morphogenesis of petunia cells cultured *in vitro*, *Biochem. Physiol. Pflanz.*, 170, 77, 1976.

376. **Sharma, A. K. and G. C. Mitra**, *In vitro* culture of shoot apical meristem of *Petunia hybrida* for mass production of plants, *Indian J. Exp. Biol.*, 14, 348, 1976.

377. **Murakishi, H. H. and Carlson, P. S.**, Regeneration of virus-free petunia, tobacco and brassica plants from green islands of mosaic-infected leaves, *In vitro*, 15, 192, 1979.

378. **Hayward, C. and Powder, J. B.**, Plant production from leaf protoplasts of *Petunia parodii*, *Plant Sci. Lett.*, 4, 407, 1975.

379. **Sink, K. C. and Power, J. B.**, The isolation, culture and regeneration of leaf protoplasts of *Petunia parviflora* Juss., *Plant Sci. Lett.*, 10, 335, 1977.

380. **Zenkteler, M.**, *In vitro* formation of plants from leaves of several species of the Solanaceae family, *Biochem. Physiol. Pflanz.*, 163, 509, 1972.

381. **Hughes, H., Lam, S., and Janick, J.**, *In vitro* culture of *Salpiglossis sinuata* L., *HortScience*, 8, 335, 1973.

382. **Lee, C. W., Skirvin, R. M., and Janick, J.**, Propagation of *Salipiglossis sinuata* L. from leaf discs, in tissue culture, *HortScience*, 11, 321, 1976.

383. **Lee, C. W., Skirvin, R. M., Solterno, A. I., and Janick, J.**, Tissue culture of *Sapiglossis sinuata* L. from leaf discs, *HortScience*, 12, 547, 1977.

384. **Nehis, R.**, Isolation and regeneration of protoplasts from *Solanum nigrum* L., *Plant Sci. Lett.*, 12, 183, 1978.

385. **Hughes, K.**, unpublished data, 1976.

386. **Knox, C. A. P. and Smith, R. H.**, A scanning electron microscopic study of healthy pecan tissues showing the presence or absence of internal fungal contamination, *In Vitro*, 16, 651, 1980.

387. **Hartman, R. D. and Zettler, F. W.**, Mericloning as a potential means for obtaining virus-free plants from aroids commercially produced in Florida, in *Proc. 4th Organic Soil Vegetable Crop Workshop*, Agricultural Research Center, Belle Glade, Fla., 1974, 60.

388. **Forsberg, J. L.**, An unexpected effect of benomyl on two gladiolus varieties, *Plant Dis. Rep.*, 53, 318, 1969.

389. **Gupta, S. D. and Hadley, G.**, Phytotoxicity of benomyl on orchid seedlings, *Am. Orchid Soc. Bull.*, 46, 905, 1977.

390. **Kohlenbach, H. W.**, Basic aspects of differentiation and plant regeneration from cell and tissue cultures, in *Plant Tissue Culture and Its Bio-Techniological Application: Proceedings,* Barz, W., Reinhard, E., and Zenk, M. H., Eds., Springer-Verlag, New York, 1977, 355.
391. **Raghavan, V., Ed.**, *Experimental Embryogenesis in Vascular Plants,* Academic Press, New York, 1976.
392. **Narayanaswamy, S.**, Regeneration of plants from tissue cultures, in *Applied and Fundamental Aspects of Plant Cell, Tissue and Organ culture,* Reinert, J. and Bajaj, Y. P. S., Eds., Springer-Verlag, New York, 1977, 179.
393. **Grout, B. and Aston, M.**, Transplanting of cauliflower plants regenerated from meristem culture. I. Water loss and water transfer related to changes in leaf wax and to xylem regeneration, *Hortic. Res.,* 17, 1, 1977.
394. **Sutter, E. and Langhans, R. W.**, Epicuticular wax formation on carnation plantlets regenerated from shoot tip cultures, *J. Am. Soc. Hortic. Sci.,* 104, 493, 1979.
395. **Hasegawa, P. W., Murashige, T., and Takatori, F. H.**, Propagation of *Asparagus* through shoot apex culture. II. Light and temperature requirements, transplantability of plants and cytological characteristics, *J. Am. Soc. Hortic. Sci.,* 98, 143, 1973.
396. **White, P. R., Ed.**, *The Cultivation of Animal and Plant Cells,* 2nd ed., Ronald Press, New York, 1963.
397. **Gamborg, O. L., Miller, R. A., and Ojima, K.**, Nutrient requirements of suspension cultures of soybean root cells, *Exp. Cell Res.,* 50, 151, 1968.
398. **Huang, Li-Chun and Murashige, T.**, Plant tissue culture media: major constituents; their preparation and some applications, *Tissue Culture Assoc. Man.,* 3, 539, 1977.
399. **Gamborg, O. L., Murashige, T., Thorpe, T. A., and Vasil, I. K.**, Plant tissue culture media, *In Vitro,* 12, 473, 1976.
400. **Nitsch, J. P.**, Experimental androgenesis in *Nicotiana, Phytomorphology,* 19, 389, 1969.
401. **Stoltz, L. P.**, Iron nutrition of *Cattleya* orchid grown *in vitro, J. Am. Soc. Hortic. Sci.,* 104, 308, 1979.
402. **Steiner, A. A. and van Winden, H.**, Recipe for ferric salts of ethylene diaminetetraacetic acid, *Plant Physiol.,* 46, 862, 1970.
403. **Tran Thanh Van, K.**, Regulation of morphogenesis, in *Plant Tissue Culture and Its Bio-Technological Application: Proceedings,* Barz, W., Reinhard, E., and Zenk, M. H., Eds., Springer-Verlag, New York, 1977, 367.
404. **Thorpe, T. A.**, Physiological and biochemical aspects of organogenesis in vitro, in *Frontiers of Plant Tissue Culture; Proc. 4th Int. Congr. Plant Tissue and Cell Culture; Proc. 4th Int. Congr. Plant Tissue and Cell Culture,* Thorpe, T. A., Ed., University of Calgary, Canada, 1978, 49.
405. **Barg, R. and Umiel, N.**, Effects of sugar concentration of growth greening and shoot formation in callus cultures of four genetic lines of tobacco, *Z. Pflanzenphysiol.,* 81, 161, 1977.
406. **Linsmaier, E. M. and Skoog, F.**, Organic growth factor requirements of tobacco tissue cultures, *Physiol. Plant.,* 18, 100, 1965.
407. **Thimann, K.**, The synthetic auxins: relation between structure and activity, in *Plant Growth Substances,* Skoog, F., Ed., University of Wisconsin Press, Madison, 1951, 21.
408. **Lam, T. H. and Street, H. E.**, The effect of selected aryloxyalkane-carboxylic acids on the growth and levels of soluable phenols in cultured cells of *Rosa damescens, Z. Pflanzenphysiol.,* 84, 121, 1977.
409. **Street, H. E.**, Embryogenesis and chemically induced organogenesis, in *Plant Cell and Tissue Culture: Principles and Applications,* Sharp, W. R., Larsen, P. O., Paddock, E. F., and Raghavan, V., Eds., Ohio State University Press, Columbus, 1979, 123.
410. **Staba, E. J., Laursen, P., and Buchner, S.**, Medicinal plant tissue cultures, in *Int. Conf. on Plant Tissue Culture,* White, P. R., and Grove, A. R., Eds., McCutchan, Berkeley, 1965, 191.
411. **Halperin, W.**, Alternative morphogenetic effects in cell suspension cultures, *Am. J. Bot.,* 53, 443, 1966.
412. **Engvild, K. C.**, Substituted indoleacetic acids tested in tissue cultures, *Physiol. Plant.,* 44, 345, 1978.
413. **Chernova, L. K., Prokhorov, M. N., and Filin-Koldakov, B. V.**, Comparison of the dedifferentiating effects of 2,4-D and 4-amin-3,5,6-Trichloropicolinic acid on tissues of legumes and cereals, *Sov. Plant Physiol.,* 22, 138, 1975.
414. **Feung, C., Hamilton, R. H., and Mumma, R. O.**, Metabolism of 2,4-D. Identification of metabolites in rice root callus tissue cultures, *J. Agric. Food Chem.,* 24, 1013, 1976.
415. **Feung, C., Hamilton, R. H., and Mumma, R. O.**, Metabolism of indole-3-acetic acid, *Plant Physiol.,* 59, 91, 1976.
416. **Verma, D. C. and Dougall, D. K.**, Biosynthetic routes for uronic acid and pentose units of cell wall in wild carrot *(Daucus carota)* suspensions, *In Vitro,* 14, 355, 1978.
417. **Halperin, W. and Jensen, W. A.**, Ultrastructural changes during growth and embryogenesis in carrot cell cultures, *J. Ultrastruct. Res.,* 18, 428, 1967.

418. **Wochok, Z. S.**, Microtubules and multivesicular bodies in cultured tissues of wild carrot: changes during transition from the undifferentiated to the embryonic condition, *Cytobios*, 7, 87, 1973.
419. **Mok, D. W. S.**, Cytokinin independence of tissue cultures of *Phaseolus:* a genetic trait, *In Vitro*, 15, 190, 1979.
420. **Meins, F. and Binns, A. N.**, Cell determination in plant development, *BioScience*, 29, 222, 1979.
421. **Skoog, F. and Tsui, C.**, Chemical control of growth and bud formation in tobacco stem segments and callus cultured *in vitro, Am. J. Bot.*, 35, 782, 1948.
422. **Pollard, J. K., Shantz, E. M., and Steward, F. C.**, Hexitols in coconut milk: their role in nurture of dividing cells, *Plant Physiol.*, 36, 492, 1961.
423. **Anderson, L. and Wolter, K. E.**, Cyclitols in plants: biochemistry and physiology, *Annu. Rev. Plant Physiol.*, 17, 209, 1966.
424. **Harran, S. and Dickinson, D. B.**, Metabolism of *myo*-inositol and growth in various sugars of suspension-cultured tobacco cells, *Planta* 141, 77, 1978.
425. **Wolter, K. E. and Murmanis, L.**, Radioautography of *myo*-inositol in cultured *Fraxinus callus, New Phytol.*, 78, 95, 1977.
426. **Wolter, K. E. and Skoog, F.**, Nutritional requirements of *Fraxinus* callus cultures, *Am. J. Bot.*, 53, 263, 1966.
427. **Kaul, K. and Sabharwal, P. S.**, Morphogenetic studies on *Haworthia:* effects of inositol on growth and differentiation, *Am. J. Bot.*, 62, 655, 1975.
428. **Van Overbeek, J., Conklin, M. E., and Blakeslee, A.**, Factors in coconut milk essential for growth and development of very young *Datura* embryos, *Science*, 94, 350, 1941.
429. **Jablonski, J. J. and Skoog, F.**, Cell enlargement and cell division in excised tobacco pith tissue, *Physiol. Plant.*, 28, 393, 1954.
430. **Shantz, E. M. and Steward, F. C.**, Coconut milk factor: the growth-promoting substances in coconut milk, *J. Am. Chem. Soc.*, 74, 6133, 1952.
431. **Letham, D. S.**, Regulators of cell division in plant tissues. XX. The cytokinins of cocnut milk, *Physiol. Plant.*, 32, 66, 1974.
432. **Van Staden, J. and Drewes, S. E.**, Identification of cell division inducing compounds from coconut milk, *Physiol. Plant.*, 32, 347, 1974.
433. **Klein, B. and Bopp, M.**, Effect of activated charcoal in agar on the culture of lower plants, *Nature (London)*, 230, 474, 1971.
434. **Anagnostakis, S. L.**, Haploid plants from anthers of tobacco-enhancement with charcoal, *Planta*, 115, 281, 1974.
435. **Ernst, R.**, The use of activated charcoal in a symbiotic seedling culture of *Paphiopedilum, Am. Orchid Soc. Bull.*, 43, 35, 1974.
436. **Wang, P.-J. and Huang, L.-C.**, Beneficial effects of activated charcoal on plant tissue and organ cultures, *In Vitro*, 12, 260, 1976.
437. **Fridborg, G. and Eriksson, T.**, Effects of activated charcoal on growth and morphogenesis in cell cultures, *Physiol. Plant*, 34, 306, 1975.
438. **Fridborg, G.**, Effects of activated charcoal on morphogenesis in plant tissue cultures, in *Abstr., 4th Int. Congr. Plant Tissue and Cell Culture*, University of Calgary, Canada, 1978, 41.
439. **Constantin, M. J., Henke, R. R., and Mansur, M. A.**, Effect of activated charcoal on callus growth and shoot organogenesis in tobacco, *In Vitro*, 13, 293, 1977.
440. **Proskauer, J. and Berman, R.**, Agar culture medium modified to approximate soil conditions, *Nature (London)*, 227, 1161, 1970.
441. **Reuveni, O. and Lilien-Kipnis, H.**, Studies of the *in vitro* culture of date palm *(Phoenix dactylifera* L.) tissues and organs, Pamphlet No. 145, Volcani Institute of Agricultural Research, Israel, 1974.
442. **Weatherhead, M. A., Burdon, J., and Henshaw, G. G.**, Some effects of activated charcoal as an additive to plant tissue culture media, *Z. Pflanzenphysiol.*, 89, 141, 1978.
443. **Fridborg, G., Pedersen, M., Landstrom, L., and Eriksson, T.**, The effect of activated charcoal on tissue cultures: adsorption of metabolites inhibiting morphogenesis, *Physiol. Plant.*, 43, 104, 1978.
444. **Murashige, T.**, Comments at round table discussion (oral communication), Tissue Culture Association Meetings, Seattle, Wash., 1979.
445. **Stone, O. M.**, The production and propagation of disease free plants, in Propagation of Higher Plants Through Tissue Culture: A Bridge Between Research and Application, Conf. No. 7804111, Hughes, K. W., Henke, R., and Constantin, M., Eds., National Technical Information Service, U.S. Department of Commerce, Springfield, Va., *1978, 25.*
446. **Buys, C., Poortmans, P., and Rudele, M.**, Serienmassige meristemkulturen und auswahl virus-freier nelken im grossem, *Gartenwelt*, 66, 305, 1966.
447. **Cheng, T.-Y.**, Adventitious bud formation in culture of Douglas fir (*Pseudotsuga menziesii* (Mirb.) Franco), *Plant Sci. Lett.*, 9, 179, 1975.
448. **Konar, R. N. and Oberoi, Y. P.**, *In vitro* development of embryoids on cotyledons of *Biota orientalis, Phytomorphology*, 15, 137, 1965.

449. **Hu, C. Y. and Sussex, I. M.**, *In vitro* development of embryoids on cotyledons of *Ilex aquifolium, Phytomorphology,* 21, 103, 1971.
450. **Pierik, R. L. M. and Steegmans, H. H. M.**, Analysis of adventitious root formation in isolated stem explants of *Rhododendron, Sci. Hortic.,* 3, 1, 1975.
451. **Murahige, T. and Nakano, R.**, The light requirement for shoot initiation in tobacco callus cultures, *Am. J. Bot.,* 55, 710, 1968.
452. **Hackett, W. P.**, The influence of auxin, catechol, and methanolic tissue extracts on root initiation in aseptically cultured shoot apices of the juvenile and adult forms of *Hedera helix, J. Am. Hortic.*
453. **Leroux, R.**, Contribution a l'etude de la rhizogenese de fragments de tiges de pois (*Pisum sativum* L.) cultives *in vitro,* These, Universite de Paris, 1971.
454. **Thorpe, T. T. and Murashige, T.**, Some histochemical changes underlying shoot initiation in tobacco callus cultures, *Can. J. Bot.,* 48, 277, 1970.
455. **Thorpe, T. T. and Meier, D. D.**, Starch metabolism, respiration, and shoot formation in tobacco callus cultures, *Physiol. Plant.,* 27, 365, 1972.
456. **Thorpe, T. T.**, Carbohydrate availability and shoot formation in tobacco callus cultures, *Physiol. Plant.,* 30, 77, 1974.
457. **Sironval, C.**, Action of day length upon the formation of adventitious buds in *Bryophyllum tubiflorum* Harv., *Nature (London),* 178, 1357, 1965.
458. **Heide, O. M.**, Effects of 6-benzylaminopurine and 1-napthaleneacetic acid on the epiphyllous bud formation in *Bryophyllum, Planta,* 67, 281, 1965.
459. **Heide, O. M.**, Photoperiodic effects on the regeneration ability of *Begonia* leaf cuttings, *Physiol. Plant.,* 18, 185, 1965.
460. **Gautheret, R. J.**, Investigations on the root formation in the tissues of *Helianthus tuberosus* cultured *in vitro, Am. J. Bot.,* 56, 702, 1969.
461. **Vasil, I. K. and Hildebrandt, A. C.**, Variations of morphogenetic behavior in plant tissue cultures. II. *Petroselinum hortense, Am. J. Bot.,* 53, 869, 1966.
462. **Heide, O. M. and Skoog, G.**, Cytokinin activity in *Begonia* and *Bryophyllum, Physiol. Plant.,* 20, 771, 1967.
463. **Naef, J. B. and Simon, P.**, Photoregulation of *Funaria* protonemas and its relation to cytokinin action, in *Abstr., 4th Int. Congr. Plant Tissue and Cell Culture,* Thorpe, T. A., Ed., University of Calgary, Alberta, Canada, 1978, 114.
464. **Kadkade, P. G., Wetherbee, P., Jopson, H., and Botticelli, C.**, Photoregulation or organogenesis in pine embryo and seedling tissue cultures, in *Abstr., 4th Int. Congr. Plant Tissue and Cell Culture,* University of Calgary, Canada, 1978, 29.
465. **Weis, J. S. and Jaffe, M. F.**, Photoenhancement by blue light of organogenesis in tobacco pith cultures, *Physiol. Plant,* 22, 171, 1969.
466. **Seibert, M.**, The effects of wave length and intensity on growth and shoot initiation in tobacco callus, *In Vitro,* 8, 435, 1973.
467. **Carew, D. P. and Staba, E. J.**, Plant tissue culture: its fundamentals, applications and relationship to medicinal plant studies, *Lloydia,* 28, 1, 1965.
467. **Puchan, Z. and Martin, S. M.**, The industrial potential of plant cell culture, *Prog. Ind. Microbiol.,* 9, 13, 1971.
469. **Bonga, J.**, Applications of tissue culture in forestry, in *Applied and Fundamental Aspects of Plant Cell, Tissue and Organ Culture,* Reinert, J. and Bajaj, Y. P. S., Eds., Springer-Verlag, New York, 1977, 93.
470. **Nakosteen, P. C. and Hughes, K. W.**, Sexual life cycle of three species of Funariaceae in culture, *Bryologist,* 81, 307, 1978.
471. **Tanaka, M. and Sakanishi, Y.**, Factors affecting the growth of *in vitro* cultured lateral buds from *Phalaenopsis* flower stalks, *Sci. Hortic.,* 8, 169, 1978.
472. **Goloff, A. A., Black, D. G., Romono, J. B., and Lawrence, R. H.**, Selective induction of adventive shoot and/or root morphogenesis in cultured lettuce leaf explants, *In Vitro,* 15, 191, 1979.
473. **Beasley, C. A. and Eaks, I. L.**, Ethylene from alcohol lamps and natural gas burners: effects on cotton ovules cultured *in vitro, In Vitro,* 15, 263, 1979.
474. **LaRue, T. A. G. and Gamborg, O. L.**, Ethylene production by plant cell cultures, *Plant Physiol.,* 48, 394, 1971.
475. **Dalton, C. C. and Street, H. E.**, The role of the gas phase in the greening and growth of illuminated cell suspension cultures of spinach (*Spinacia oleraceae* L.), *In Vitro,* 12, 485, 1976.
476. **Thomas, D. S. and Murashige, T.**, Volatile emissions of plant tissue cultures. I. Identification of the major components, *In Vitro,* 15, 654, 1979.
477. **Thomas, D. S. and Murashige, T.**, Volatile emissions of plant tissue cultures. II. Effects of the auxin 2,4-D on production of volatiles in callus cultures, *In Vitro,* 15, 659, 1979.
478. **Tisserat, B. H. and Murashige, T.**, Probable identity of substances in citrus that repress asexual embryogenesis, *In Vitro,* 13, 785, 1977.

479. **Gautheret, R. J.**, Recherches sur la polarite des tissus vegétaux, *Rev. Cytol. Cyrophysiol. Veg.*, 7, 45, 1944.
480. **Bloch, R.**, Polarity in plants, *Bot. Rev.*, 9, 261, 1943.
481. **Jacobs, W. M. D.**, Auxin-transport in the hypocotyl of *Phaseolus vulgaris* L., *Am. J. Bot.*, 37, 248, 1950.
482. **Takatori, F. H., Murashige, T., and Stillman, J. I.**, Vegetative propagation of *Asparagus* through tissue culture, *HortScience*, 3, 20, 1968.
483. **Seabrook, J. E. A. and Cumming, B. G.**, The *In Vitro* propagation of Amaryllis (*Hippeastrum* spp. Hybrids), *In Vitro*, 13, 831, 1977.
484. **Tamura, S.**, Adventitious bud formation from excised *Hyacinth* bulb scale *in vitro*, *J. Jpn. Soc. Hortic. Sci.*, 46, 501, 1978.
485. **Vasil, I. K., Hildebrandt, A. C., and Riker, A. J.**, Endive platlets from freely suspended cells and cell groups grown, *in vitro*, *Science*, 146, 76, 1964.
486. **Reinert, J. and Backs, D.**, Control of totipotency in plant cells growing *in vitro*, *Nature (London)*, 220, 1340, 1968.
487. **Wochok, Z. S. and Wetherell, D. F.** Restoration of declining morphogenetic capacity in long-term cultures of *Daucus carota* by kinetin, *Experientia*, 28, 104, 1972.
488. **Sheridan, W. W.**, Plant regeneration and chromosomal stability in tissue cultures, in *Genetic Manipulation with Plant Material*, Ledoux, L., Ed., Plenum, New York, 1974, 263.
489. **Torrey, J. G.**, Cytological evidence of cell selection by plant tissue culture media, in *Int. Conf. on Plant Tissue Culture*, White, P. R., and Grove, A. R., Eds., McCutchan, Berkeley, 1965, 473.
490. **Orton, T., J. and Nelson, J. L.**, Chromosomal variability in plants regenerated from callus cultures of an interspecific *Hordeum* hybrid, *Genetics*, Suppl. 91, 91, 1979.
491. **Flashman, S.**, Oral communication, 1979.
492. **Shimada, T., Sasakuma, T., and Tsunewaki, K.**, *In vitro* culture of wheat tissues. I. Callus formation, organ redifferentiation and single cell culture, *Can. J. Genet. Cytol.*, 11, 294, 1969.
493. **Torrey, J. G.**, Morphogenesis in relation to chromosomal consitution in long-term plant tissue cultures, *Physiol. Plant.*, 20, 265, 1967.
494. **Murashige, T. and Nakano, R.**, Morphogenetic behavior of tobacco tissue cultures and implications of plant senescence, *Am. J. Bot.*, 52, 819, 1965.
495. **Murashige, T. and Nakano, R.**, Chromosomal complement as a determinant of the morphogenetic potential of tobacco cells, *Am. J. Bot.*, 54, 963, 1967.
496. **Nishi, T., Yamada, Y., and Takahashi, E.**, Organ differentiation and plant restoration in rice callus, *Nature (London)*, 219, 508, 1968.
497. **Malnassy, P. and Ellison, J. H.**, Asparagus tetraploids from callus tissue, *HortScience*, 5, 444, 1970.
498. **Heinz, D. J. and Mee, G. W. P.**, Morphologic, cytogenetic and enzymatic variation in saccharum species hybrid clones derived from callus tissue, *Am. J. Bot.*, 58, 257, 1971.
499. **Bennici, A. and D'Amato, F.**, *In vitro* regeneration of *Durum* wheat plants. I. Chromosome numbers of regenerated plantlets, *Z. Pflanzenzuecht.*, 81, 305, 1978.
500. **Sacristan, M. D.**, Karyotypic changes in callus cultures from haploid and diploid plants of *Crepis capillaris* (L.) Wallr., *Chromosoma*, 33, 273, 1971.
501. **Bennici, A.**, Cytological analysis of roots, shoots, and plants regenerated from suspension and solid *in vitro* cultures of haploid *Pelargonium Z. Pflanzenzwecht.*, 72, 199, 1974.
502. **Matern, U., Strobel, G., and Shepard, J.**, Reaction to phytotoxins in a potato population derived from mesophyll protoplasts, *Proc. Natl. Acad. Sci. U.S.A.*, 75, 4935, 1978.
503. **Izhar, S. and Power, J. B.**, Genetical studies with petunia leaf protoplasts. I. Genetic variation to specific growth hormones and possible genetic control on stages of protoplast development in culture, *Plant Sci. Lett.*, 8, 375, 1977.
504. **Hanson, M. R., Skvirsky, R. C., and Ausubel, F. M.**, Genetic analysis of cytokinin response in *Petunia hybrida, Plant Physiol. Suppl.*, 61, 46, 1978.
505. **Nitsch, C. and Hughes, K.**, Roundtable on Anther culture, haploids and chromosome doubling, in *Frontiers of Plant Tissue Culture: Proc. 4th Int. Congr. Plant Tissue and Cell Culture*, Thorpe, T. A., Ed., University of Calgary, Canada, 1978, 482.
506. **Bernhard, S.**, Développement d'embryons haploïdes a partir d'anthères cultivées *in vitro*. Etude cytologique comparée chez le tabac et la pétunia, *Rev. Cytol. Biol. Vég.*, 34, 165, 1971.
507. **Sangwan, R. S. and Norreel, B.**, Induction of plants from pollen grains of Petunia cultured *in vitro*, *Nature (London)*, 257, 222, 1975.
508. **Binding, H.**, Cell cluster formation by leaf protoplasts from axenic cultures of haploid *Petunia hybrida* L., *Plant Sci. Lett.*, 2, 185, 1974.
509. **Durand, J., Potrykus, I., and Donn, G.**, Plants issues de protoplastes de *Pétunia Z. Pflanzenphysiol.*, 69, 26, 1972.
510. **Binding, H.**, Regeneration von haploiden und diploiden pflanzen aus protplasten von *Petunia hybrida* L., *Z. Pflanzenphysiol.*, 74, 327, 1974.

511. **Vasil, V. and Vasil, I. K.,** Regeneration of tobacco and petunia protoplasts and culture of corn protoplasts, *In Vitro,* 10, 83, 1974.
512. **Power, J. B., Berry, S. F., Frearson, E. M., and Cocking, E. C.,** Selection proceedures for the production of inter-species somatic hybrids of *Petunia hybrida* and *Petunia parodii.* I. Nutrient media and drug sensitivity complementation selection, *Plant. Sci. Lett.,* 10, 1, 1977.
513. **Cocking, E. C., George, D., Price-Jones, M. J., and Power, J. B.,** Selection proceedures for the production of inter-species somatic hybrids of *Petunia hybrida* and *Petunia parodii.* II. Albino complementation selection, *Plant Sci. Lett.,* 10, 7, 1977.
514. **Power, J. B., Frearson, E. M., Hayward, C., George, D., Evans, P. K., Berry, S. F., and Cocking, E. C.,** Somatic hybridization of *Petunia hybrida* and *P. parodii, Nature (London),* 263, 500, 1976.
515. **Davidson, E. H. and Britten, R. J.,** Regulation of gene expression: possible role of repetitive sequences, *Science,* 204, 1152, 1979.
516. **Verma, D. C. and Dougall, D. K.,** DNA, RNA, and protein content of tissues during growth and embryogenesis in wild carrot suspension cultures, *In Vitro,* 14, 183, 1978.

Chapter 3

FRUIT CROPS

Robert M. Skirvin

TABLE OF CONTENTS

I. INTRODUCTION

Both reproductive and vegetative portions of many fruit crops have been studied in tissue culture.* Most investigations have dealt with three primary subjects: (1) the production of callus, (2) proliferation of axillary and adventitious buds, and (3) rooting. Since all plant tissue cultures are derived from small portions of a parent plant, a primary concern of the tissue culturalist is to develop a medium that contains all essential components (both organic and inorganic) that would have been provided by the roots, leaves, storage organs, and tissues of the parent plant.

II. MEDIA AND MEDIA COMPONENTS

In a recent survey of tissue culturalists, Zimmerman** reported that most investigators working with fruit crops used a modification of the Murashige and Skoog[255] (MS) high mineral salts medium at some stage of culture. This fact is obvious when one examines the literature of Tables 1, 2, and 3.*** The large number of systems utilizing MS medium is particularly remarkable since the medium was not published until 1962. It is generally agreed that the development of the MS medium constituted one of the most important breakthroughs in all plant tissue culture. Zimmerman warns, however, that other media should not be ignored, and researchers should strive for an optimal medium for each crop. The components of the tissue culture medium are discussed in considerable detail below.

* The subject has been briefly reviewed by Murashige,[250-252] Hedtrich,[144] Skirvin,[344] and Zimmerman.[404]

** Results of survey mailed to respondents by R. H. Zimmerman, USDA-SEA, Beltsville, Md.

***Tables 1, 2, and 3 will be found at the end of the text.

There are many media which are reported to have value for tissue culture of fruit crops (Table 1), and certain plants may be medium-sensitive. It is common for a crop to perform well on one medium and poorly on another. Caponetti et al.,[62] for example, reported that cultures of black cherry callus grew best on Wetmore and Rier[386] medium and performed more poorly on MS media. Both negative and positive information related to media is of value to colleagues interested in working with a particular crop and should be included in research reports. Unfortunately, information of this nature is seldom included, and it is, therefore, arduous to determine whether differences in plant growth on various media are due to osmotic potential (due to salt and sugar level) or are the result of specific components in the medium.

A. Inorganic Components of the Medium

In the early days of tissue culture it was difficult for researchers to identify and provide all constituents necessary for good growth of tissues and organs in vitro. Some researchers successfully ignored the problem. Knudson,[182] for instance, did not use microelements in his now famous medium for orchid seed germination because ". . . these minor elements are always present as impurities in the salts and other substances which are used for [the medium]." Knudson's medium has been useful for the culture of some plants but certainly not for all.

In general, the tissue culture medium contains the 16 essential elements for plant growth as well as a few possible extra compounds such as aluminum and nickel. Some media are more simple than others; Knop's medium, for instance, contains only four compounds. The most important difference among media may be the overall salt level. There seems to be basically three different media types by this classification: high salt (e.g., Murashige and Skoog[255] medium), intermediate levels (e.g., Nitsch and Nitsch[275]), and low salt media (e.g., White[389]). The choice of inorganic salts and salt levels is basically dictated by the resources of the researcher and the particular plant in question.

Boxus and Quoirin[48] and their allies have had considerable success propagating many Rosaceous plants in vitro (including numerous *Prunus* species and strawberry) on their basic medium (Table 1). Throughout the world, this medium has been considered acceptable and has performed well for most researchers. One of the most peculiar features about their media, however, was the very high level of microelements that they used in comparison to most tissue culture media (they used 100 times the recommended rate of Heller[149]). Quoirin and Lepoivre[302] studied this in more detail and found that manganese was limiting the growth of *Prunus* shoot tips and the formation of axillary buds. The main advantage of the 100× Heller's microelements had been to provide additional manganese. The "improved" media that resulted from these investigations reflects this higher rate of manganese.

A common problem associated with fruit tissue cultures is chlorosis or loss of green pigmentation. Researchers have assumed that the problem is due to iron deficiency and iron has been added to the media in a variety of forms. The most common method of adding Fe is using the procedure described by Murashige and Skoog[255] (i.e., 7.45 g/ℓ Na_2EDTA + 5.57 g/ℓ $FeSO_4 \cdot 7H_2O$ dissolved in 1 ℓ of water and dispensed as a stock solution at 5 mℓ/ℓ of medium). In our laboratory we have found that it is sometimes difficult to solubilize this mixture so we have developed modified procedures as follows:*

1. Add 800 mℓ of water plus 3 pellets of NaOH to a beaker.
2. Heat and mix to boiling and reduce heating level.

* Courtesy of Caryn L. Carlson, Hurst Plant Science Department, University of Wisconsin at River Falls.

3. Add Na_2EDTA and mix until dissolved with heat off.
4. Slowly add $FeSO_4 \cdot 7H_2O$ and mix until dissolved.
5. Dilute to 1000 mℓ.

Even extra iron may have little effect on chlorosis: Seirlis et al.[338] reported that they grew their *Prunus* cultures on medium supplemented with 4× iron levels for several subcultures with little effect. Seirlis suggested that the effect could be related to high cytokinin in the medium.

B. pH of the Medium

Most fruit tissue cultures are grown at pH 5.6 to 5.8. Several researchers have reported growing their cultures at pH 5.0. Sommer et al.[356a] grew peach cells at pH 7.0, but the use of such a neutral pH is rare in the literature. Members of the acid-loving heath family (Ericaceae) such as blueberries and rhododendrons grow best on the medium of Anderson[19] at pH 4.5

C. Organic Components of the Medium

1. Organic Complexes

The problem of determining the organic compounds essential for plant growth in tissue culture was particularly difficult for early investigators since few natural plant constituents were known when plant tissue cultures began. Researchers quickly found that the addition of "complexes" to the basic medium frequently resulted in successful growth of tissues and organs. Some of these complexes have included green tomato extract, coconut milk, orange juice, casein hydrolysate (called "edamin" in the original MS medium), yeast and malt extract. The use of these complex substances in tissue culture medium results in an "undefined" medium.

Most researchers feel that they have little control of their experiments when they used undefined media; therefore, many researchers avoid the use of organic complexes and prefer "defined" media. However, many plants are still grown on undefined media for lack of a suitable, defined medium. *Citrus* cultures, for instance, are commonly grown on Murashige and Tucker medium[256] supplemented with 500 mg/ℓ malt extract.

a. *Agar*

One of the most common organic complexes used in tissue culture is *agar*. Since agar is produced from a red algae that is harvested from the ocean, each batch is slightly (or drastically) different and may represent a source of variation in an experiment that could be especially pronounced between batches of media. For this reason many researchers are using more purified forms of agar (e.g., "phytagar") and/or are "washing" their agar prior to use.

The amount of agar used in the medium varies considerably. MS medium calls for 10 g/ℓ (1%), but many researchers have reduced this agar level to 8 g/ℓ or less. The osmotic potential of high agar concentration, and the concomitant reduction in nutrient and organic matter availability to the tissue is the common reason for reducing the agar concentration.

There is an increased water loss due to evaporation associated with reduced agar concentration. For this reason, individuals interested in reducing agar concentration in their media may have to compromise and choose an agar level that is hard enough to minimize water loss, but soft enough to allow good nutrient diffusion. Excessive water loss can be reduced by raising the humidity of the culture room or by sealing the culture tubes with an impermeable (e.g., "Parafilm"®) or semipermeable (e.g., commercial plastic wraps such as "Saran Wrap"®) seal. Ma and Shii[218] found that tissue cultures of banana formed callus and corm-like structures readily on Smith and Murashige medium[355] supplemented with 160 mg/ℓ adenine sulfate and 10 g/ℓ agar. These corms remained "dormant" on the medium until the agar concentration was either reduced by one half or eliminated, whereupon the corms grew to whole plants.

Many researchers have completely eliminated the use of agar from their cultures by using shake or roller drum cultures and/or by using the so-called "Heller bridges"[148] made of filter paper standing in liquid medium.

b. Coconut Milk

Of all the complexes added to plant tissue culture medium, coconut milk is by far the best known. It is usually added in concentrations of 3 to 15% by volume. The chemical composition of coconut milk has been studied by a number of researchers over the years, but the exact components of the milk remain unknown. Raghavan[409] has examined the entire literature related to coconut milk and compiled an impressive list of its known components and their concentrations.

Coconut milk is generally believed to contain cytokinin-like substances as well as some reduced nitrogen, all of which may have value for certain tissue cultures. Typical effects of coconut milk are reported by Hawker et. al.[142] who determined that replacement of casein by coconut milk could double the growth rate of grape berry callus. De Guzman,[85] however, reports the paradox that coconut milk inhibits root formation by coconut embryos.

Not all coconut milk is equally useful for tissue cultures. Young coconuts give a higher quality product while milk from old coconuts may actually inhibit growth.[78] Most investigators in temperate zones of the world do not have a year-round source of green coconuts and must settle for milk from mature coconuts.

2. Carbohydrates

Carbohydrates serve two principal functions in the tissue culture medium: (1) they provide an energy source to the tissues, and (2) they maintain a minimal osmotic potential within the medium. There are a number of carbohydrates that are used in fruit tissue cultures, but the most popular and versatile of these has been sucrose.

The level of carbohydrates in most media runs about 20 to 30 g/ℓ, but in some instances the level of carbohydrates is much higher. Zwagerman and Zilis,[407] for instance, found that axillary buds of "Thundercloud" (an ornamental plum) proliferated well on a modified MS medium supplemented with benzyladenine (BA) (1.0) and naphthalenacetic acid (NAA) (0.1), but when the sucrose level was raised to 60 g/ℓ, the young shoots no longer required NAA for proliferation, and the shoots developed the purple-red coloration characteristic of the cultivar.

Arya et. al.[21a-23] reported a very interesting study of grape stems in vitro. They compared the growth characteristics of normal and Phylloxera gall infected grape cells on a variety of media containing many different carbohydrate sources. They were able to isolate many variant lines as a function of the ability of cells to utilize particular sugar sources. Some cell lines, for instance, grew well on galactose while others did not.

Researchers studying protoplasts derived from fruit tissues make particular use of carbohydrates to control osmotic potential. Without a careful control of the osmotic environment of the medium, the protoplasts would burst. Skene[342] uses either sorbitol or sucrose to maintain this protoplast medium at about 0.23 *M*. He also reports using four other sugars in this medium.

Coffin et al.[76] reported that sorbitol (D-glucitol) is a common carbohydrate of the Rosaceae family. They also determine that many Rosaceous plants can grow well on medium supplemented with either sucrose or sorbitol. Some plants even grew better with sorbitol than sucrose (e.g., "Reliance" peach).

It has been suggested that the beneficial effect of adding certain organic complexes (e.g., fruit juices and extracts) to the medium is due to increased sugar level. Einset,[93] for instance, reported that *Citrus* cultures grew about six times more than the control with the addition of 10% orange juice to the medium. This was equivalent to adding about an extra 5% sucrose or 10% total sugars to the medium. When they added additional sugar to the medium instead of orange juice, the beneficial effect of the orange juice was not noted. The effect could not be completely mimicked by the addition of citric acid.[93,100]

3. *Nitrogenous Compounds*

It is generally agreed that reduced forms of nitrogen have value for certain plant tissue cultures. Nitsch et al.[274] confirmed this using an apple and pear callus. They found that NH^{+}_{4} ion (provided as either NH_4Cl or NH_4NO_3) gave good growth, but the addition of asparagine, glutamine, or other amino acids to the medium in the presence of NH^{+}_{4} gave no particular increase in growth.

Several media listed in Table 1 are routinely supplemented with amino acids or casein hydrolysate (a partially digested mixture of protein), but it remains unclear whether these compounds are absolutely essential for growth. Nitsch[273] reported that no organic compounds or vitamins, other than sucrose and agar, were required to stimulate the formation of haploid tobacco plants on his medium.

4. *Vitamins*

Almost every medium used for plant tissue cultures contains at least some vitamins. The most complex vitamin mix is that used by Skirvin and Chu[346] (Table 2) that they call "Staba" vitamins [since the formulation for this mix was provided by Dr. Staba of the University of Minnesota]. The mix contains ten different vitamins. At the other extreme, the Linsmaier and Skoog medium[210] uses only thiamine·HCl in its formulation. Most tissue culturists will find that the number of vitamins suitable for their crop lies between these two extremes. Most researchers report that thiamine, nicotinamide (niacin), and pyridoxine are essential for their cultures, but the relative concentrations of these compounds vary considerably.

5. *Antioxidants*

Many fruit tissue explants when placed on culture-medium oxidize the medium to yield a diffusion shell of a black substance. Frequently the formation of this oxidized material is toxic or inhibitory to further development of the tissue. Such oxidation can sometimes be inhibited by the addition of certain compounds to the medium. Reynolds and Murashige[315] reported that the addition of activated charcoal to the medium at 3 g/ℓ could solve this problem. Lee and de Fossard,[205,206] as well as Skirvin and Chu,[346] routinely add ascorbic acid to their media to prevent this oxidation. Sometimes the transfer of tissue to fresh nonoxidized medium will suffice to cease the oxidation, in other cases it will not.

6. *Hormones*

An examination of the various media and hormone combinations utilized by researchers for fruit crops (Tables 1 and 3) reveals some interesting facts: (1) callus formation can be stimulated by a variety of auxins and kinins; (2) shoot proliferation (both as adventitious buds from callus and stems as well as growth of axillary buds) generally requires the presence of both an auxin and a kinin, but in subculture, a kinin alone is often sufficient; (3) rooting is frequently inhibited by the presence of kinin; (4) the most common auxins used in tissue culture are 2,4-dichlorophenoxyacetic acid

(2,4-D), indoleacetic acid (IAA), indolebutyric acid (IBA), and α-naphthalenacetic acid (NAA) (Table 3); and (5) the most common cytokinins are kinetin (6-furfurylamino-purine) and 6-benzylaminopurine (BA).

There have never been standardized procedures for the manufacture of hormone stock solutions. Miller[230a] reports that kinetin can be dissolved in water by autoclaving for 10 min at 15 psi. In our laboratory, we routinely dissolve kinetin and BA in a small quantity of 0.1 *N* HCl while auxins 2,4-D, and NAA are dissolved in small quantities of KOH.* The acid or base solutions are then diluted to volume with water. Some laboratories routinely dissolve their hormones in dimethyl sulfoxide (DMSO).[408] As far as this author knows, however, a booklet or article which compiles and compares various methods of stock solution preparation has never been published. The author suggests that this would be good topic for a future article.

In general, the classic Miller and Skoog[233] paper that first reported the effects of various auxin to kinin ratios on differentiation of tobacco plants in vitro applies directly to all fruit tissue cultures, i.e., differentiation of organs from callus is largely a function of the correct ratio of hormones. It appears that researchers interested in studying a particular crop should investigate interactions of different auxins and kinins with respect to concentration, salt formulation of the medium, vitamin levels, photoperiod, light intensity, agar concentration, and the addition of the organic complexes to the medium. Ghugale et al.,[132] for instance, studied the effects of several auxins on rooting of mulberry (*Morus alba*) tissues. The results of their studies are summarized below:

Kinetin (1 ppm)	% Rooting
+ NAA	20
+ IAA	50
+ IBA	70
+ IPA	80

James and Newton[165] reported a very complicated factorial experiment with "Gento" strawberry in which they determined a range of concentrations of IBA and BA that would ensure good production of adventitious bud proliferation in culture. They found that BA at a concentration of 0.25 to 2.5 μmol ℓ^{-1} + IBA at 0.25 to 1.0 μmol ℓ^{-1} yielded an optimum number of buds with a minimal amount of undesirable morphological problems such as curled leaf and callus formation.

It is generally agreed that callus formation is antagonistic to root and shoot proliferation. For this reason most fruit tissue culturalists strive to minimize callus. It is also well known that the use of callus cultures increases the likelihood of cellular variation[345] and, hence to ensure clonal stability and maximize shoot and/or root proliferation, callus should be minimized.

The concentration of hormones used for fruit tissue cultures varies considerably from report to report. Apparently there is an optimum hormone concentration for each plant on each medium. In general, the concentration of any single hormone used in tissue culture is less than 10.0 mg/ℓ. Reynolds and Murashige[315] reported using 100 mg/ℓ 2,4-D in their medium for date palm. Such a high concentration of 2,4-D would be toxic under normal conditions, but the authors reported that palm cells tended to produce "auto-intoxicating" substances that were leached into the medium inhibiting or killing the tissues. The addition of activated charcoal (3 g/ℓ) solved this problem,

* Special care should be taken with IAA. Epstein and Lavee[97] report that autoclaving of IAA resulted in a 3% loss of their ^{14}C IAA, and photolysis destroyed another 19% in the medium.

but it also severely reduced the auxin available to the plant; therefore, it was necessary to use very high levels of 2,4-D.

III. EXPLANTS FOR TISSUE CULTURE

Tissue cultures can be established from any portion of a plant, but certain sections of a plant have proved to be particularly amenable to tissue culture conditions.

A. Age of Explant

The youngest tissue from a plant frequently is the best source of explant material. Asahira and Kano,[24] for instance, reported that strawberry fruit tissue gathered either 2 or 7 days after anthesis formed a callus that was capable of differentiating adventitious buds; whereas fruits that were 12 days past anthesis formed little callus, and none of the callus differentiated plantlets. Similarly, Hawker et al.[142] found that grape berries less than 0.55 mm in diameter would form callus most easily on their medium. Zatykó et al.[401] reported that *only* 3- to 4-week-old fertilized ovules taken from 7- to 9-mm red currant berries were capable of developing embryoids. Similar reports have been filed for strawberry[25] and *Annona squamosa.*[27]

Juvenile tissue performs well in tissue culture. Trippi[374,375] showed that suckers from mature chestnut trees possessed the ability to produce callus equal to juvenile trees; apparently suckers of mature trees had returned to juvenility. Eichholtz et al.[92,92a] succeeded in producing adventitious shoots and embryoids from young apple fruits. The authors reasoned that tissues associated with reproduction and seed production would be more likely to behave in a juvenile-like fashion that mature wood.

In some cases, age and size of the subculture may determine the success of culture. Quoirin et al.,[303] for instance, reported that elongated shoots rooted more easily on their medium than small shoots. The fact that explant age and size influences differentiation cannot be disputed and researchers should consider this when beginning a study. Further, this specificity suggests that the frequent report that particular tissues have not been successfully cultured may be due to unsatisfactory tissue culture explant material.

B. Source of Explant

Letham[208] found the pear fruit tissue that contained a section of vascular bundles grew best in vitro. Harada[138] reported that adventitious shoots formed more easily from root tissue than from shoots of *Actinida.*

The choice of organs for explant should be carefully and systematically explored by the fruit tissue culturalist. Not all tissues on a plant are equally capable of differentiating. Scorza and Janick[337] demonstrated that intact plants could be derived from leaf discs of a number of *Passiflora* species. They noted, however, that discs of young leaves of *P. suberosa* (less than five leaves from the apex), in addition to forming adventitious shoots, developed flowers directly from the discs on the same medium.

C. Explant-Hormone Interaction

Different organs can respond to ingredients in the medium in various ways. This has been most clearly demonstrated by Sharp and colleagues[339] using coffee. Part of this impressive research is given below.

Plant material	Growth factor	Growth pattern
Seeds	NAA or 2,4-D	Normal seed germination
Fruits	2,4-D	Clumps of slow growing friable brown callus

Orthotropic shoots (upright growth)	2,4-D	Rapidly proliferating, firm white or yellow callus*
Plagiotropic shoots (lateral growth)	2,4-D	No growth or slow growing, firm, white callus*
Leaves	2,4-D	Slow growing, firm, white callus
Leaves	NAA	Massive adventitious root formation
Anthers	2,4-D ± coconut milk	Slow growing, white callus, ranging from very friable to firm

It is obvious that different organs can give different results. The subject has been discussed by Hildebrandt.[153] Sometimes portions of a single explant may behave differently. Powell,[298] for instance, reports the basal portion of an apple stem is more dormant than the upper portions and, hence, may produce different results in vitro. Feucht[107] reported that the cortex and cambium from stem explants of *Prunus* showed different growth rates as a function of hormones. NAA at 10 mg/ℓ stimulated good cortex proliferation while NAA at 1.0 mg/ℓ produced good cambial proliferation. They also found that cortical parenchyma was more inhibited by abscisic acid (ABA) than cambial tissue in vitro. Nitsch[272] reported that 10 mg/ℓ of kinetin had very little effect on the growth of peach fruit mesocarp, but strongly stimulated the endocarp. Similarly, he reported that 1-naphthalene acid amide exhibited strong effects on the mesocarp and only moderate effects on the endocarp.

D. Preparing the Explant for Culture

Abbott[1] clearly stated the problem of tissue culture of woody plants, "For some years . . . woody plants were assumed to be intractable in sterile culture, so that research workers have given these plants only scant attention compared with herbaceous species."

There are two reasons for the preconceived notion that woody plants were not well-suited to tissue culture: (1) the woody plant is often slow to propagate, has complicated dormancy cycles, and frequently exists in various forms of maturity and juvenility and (2) woody plants are grown in the soil for many years and, hence, they are routinely infected with microorganisms both internally and externally which are often difficult to control in vitro.

The usual method of tissue sterilization is by using a known concentration of sodium hypochlorite (usually supplied as diluted commercial bleach, e.g., Clorox®). A dilution of 10% v/v bleach normally does a good job of surface-disinfecting the tissue, particularly when it is mixed with a surfactant (e.g., 0.1% Triton® X-100, Tween 20®, etc.). Even careful use of bleach followed by good rinsing can produce damage, however, and Fisher and Tsai[113] reported the growth of their coconut cultures was inhibited by normal levels of bleach. Only when the bleach concentration was reduced to 0.5 to 1.5% was good growth restored. Tissue that is pubescent present particular problems, and it is sometimes of benefit to use a vacuum system to ensure good disinfection. Since bleach is toxic to plant cells, it is necessary to wash the tissue two to three times with sterile, distilled water to ensure dilution of the bleach.

Many researchers have found that a single bleach treatment is not sufficient to ensure contaminant-free cultures. In our laboratory, for instance, we have found that it is very difficult to obtain clean cultures of peach from field-grown plants during the summer. Our normal disinfection procedures include 10% Clorox® + 0.1% Triton® X-100 for 10 min, followed by two 5-min rinses in sterile, distilled water. This treat-

* This is particularly upsetting to those individuals working with new cultivars or other plants in limited supply since the researcher is forced to use all tissue and it may not necessarily be uniform — particularly notice the orthotropic vs. plagiotropic shoots.

ment works well with most plants, but peaches frequently are 100% contaminated with this treatment. We have developed an improved method: first, peach tips are placed into 25% Clorox® + 0.1% Triton® X-100 for 10 min, followed by two 5-min water rinses; then the tips are soaked in 70% ethanol for 5 min followed by two more 5-min water rinses. By this procedure we have found that the contamination rate can be reduced by 50% with only a 10% kill rate.

Marõti and Lévi[222] reported that it was better to rinse with ethanol (45%) first for 3 min followed by a 10-min bleach treatment (5 to 10%) and three rinses with sterile water.

Nekrosova[262] reported high contamination rates with several fruit crops until optimal sterilization times were determined using 0.1% $HgCl_2$. His results are summarized below:

Species	Optimal period of sterilization (min)
Prunus armeniaca (apricot)	2
P. persica (peach)	
Shoot tips	6
Lateral buds	20—25
P. cerasus (sour cherry)	6
P. cerasus var Besseyi	6
Malus sylvestris	8
Citrus limon (lemon)	
Shoot tips	4
Lateral buds	35—40
Ribes nigrum (black currant)	18

Jones et al.[171] have developed a complicated sterilization procedure that is summarized below:

1st stage of sterilization

1. Tissue is dipped in 0.01—0.1% surfactant for 20 sec
2. Then the tissue was soaked in 0.14% available Cl solution for 1 min
3. The tissue was given three rinses in sterile distilled H_2O
4. The tissue was then transferred to media with no hormones or vitamins for 24 hr

2nd stage of sterilization

1. Tissue was removed from medium and dipped in 0.01—0.1% surfactant for 20 sec
2. Then the tissue was soaked in 0.1% benomyl (Fungicide®) for 15 min.[140]
3. Then the tissue was soaked in 0.42—0.50% available Cl for 30—40 min
4. The tissue was given three rinses in sterile distilled H_2O
5. The tissue was transferred to a fresh complete medium

Although this system may be tedious, it has been successfully utilized by Broome and Zimmerman[52] and Harper.[140]

Bacterial contamination in strawberry cultures has been successfully controlled using

Difco® brand antibiotic discs according to Skirvin and Larson.[352] Unfortunately, certain antibiotics also inhibit the growth of the plant cultures.

Cheng[70a] reported a slightly different procedure for sterilizing her fruit tree cultures. She routinely washes her cultures with water containing Alconox®, and then the cultures are sterilized with 6 to 20% Clorox® + 0.2% Alconox®. Interestingly, Cheng then explants her cultures onto a basal medium without hormones for one week to "condition" the cultures; whereupon they are transferred to normal medium. Sondahl and Sharp[357] reported that their coffee cultures gave best results when they were "conditioned" on a special medium for 7 weeks prior to movement to differentiation medium. Jones and Vine[173] use an induction medium (Medium "A") for 3 to 4 weeks where shoot tips are induced to produce axillary buds after which they are moved to Medium "B" for further development.

Herman and Haas[151] reported that they had to use two different media to allow "organoids" that had formed on the LS medium[210] to develop into plants on Gresshoff and Doy No. 4 medium.[134]

IV. DIFFERENTIATION

A. Shoots

Many researchers do not allow enough time for differentiation to occur. Boxus[45] reports that his strawberry cultures could remain on medium for 6 to 18 months with little or no sign of growth, but when explanted onto medium with BA (1.0 mg/ℓ) the shoots began to grow.

Shoot proliferation does not seem to be a particular problem in any genera investigated to date. Many researchers report problems with particular species and cultivars, but success seems to be within the grasp of a patient investigator.

B. Roots

Rooting of shoots may be a different matter. Many fruit species are normally difficult-to-root in the mature state. For this reason, much work has been done with seedlings and easy-to-root species. Success with difficult-to-root species has been more sparse. Feucht and Dausend[110] studied *Prunus avium* (a difficult-to-root species) and *P. pseudocerasus* (an easy-to-root species) in tissue culture. They found that rooting of stem sections in vitro showed the same relationship observed in cuttings. Calli from both species rooted readily on appropriate media. They found that ABA stimulated rooting of these species. Favre[103] reported that a hybrid clone of grape that was hard-to-root by normal procedures rooted 100% in vitro.

There is, of course, an interaction of clones and hormones in rooting. GA_3, for instance, is reported by Button and Bornman[56] to give enhanced rooting of "Washington Navel" orange in vitro, but Putz[299] reports that GA_3 inhibits rooting of raspberry. Chaturvedi and Mitra[69] even report an interaction of hormone and shoot growth vigor. They found that "low vigor" shoots rooted with 0.1 mg/ℓ NAA while "high vigor" shoots required either 0.5 mg/ℓ NAA or 0.25 mg/ℓ NAA + 0.25 mg/ℓ IBA.

Abbott and Whiteley[3] successfully rooted up to 80% of their apple cultures by dipping the young shoots in IBA (1.0 mg/ℓ) for 15 min. Then the shoots were planted upside down on fresh medium with Heller[148] bridges where rooting occurred.

Hedtrich[143] reported with *Prunus mahaleb* that neither leaf discs with mid ribs, petioles, nor internode segments would produce roots on their medium while leaf lamina readily rooted. Further, they found that roots that developed from leaves were geotropic and had hairs on them while roots from callus were nongeotropic and had no hairs.

C. Grafting

It has been difficult to decide whether it is more sensible to produce fruit crops on their own root systems or to provide them to growers and researchers in the normal grafted manner. A number of researchers have decided that it is more sensible to graft plants that are normally grafted for a variety of reasons: (1) grafted plants by-pass the juvenile stage that is associated with nucellar and sexual seedlings, (2) many cultivars have been commercially propagated by grafting for many generations, and growers and researchers know very little about the behavior of such plants on their own roots with regard to plant size, pathogenic and insect pests, cold-hardiness, etc., and (3) shoot tips grown in tissue culture in conjunction with heat treatments can be useful for production of virus-free plants.

The system of grafting shoot tips to rootstocks in vitro has been best developed in *Citrus*. The system consists of grafting virus-free *Citrus* scions onto germinated nucellar or sexual seedlings followed by movement to soil. The system has been detailed in several papers.[253,260,261,399] Successful in vitro grafts have also been reported for *Prunus*[11,223] and apple.[13,159,216]

The behavior of many perennial fruit crops on their own roots is unknown since these plants have almost always been propagated by grafting. Self-rooted plants obtained through tissue culture could be compared to grafted plants in the field.

D. Movement of Young-Rooted Plants to Soil

The literature contains reports of rooted cultures which were subsequently lost upon transfer to soil. Since the young tissue-cultured plant is grown in a sheltered environment at 100% relative humidity, the shock of transfer to a nonfavorable atmosphere (such as the greenhouse or laboratory) results in immediate desiccation. Broome and Zimmerman[52] report that they can successfully root thornless blackberries in Jiffy 7®peat pellets under mist. This mist treatment helps the plant harden and increases its chances of survival in soil. Many researchers pot their rooted cultures in sterile soil or vermiculite. The cultures are covered by a glass vessel or plastic bag to maintain humidity. At the end of 10 to 21 days, most plants can survive in the natural atmosphere.

The soil mixes used for initial potting vary with the researcher. Stone[364] reported that carnation shoots grew well in a 1:1:1 mixture of peat: sand:loam covered with a layer of peat or sand. They claimed that this mix was superior to perlite or vermiculite mixes. Rosati et al.[321] potted their Japanese plum plants in a 1:1:1 mix of peat:perlite:sand. Popov et al.[297] reported that either a 1:1:1 soil:peat:sand or 3:1 peat:sand mix was satisfactory for "Shubinka" sour cherry.

V. ENVIRONMENTAL EFFECTS ON TISSUE CULTURE DEVELOPMENT

A. Time of Year

It has been fairly well documented that tissues taken from field grown plants are not equally amenable to tissue culture conditions throughout the year. Monsion and Dunez[239] determined that the optimum time to gather *Prunus mariana* shoot tips and buds for explants in France was between May 25 and August 14 to ensure shoot and root formation. Borrod[38] determined that wood from *Castanea sativa* formed callus the best in March and the worst in December; interestingly, another chestnut clone was found to grow best in March, but worst in July.

Litz and Conover[214] reported "the time of the year that cultures were established [for papaya], and the cultural conditions in the field were crucial for success of this procedure. Explants responded better early in the growing season. Consistent applications of fertilizer and water to the plants in the field also encouraged a more sensitive in vitro growth response." They explained that bacterial and fungal attack was a serious problem during the winter.

Skirvin and Chu[348] reported that peach scions gathered in the early spring and grown in vases in the laboratory represented a good source of explants which were consistently low in contamination. As the season progresses and the outside temperatures increase, the rate of contamination shows a concomitant increase.

It would seem likely that on the basis of this information a good source of tissue culture explants would be stored scion wood that could remain refrigerated in ethylene-free cold rooms until needed. The dormant bud sticks could then be placed in water and allowed to grow. Even those procedures may have problems. Quoirin et al.[304] have shown that *Prunus accolade* twigs stored at 4°C from December until April produced abnormal rosetted shoot tip cultures in vitro.

An examination of the literature has convinced this author that many tissue culture studies have been performed during winter months by investigators who find their field projects terminated until Spring. It has been fortuitous that this period of time generally coincides with the period of minimal contamination.

B. Photoperiod Effects

Since many fruit crops are photoperiod sensitive, most researchers grow their tissue cultures with some day-night regime, the most common light cycle being 16 hr of light followed by 8 hr of darkness. Alleweldt and Radler[8,9] investigated the effect of photoperiod on growth of a short-day-sensitive grape. They found that the cultivar produced roots under short days and callus under long days. Pence et al.[291] reported that Cacao embryoids rooted better under a 16-hr photoperiod than in the dark.

Some plants grow better in total darkness. Blueberry callus, for instance, grows much better in darkness than in light,[266] while callus of "Robusta" coffee grows only in darkness.[362] Chong and Taper[75] studied callus of three apple cultivars in vitro and found that "Cortland" grew best in darkness, *M. Robusta* No. 5 grew well in either light or dark, and "McIntosh" showed no response to light of 0 to 3350 lx, but 7800 lx slightly stimulated growth.

C. Temperature

Most tissue cultures are grown at temperatures between 23 and 32°C. Saad and Boone,[322] for instance, report that apple callus grows best between 28 and 32°C, but grows poorly at temperature less than 20°C. Although Schroeder and Kay[332,333] report that tissue cultures of avocado can survive temperatures of 55°C following a preexposure to 50°C, it is rare that temperatures above 30°C are utilized in vitro.

A notable exception to this rule is reported by Galzy[119,120] who grew grape meristem cultures at 35°C for 3 months to produce virus-free stocks of several grapes. Belkengren and Miller[31] report treating *Fragaria vesca* meristems to 38°C for 3 to 5 days to produce virus-free cultures. They reported that this treatment yielded about 30% survival.

Startisky[362] reports that coffee cultures are very sensitive to temperature changes. "When the dark growing chamber was out of order for a week — at a temperature of about 30°C, all the callus tissues ceased growing. None of them recovered after a repair of the cabinet and new isolations had to be made. The same is true of temperature rises. When a photograph is made, heat of the light kills tissues and especially green plantlets."

Rosati et al.[320] report that chilling the flowers of "Rabunda" strawberry for 48 hr prior to explant served to condition the anthers to tissue culture conditions and was beneficial for eventual differentiation.

D. Culture Vessels

Surprisingly, the size of the culture vessel is also reported to affect differentiation and growth. Adams[6] reported that strawberry shoots rooted within 2 months in small containers; whereas roots developed much faster on the same medium in large vessels.

VI. STUDIES WITH WHICH FRUIT TISSUE CULTURES HAVE BEEN UTILIZED

A. Study of the Graft Compatibility In Vitro

When compatible scions are grafted to rootstocks, the first step in the completion of the graft union is callus formation. If the graft is strong and complete, then the likelihood of a successful graft is high. Several researchers have studied various rootstocks and graft-scion relationships in vitro. Feucht[108] cultured stem explants from dwarfing, semidwarfing, and invigorating *Prunus* species in vitro. The callus fresh weights showed no relationship with growth vigor of original trees. The growth potential therefore is quite similar in vigorously growing and dwarfing trees when cultured on the same hormone levels. It was found, however, that the beginning of cell division of the explants is in relationship with the growth vigor of the *Prunus* species: the more vigorous the growth, the shorter the lag phase is for cell division. The difference between vigorous *P. avium* and dwarfing *P. fruticosa* is about 2 days.

Saavedra and Feucht[324] have found that neither callus growth, cell division, nor cell expansion was related to tree vigor in apple root stocks. They did report that there was a negative correlation between the vigor of the original tree and the starch content of the cells.

Fujii and Nitro[118] studied the compatibility of callus derived from grape, peach, chestnut, persimmon, pear, and apple in all combinations. The study is fascinating, and it was found that even members of different families (e.g., persimmon and chestnut; grape and apple) could grow together on the authors' medium. Some combinations would grow side-by-side without ever touching, and some combinations inhibited each other. Schneider et al.[328] have noted that callus from certain apple cultivars exhibits growth inhibition when grown side-by-side. The authors suggested that such studies may aid investigators interested in graft compatibility in vivo. A well-known cause of graft scion incompatibility in certain plants is due to cyanide production by either the scion or stock. Heuser[152] has studied this phenomenon with three dwarfing rootstocks of *Prunus* by adding known levels of cyanide to his medium and studying growth rates.

B. Biochemical Studies

Rubus suspension cultures have been utilized by several researchers to study enzyme production and induction. Rekoslayskaya and Gamburg,[312] for example, compared habituated and nonhabituated callus with respect to indolyacetyl asparate synthetase induction by auxin. They found that, following induction, habituated callus exhibited a 6 to 12 hr lag period before the enzyme appeared. Nonhabituated cells, on the other hand, exhibited no lag period. Elder et al.[94] demonstrated that microsomes from suspension cultures of *Rubus fruticosus* could convert squalene 2,3-epoxide to alpha-amyrin. Similarly Lobov et al.,[215] Pilet and Roland,[295] and Bĕjaoui and Pilet[30] have utilized *Rubus* cells to study the effects of the growth inhibitors maleic hydrazide, abscisic acid, and fluoride, respectively, on growth in vitro.

Growth-inhibiting substances have been isolated from tissue cultures of *Citrus*.[370,371] Growth-promoting substances from *Prunus* species have been identified and investigated by several authors.[112,112a,356] Proteolytic enzymes have been extracted from tissue cultures of pineapple.[21]

The rate of ^{14}C labeled sucrose uptake in tissue culture has been studied[211] using strawberry fruit callus. When old (3 months) apple callus is grown with ^{14}C IAA within 4 hr, 14% of the IAA is decarboxylated and only 1.4% is absorbed into the callus. Young apple callus, on the other hand, decarboxylates only 20% of the IAA and 35% is absorbed.[97]

C. Developmental Studies

Because the development of organs in vitro frequently proceeds at a slower rate than in vivo, tissue culture has been of value to scientists interested in detailed development studies. Kordan[193,195] compared the in vivo and in vitro development of juice vesicles of "Eureka" lemon and concluded that development was identical in both cases. Similarly, the effect of various growth regulators on strawberry fruit development has been studied in some detail by Kano and Asahira.[178]

Flower differentiation and development in vitro has been studied in *Passiflora*[336] Srinivasan and Mullins[361] found that tipe of grape tendrils could be induced to form flower buds on either liquid or solid medium supplemented with BA.

Slabaeck-Szweykowska[354] utilized callus cultures of "Healthy" grape (that had been obtained from Morel's lab and was a subculture from the Morel[240] study) to study the effect of various chemical and environmental factors in anthocyanin formation. Many interesting observations were made including the observation that increased sucrose levels in the medium resulted in more anthocyanin development.

Latché et al.[201] have utilized fruit discs of pear grown in vitro to study ripening characteristics in vivo. They found that pear fruit ripened in vitro the same as it did in vivo.

D. Production of Virus-Free Plants

Quak[300] has recently published an impressive and detailed report concerning meristem cultures and the production of virus-free plants. Morel and Martin[244] were the first individuals to demonstrate that virus-free plants could be produced through culture using the meristematic dome of a Dahlia. Galzy[122-124] has demonstrated that virus-free grape plants can be produced from meristems in vitro when cultures are treated to high temperatures (35°C) for longer than 21 days. Mullin et al.[247,248] demonstrated that strawberries could be freed of at least three viruses through meristem culture. These authors were able to store over 50 cultivars of meristem-derived plants in a virus-free condition for 6 years or more on special medium at 4°C in the dark. Lundergan and Janick[217] reported a similar study with apple. These systems may represent the first true germ plasm banks established for fruit crops.

E. Protoplasts

The rigid cell wall of certain fruit crops has been removed enzymatically to produce naked cells called protoplasts. These protoplasts are grown in special medium[343] [discussed earlier] and can be induced to regenerate a new cell wall and divide to form cell colonies. The use of protoplasts for fusion to form nonsexual hybrids is common in the Solanaceae family, but has not been reported for fruit crops. Vardi[379] has reported the production of intact plants from protoplasts of callus from embryogenic lines of "Shamouti" orange. It was found that with a minimum plating rate of 10^4 cells per milliliter that cell division would occur in vitro on Murashige and Tucker medium.[256] Much better growth was obtained at 10^5 cells per milliliter, but colony formation from these cells frequently coalesced to not yield true clones.

Smaller cell numbers could be grown successfully on medium which contained non-dividing "feeder" cells of either orange or tobacco using a system described by Raveh et al.[310a] Under this system, nondividing feeder cells were produced by irradiation of protoplasts to 15 KR. This exposure prevented cells from forming colonies, but did not prevent them from serving as feeders. The desired protoplasts were plated above the irradiated ones, and even at very low densities (5×10^2 cells per milliliter) growth rates of 4 to 8% were obtained.

Protoplasts of *Prunus persica* and *P. brigantica* have been treated with a hypotonic shock in suspension culture to cause chromosome scattering and banding.[246,326a] Such studies may represent the beginning of a better understanding of the cytogenetics of fruit crops.

F. Production of Compounds of Economic Importance

Callus cultures of cacao grown in suspension cultures have been shown to give off an "aroma" that mimicks natural chocolate.[372] Suspension cultures of coffee, in addition to producing caffeine,[179] when harvested and "roasted" produce a coffee-like substance which smells and tastes like coffee derived from green coffee beans.[373] Townsley reports that his coffee cultures were derived from coffee stems rather than young beans, and therefore, the coffee cultures contain linolenic acid that is not found in green coffee beans,[378] but in other respects, the cultures are similar to coffee derived from beans. The process has now been patented. The use of coffee tissue cultures has been reviewed.[237]

G. Anther Cultures

The use of anthers and pollen grains to produce haploid plants of fruit crops have been reported by several authors. A haploid cell line of *Vitis vinifera* has been reported by Gresshoff and Doy.[135] Jordan[174] and Jordan and Feucht[177] have succeeded in producing multicellular pollen grain which burst the exine from *Prunus* × *pandora*. "Jonathan" apple pollen grains have been reported to develop torpedo-stage embryoids in vitro.[197,230] Fowler et al.[116] reported a callus from strawberry anthers; later this callus proved to be of maternal rather than androgenic origin.[115] *Citrus limon* anthers have produced androgenic callus; while *C. medica* callus has yielded callus of maternal origin.[88]

H. Propagation

In a recent review article concerning the use of in vitro procedures for propagation of shrubs and trees, Pierik[294] suggested that there were four principal reasons that tissue culture procedures have been important for propagation and will remain so:

1. When the existing methods of vegetative propagation in vivo are too slow and/or sometimes nonprofitable
2. When vegetative propagation in vivo is not possible
3. When plant breeders need several plants of a selected genotype without delay
4. When production of plant material free from pathogens (viruses, bacteria, molds, nematodes) is necessary

Most fruit tissue culture studies to date have been concerned with propagation to some extent. In some crops, tissue culture may represent the only potential mode of propagation. Eeuwens,[90] for instance, states that excepting the rare instances that date palms can be propagated from root suckers, there is no way to commercially asexually propagate coconut and date palms.

I. Pathogen-Host Interactions

Arya and Hildebrandt[21a-b] and Arya et al.[22,23] have grown crown-gall induced galls of grape in vitro. Caporali and Weltzien[63] have actually infected callus cultures of peach with *Taphrina deformans* in vitro and studied the host pathogen interactions.

J. Embryo Cultures

One of the earliest uses for tissue culture was growing immature embryos to maturity in vitro. Tukey[377] developed a very simple medium for growing early aborting cherry embryos to maturity. Other authors have reported similar studies.[4,402]

Embryo cultures have also been used as an aid in obtaining viable offspring from wide crosses. The most famous example of this has been associated with the development of Triticale, a hybrid between wheat (*Triticum*) and rye (*Secale*). The Triticale literature has been reviewed by Gustafson.[137] The author found no literature relating to such wide crosses in fruit crops.

Williams[396] has reported an exciting new method that utilizes a modified embryo culture for producing wide crosses. Caucasian clover (*Trifolium ambiguum*) and white clover (*T. repens*) were hybridized by standard procedures. After 14 to 16 days, pods from the female parents were removed, surface-sterilized, and dissected. If an ovule was found to contain an embryo large enough for removal, it was gently transplanted into normal cellular "nurse endosperm" that had been obtained from a fertile cross. The entire structure, hybrid embryo and nurse endosperm, was then transplanted onto tissue culture medium for further development. In this manner, hybrids of this sexually infertile cross were obtained. The hybrid nature has been confirmed by chromosome counts and by morphological characters. Williams' system will undoubtedly have value for fruit breeders interested in wide crosses.

VII. VARIATION IN TISSUE CULTURE

Variation has been an ubiquitous phenomenon associated with tissue culture of plants.[345] Variants associated with in vitro plant tissue culture may be classified as follows: (1) physical and morphological changes in undifferentiated callus; (2) variability in organogenesis; (3) changes manifested in differentiated plantlets; and (4) chromosomal changes. Each will be discussed briefly.

A. Physical and Morphological Changes in Unorganized Callus

A number of different expressions of physiological and morphological changes have been reported in unorganized tissue including (1) habituation (loss of exogenous requirement for some growth factor, usually auxin), (2) changes in biochemical sensitivity and requirements, (3) alterations of growth habit, and (4) modifications of cellular constituents.

1. Habituation

The disappearance of the auxin requirement received considerable study in early investigations.[129-131] Habituation can be detected by several different procedures: (1) following long-term culture of various plants,[129] (2) progressively lowering the quantities of auxin in the medium,[117] and (3) isolating callus from tumors of virus- or bacterial-infected plants.[33] For example, Fox[117] isolated two strains of tobacco callus, one of which had no requirement for auxin and another which had no requirement for either auxin or kinetin from pith explants obligate for both auxin and kinetin indicating that various types of habituation can be independent of each other.

Kochba et al.[191] succeeded in producing embryogenic and nonembryogenic callus lines of "Shamouti" orange. Some of these lines were habituated in subsequent years. This line has been utilized for many studies of embryogeny and the effects of media components on this embryogeny. It has been reported,[190] for instance, that both embryogenic and nonembryogenic callus lines of "Shamouti" orange (*Citrus sinensis* Osb.) grew as well on media containing 1 to 5% galactose, lactose, or raffinose as on medium containing 5% sucrose. Further, the authors noted that nonembryogenic callus produced embryoids on medium with 7 to 10% galactose while the presence of sucrose in any concentration suppressed the formation of embryoids. They found that increasing the levels of IAA in the medium suppressed this galactose-induced embryoid formation. They postulate that these sugars inhibit auxin synthesis and that reduced auxin level is conducive to embryogenesis in *Citrus* callus. This hypothesis was supported by using labeled IAA. It was found that embryogenic callus lines were capable of disposing of IAA by forming an IAA-asparate complex, and this was not detected in the nonembryogenic lines.[188,190] Epstein et al.[99] reported that the rate of this IAA degradation by embryogenic lines occurred faster in the dark than in the light.

The phenomenon of habituation was reported to be so frequent in "Shamouti" orange that it could be utilized as an aid in the isolation of single cell clones derived from protoplasts.[360] These Citrus lines have also been used to study the effects of hormones,[190] sucrose,[133] radiation,[186,187] and transfer interval[183] on differentiation. The literature of Citrus tissue culture has been reviewed by Button and Kochba.[58] Habituated callus have also been reported for *Rubus* tissue[312] and apple.[287] The cause of habituation remains unclear. Syono and Furuya[365] have shown that it is reversible in tobacco if the callus is treated with 1.0 mg/ℓ of IAA at an early stage of subculturing, but not after prolonged transfers.

2. Biochemical Sensitivity and Requirements

This subject has been reviewed by Chaleff and Carlson.[66,67] Although there have been no reports of utilizing fruit cultures for such studies, plant cells in culture have demonstrated biochemical sensitivities similar to that observed in microbial genetic studies. For instance, isolation of artificially produced and spontaneous mutant cells in culture have yielded strains of carrot resistant to acriflavine,[37] tomato roots which tolerate low pyridoxine levels,[395] streptomycin-resistant cells of haploid Petunia[33a-b] and plants of tobacco,[220] and 8-azaquanine resistance has also been reported in tobacco,[206a] The ability of plant cells to grow in normally inhibitory levels of various amino acid analogs[67,285,390-394,403] has yielded strains of cells which vary in their content of various nutritionally important amino acids. Resistance to threonine[145] and sodium chloride[257] has also been selected. While 5-bromodeoxyuridine resistant strains of soybean[280,281] and tobacco[219] have been found, its incorporation into haploid cells of tobacco and fern has yielded presumptive mutants which may require amino acids, vitamins, or purines.[64,65]

3. Changes in Growth Habit

Variation in plant tissue culture also includes changes in the growth habit, rates, appearance, and requirements. Callus cultures of almond,[227] chestnut,[381] grape,[21a-23] *Citrus*,[68] and apple[384] have exhibited callus variability. The *Citrus* cultures, for instance, have yielded two types of callus, types "A" and "B". Type A was characterized as compact, slow growing, while Type B was friable, spongy, and fast growing.

Single cell lines derived from normal and *Phylloxera* incited-galls produced lines that varied in their growth rate, texture, color, and ability to utilize various carbohydrate sources. A fast growing line, for instance, was whitish grey, a medium growth rate form had a whitish yellow culture, and their slow growing clone was lemon yellow and firm.[21a-23] Townsley[372] reported that tissue cultures of cacao contained two cell types, dark brown cells and clear cells. These two types could be separated and subcultured as different forms.

4. Modification of Cellular Constituents

The most obvious change in cellular content has been changes in pigmentation. This was discussed above. Other authors have reported other changes within tissue cultured fruit cells. Mehra and Mehra,[227] for instance, reported oil bearing cells among their suspension cultures of Almond. Chaturvedi et al.[68] reported *Citrus grandis* cultures that, in addition, to exhibiting differences in growth rate, also differed in their content of amino acids, total protein, nitrogen, and sugars.

B. Variability in Organogenesis

A fairly general phenomenon of tissue culture is variation in ability to produce embryoids, organs, or tissues. One manifestation is the varying ability of certain cells to form embryoids. In *Citrus* callus of integument tissue could not be induced to form embryoids whereas nucellar tissue produces embryoids with relative ease. The rate of embryogenesis tends to decline with age.[235] Yokoyama and Takeuchi[398] reported that only callus derived directly from hypocotyl explants of persimmon were capable of forming shoots, and that within five generations of subculture, root-forming capacity was also lost from the callus.

Eeuwens and Blake[91] reported that coconut endosperm grown in vitro produced a callus that could be subcultured while stem, leaf, and inflorescence tissue yielded a callus that would not subculture. Trippi[375] reported that a similar loss of callus-forming capacity in Chestnut was related to the age of the parent plant from which the explant came.

Grinblat[136] reports that *Citrus madurensis* cultures derived from seedling stems varied in their capacity to differentiate. Some cultures formed only roots, some only shoots, and others did not differentiate at all.

Hedtrich[143] induced root formation from leaf discs and callus of *Prunus mahaleb.* He found that roots derived from leaves were geotropic and possessed root hairs while callus-derived roots were nongeotropic and had no root hairs.

Mullins and Srinivasan[249] found that a *Vitis vinfera* clone in vitro produced abnormal-looking embryoids with lobed cotyledons and pluricotyly, but when these plants grew to intact clones, all appeared normal. Krul and Worley[196] reported a similar phenomenon in which some embryoids had variegated or malformed primary leaves. The authors reported that the variegated-leaf forms did not develop into intact plants, but the other types developed plants that were all of normal type. Takatori et al.[367] reported a similar phenomenon in asparagus where abnormal appearing plantlets failed to develop into intact plants.

Holsten et al.[158] produced callus lines from cotyledons and roots of seedling soybean. They found that *Rhizobium* exhibited no propensity to form nodules on cotyledon tissue, while root tip callus readily supported the bacteria to form nodules in vitro.

C. Changes Manifested in Differentiated Plantlets

Intact plants isolated from tissue culture have been reported to be variants in many cases (this area has been briefly reviewed by Skirvin[345]). Plantlets of sugarcane isolated through tissue culture techniques have shown variation in morphology, chromosome number, and enzyme systems. Heinz and Mee,[147] for instance, have isolated plants with heavier tillering, slower growth rate, and increased erectness. One plant in particular had a dwarf-growing habit with stiff leaves and high silicate deposits on leaf epidermis. At the present time many hundreds of tissue culture-derived plants are in selection fields in Taiwan and Hawaii.[263]

Widholm[390,391,393] has succeeded in selecting both callus and intact plant lines of carrot which are resistant to high levels of 5-methyl-tryptophan. This selection was accomplished by growing cells in media with high levels of this inhibitor.

Skirvin and Janick[349] developed in vitro techniques to regenerate plantlets (calliclones) from callus of scented geraniums. Calliclones were compared to plants derived from stem, root, and petiole cuttings of five cultivars. Plants from stem cuttings of all cultivars were uniform and identical to the parental clone. Plants from root and petiole cuttings were more variable, the amount of variation being dependent upon cultivar. High variability was associated with calliclones. Aberrant types included changes in plant and organ size, leaf and flower morphology, essential oil constituents, fasciation, pubescence, and anthocyanin pigmentation. Calliclone variation was dependent upon clone and age of callus. When variable calliclones were transferred back to tissue culture for a second generation, the variation decreased, and certain lines apparently had stabilized in their mutated phenotype. Variability in calliclones was due to segregation of chimeral tissue, euploid changes, and heritable changes which may involve individual chromosomal aberrations or simple gene mutations. One of the calliclones developed in this study has been released as "Velvet Rose"[350] and is believed to be the first named cultivar produced from tissue culture of any type. Polyploid calliclone production has also been reported in these studies.[166,351] Few fruit tissue cultures have developed to the point that variation can be observed. Lane[199] has reported the sole variant in this area. He found that in vitro propagated apples lacked trichomes. It was unclear whether this smooth state persists into maturity. Grape plants produced in vitro frequently revert to a condition of altered phyllotaxy and lack of tendrils.[196,277-279] It has been determined that these plants have reverted to juvenility. Similar variability is undoubtedly to be encountered in *Citrus* plantlets derived in vitro since they always revert to juvenility, and certain *Citrus* cultivars exhibit thorns during juvenility. Researchers investigating in vitro-derived variability should be wary of this form of variation and not assume these changes to be permanent. Interestingly, Boxus and Quoirin[49] reported no variation among *Prunus* and *Malus* plantlets potted and grown in the field over an 11-year period.

D. Chromosomal Changes

Although the chromosomal constitution of certain plants seems to be highly stable in vitro[340a] much of the variability found in tissue culture can be directly or indirectly attributed to gross chromosomal changes and chromosomal abnormalities. Chromosomal variation is a common feature of plant tissue cultures. The most common changes are increased in ploidy levels. Murashige et al.[254], for instance, reported that within one passage in culture, lemon callus had developed about 30% tetraploid (4n) cells; by the fourth transfer, 70% of the cells were 4n; and within 1 year the entire culture was 4n and 8n. Anther cultures of *Prunus* produce n, 2n, and 3n callus in vitro, but no organogenesis has been achieved.[229] Diploid callus derived from anthers, as well as tetraploid callus derived from 2n somatic tissue of coffee, has been reported.[339] Ravkin and Popov[311] have reported chromosomal variation apple cultures.

Vardi[379] reports that both a 0.2% ethyl methane sulfonate (EMS) and 7.5KR of X-irradiation have been applied to "Shamouti" orange protoplasts. Regenerated plants have been mainly diploid ($2n = 18$), but at least one plant is tetraploid ($2n = 36$).

VIII. FRUIT IMPROVEMENT THROUGH TISSUE CULTURE

Many fruit cultivars, although widely grown and popular with the public, are found to be deficient for certain characteristics such as disease and insect resistance, pigmentation, time of harvest, fruit set, nutritive value, storage ability, etc. Cultivar improvement for such characters has traditionally been approached from two directions: (1) using the sexual system to reproduce variable populations from which improved forms of the original cultivar can be selected and (2) the use of mutagens (both chemical and physical) on clonal plants to produce mutants which will represent an improved form of the parental plant.

The sexual approach has been remarkably efficient, but as unique cultivars have been developed, the public's conception of particular fruits may become so prejudiced that later cultivars must resemble the older, standard type or face consumer rejection. For example, the "Bartlett" pear, introduced about 1770, is still the standard of quality in pear production although it is extremely sensitive to fireblight and, hence, is unsuited for culture in the humid portions of the world (including much of the U.S.).

Due to heterozygosity, polyploidization, and outcrossing, most fruit crops do not breed true from seed. Hybridization of fruit crops may yield improved plant types, some of which may superficially resemble the parental cultivar, but the likelihood of developing a clone sexually that is identical to the parental cultivar with only a single improvement (e.g., "Jonathan" apple with fireblight resistance) is essentially nil. The sexual system imposes an even more severe restraint to genetic improvement for fruit crops which are partially or completely sterile (e.g., "Thompson seedless" grapes).

The use of mutagens on the parental cultivar is, in theory, a good procedure to obtain improved forms of a clone, but in practice has usually proved to be less than satisfactory. The problem is that when mutations are induced in a multicellular organism, the mutated cells are immediately placed in competition with all of the nonmutated cells which surround it.* The mutated cell may develop into a mass of cells and perhaps into a cell layer provided that the mutation does not reduce the division rate or limit the cell's ability to grow outward and express its own phenotype. In most cases the mutated cell's ability to duplicate is so reduced that only a limited number of mutated plants and plant parts are observed.

In addition, because most mutations occur in rapidly dividing cellular layers at the apices of plants (the histogenic layers), often only a single layer is affected. If the mutation occurs in the epidermis (L_{I}) layers, for instance, then the mutation may be expressed directly, but if it occurs in the internal layers (L_{II}, L_{III}), mericlinal or periclinal chimeras of various complexity may arise. If the chimera is mericlinal, a segment of the plant may exhibit the mutant character either directly or with adventitious bud formation. If the chimera is periclinal ("hand-in-glove"), the mutation may be completely masked within the epidermal layer.

Inability to either recognize or separate fruit crops chimeras into pure types has limited the usefulness of mutation induction to improve particular clones. This unfortunate situation could be avoided by growing shoots or even complete plants from single cells. When a mutation occurs in a single cell, the resulting individual will be either completely normal or completely mutant in genotype. In addition, with plants derived from single cells, the number of mutated plants produced should be considerably higher and the range of mutation types wider. It therefore seems to be of utmost importance in fruit crops (as well as other vegetatively propagated plants) to look for or to develop methods by which shoots or plants can be obtained from single cells.

There are several methods which might be useful in the production of large numbers of plants of single-cell origin; some of these include: (1) tissue culture; (2) production of adventitious shoots from internodal portions of plants; (3) leaf-petiole cuttings; and (4) root sucker formation.

Kochba et al.[191] suggest "the use of our technique, namely the culture of plants from nucellar and ovular explants, allows induction of mutations in cells, giving subsequently rise to entire plants . . . These results may be of aid in mutation research, allowing irradiation at stages prior to embryonic development."

* Some of this discussion has been paraphrased from the fascinating paper by Broertjes et al.[51]

The propensity of certain fruit crops to produce adventitious shoots from either stem or root portions (e.g., brambles and apple rootstocks) may be useful to produce large numbers of pure-type individuals which vary only slightly from the parental clone.[80,81] The intraclonal variation sought in such studies may either be induced by mutagens or appear spontaneously.

Since a clone is ultimately derived from the union of a single egg and sperm, one would expect that all cells of a clone are genetically identical, and, therefore, single cells of a clone which are induced to yield whole plants should exhibit extreme uniformity. Such is not the case, and variability, not uniformity, seems to be the rule with nearly all plants cultured in tissue culture exhibiting some form of variation such as: (A) physical and morphological changes in undifferentiated callus (habituation, biochemical sensitivity and requirements, change in growth habit, modification of cellular constituents); (B) variability in organogenesis; and (C) changes manifested in differentiated plants (growth rate, growth form, morphology, leaf morphology, increases or decreases in production of certain biochemicals, phyllotaxy, partial disease resistance, etc.).

Variation of this nature has been thoroughly documented for many crops. The utilization of such tissue culture derived intraclonal variants for fruit crop improvement is presently blocked by a few technical barriers which include: (1) the routine production of contaminant-free cultures from mature fruit cultivars; (2) the development of suitable media for callus proliferation; (3) development of optimal media for plantlet production from single cells; and (4) production of independent plants from callus in quantity.

The natural variability associated with callus-derived clones (calliclones) represents a pool upon which selection can be imposed. The amount of variability that can be expected will vary with the clone. This system of exploiting natural variation seems especially applicable to old cultivars which could be expected to have accumulated large numbers of mutant cells that may have stabilized into chimeras of various complexity.

Variability of callus and calliclones may also be increased by aging and repeated transfer. Other techniques useful to increase the percentage of variability will, of course, include the use of mutagenic agents in the callus proliferation stage and selection applied to single cell clones for stress conditions and/or ability to resist or utilize specific metabolites.

The association of polyploidy with tissue culture may be utilized as a technique to obtain either increased or reduced chromosome number. However, affinity of cells to double in culture appears to be a disadvantage in most cases, and a screening technique may be necessary to eliminate this type of change.

There are a number of immediate and potential situations in which such variation could be exploited. These include: (1) old vegetively-propagated adapted clones (e.g., "Bartlett" pear, "Concord" grape, "Russet Burbank" potato, and clonal rootstocks such as the Malling series); (2) seedless or apomictic lines (e.g., "Thompson Seedless" grape, "Washington Navel" orange, certain brambles, and Kentucky bluegrass); and (3) sterile or nonflowering lines (e.g., certain sweet potatoes).

The ability to obtain whole plants from single cells of fruit cultivars will hold special interest for the plant breeder, and will undoubtedly provide an additional method to increase intraclonal variation. Its most important use will be to obtain mild changes in unique, highly-adapted clones or to obtain variability in clones in which the sexual apparatus is disturbed. This system will probably never completely replace conventional breeding methods for most crops, but it does appear to offer significant benefits in fruit improvement.

IX. FRUIT TISSUE CULTURES IN THE FUTURE*

The major thrust of fruit tissue culturalists to date has been related to propagation of fruit crops. This research has resulted in important procedures for propagation, but, more importantly, this research provides tools for more basic scientific studies related to gene stability, differentiation, and genotype-environment interactions. In 1963 Nitsch[270] predicted that fruit tissue cultures would have value in answering questions in four areas:

1. Finding of the factors which control fruit growth and which are supplied to the ovary by the rest of the plant
2. Physiology of ripening
3. Metabolism of certain compounds peculiar to fruits
4. Fruit diseases of physiological and microbiological origins

The use of fruit tissue cultures to study these areas has begun, but there are other important areas of research in which fruit tissue cultures will be valuable. Some of the more significant studies to research for the future are related below:

1. Embryogenic callus of "Shamouti" orange and suspension cultures of *Rubus* have been utilized to study the complex interactions of media components, hormones, and environmental factors on enzyme induction, differentiation, and metabolism. Continued studies with these species, as well as other plants, will undoubtedly help us understand factors related to differentiation.
2. Meristem cultures have been utilized to produce virus-free stock plants. Some of these plants have been proliferated and stored in the cold for several years. The expansion of these procedures will facilitate the establishment of germplasm banks for asexually propagated crops. The use of such procedures will also reduce the work load on the tissue culturalist who now must spend large portions of his research time subculturing old, but important cultures.
3. Anther cultures of fruit crops are becoming more common. Apple pollen is capable of forming torpedo-stage embryoids. It is only a matter of time until intact haploid plants are produced with regularity for many fruit crops.
4. Procedures have been developed to produce wide crosses of plants through protoplast fusion, embryo culture, and embryo transplant. These procedures will be valuable to plant breeders in the future for studies of cellular fusion, plasmid and organelle uptake, as well as recombinant DNA.
5. Variability encountered in tissue culture is a definite problem to individuals interested in maintaining the genetic integrity of a clone, and future studies may develop methods to minimize or eliminate such variability. On the other hand, variability in tissue culture (both natural and induced) has been demonstrated to be of value for improvement of certain clones through the screening of intraclonal variation. The use of this variability seems to be particularly well-suited to asexually propagated fruit crops and will be of importance in the future.
6. Propagation of certain difficult-to-propagate or difficult-to-root plants will continue to be of major importance in tissue culture. In some crops, tissue culture will become the principal propagation method.

* Readers interested in this topic should consult the interesting paper by F. A. de Fossard.[84]

X. CONCLUSION

Researchers working with fruit crops can be proud of the remarkable progress that has occurred since Tukey's first successful embryo cultures in 1933. There appears to be no fruit crop that cannot be grown in tissue culture provided the researcher has sufficient patience and resources. The key to successful cultures of difficult-to-propagate plants may lie with the correct combination of medium components, environmental conditions, and explant source. The future is bright for fruit crops in tissue culture and advances related to propagation, genetics, differentiation, and plant-medium interactions are inevitable.

Table 1
FRUIT CROPS GROWN IN TISSUE CULTURE

Family (Genus and/or species)	Explant source[a]	Type of differentiation encountered[b]	Medium: Salts[c] Macro	Medium: Salts[c] Micro	Medium: Vitamins[d]	Medium: Hormones[e]	Medium: Sugar[f]	Medium: Other[g]	Ref.
Actinidiaceae (Dilleniaceae)									
Actinidia chinensis Pl. (Chinese gooseberry)	S	Ad	H	H	H	Z(1.0)	S(20)	Cs(500) Glu(500) Ad(40)	138
	S	R shoots soaked 24 hr in solution of IAA (1.0) *or* IBA(1.0)					—	—	
	S	Then to medium	H	H	H	—[h]	S(10)	C(4,000)	
No name given	S	C	H-53	H-53	H-53	Kn NAA	G(40)	—	155—157, 238
	S	C, Ad, R	H-53	H-53	H-53	—	G(40)	—	
	Fr	C	MS	MS	MS	2,4-D(0.3)	G(40)	Cc(8%)	
	F	C	Monnier		Thia(1.0)	—	G(50)	—	
	Stamens	C	Monnier		Thia(1.0)	NAA(5.0)	G(50)	Cc(10%)	
Annonaceae									
Annona squamosa L. (custard apple)	E + En, Fr	C	MS (½ MS Fe)	MS	—	2,4-D(2.0) Kn(5.0)	S(20)	Cc(15%) Cs(400)	27
A. cherimola Mill. (cherimoya)	Fr	C	N-51	N-51	N-51	±IAA(10)	S(20)	—	330
Araceae									
Monstera deliciosa Liebm. (ceriman)	Fr	C	N-51	N-51	N-51	±IAA(10)	S(20)	—	330
Betulaceae									
Corylus sp.	S	C	G	G	G	(—)[h]	(—)	—	60
"European filbert" No name given	—	Protoplasts	—	—	—	—	—	—	246
Corylus avellana L.	E	C, Em(?)	MS	MS	Myo(100) CaP(10) Thia(2.0) Nia(5.0)	2,4-D(1.0) Kn(1.0)	S20	Cs(200) Ad(2.0)	306
	E	G	MS	MS	Myo(100) CaP(10) Thia(2.0) Nia(5.0)	Kn(1.0)	S20	Cs(200) Ad(2.0)	
Bromeliaceae									
Ananas sativus (*A. Comosus*) (pineapple)	C	Enz	(—)	(—)	(—)	(—)	(—)	(—)	21
"Kew"	St	Ax	KN	N	—	NAA(1.0)	S(20)	—	198

Table 1 (Continued)
FRUIT CROPS GROWN IN TISSUE CULTURE

Family (Genus and/or species)	Explant source[a]	Type of differentiation encountered[b]	Medium: Salts[c] Macro	Medium: Salts[c] Micro	Medium: Vitamins[d]	Medium: Hormones[e]	Medium: Sugar[f]	Medium: Other[g]	Ref.
No name given	St	Ax (buds swell)	NN-69	NN-69	NN-69	IAA(0.1) BA(0.1)	S(20)	Ad(40)	224
	St	Ax (buds unfold)	MS	MS	MS Myo(100) Gly(2.0)	NAA(1.8) IBA(2.0) Kn (2.1)	S(20)	Ad(40)	
	St	Ax	MS	MS	MS Myo(100) Gly(2.0)	NAA(1.8) IBA(2.0) Kn(2.1) GA_3(—)	S(20)	Ad(40)	
	St	Ax (multiple shoots)	MS	MS	MS	NAA(1.8) IBA(2.0) Kn(2.1)	S(20)	Ad(40)	
	St	R	MS	MS	MS	NAA(0.1) *or* IBA(0.4)	S(20)	Ad(40)	
Ananas comosus L. Merr. "Cayenne Lisse"	St, B	G, Ax	MS	H-53	Mo	BA(0.2) NAA(0.1)	S(30)	—	286
	St, B	Ax (multiple shoots)	MS	H-53	Mo	BA(0.1) IBA(1.0)	S(20)	—	
No name given	St, B	Ad(—), Ax(—)	—	—	—	—	—	—	368, 369
Others	—	— —	—	—	—	—	—	—	7, 221
Byttneriaceae *Theobroma cacao* L. (chocolate)	E(2.5—10 mm)	Em	MS	MS	MS Myo(100) Gly(2.0)	NAA(1.5)	S(30)	Cs(1,000) Cc(10%)	291, 292
No name given	S, P	C (+ chocolate aroma)	GE-PRL-4	GE-PRL-4	GE-PRL-4	—	—	—	372
Others	—	—	—	—		—	—	—	164a
Caricaceae *Carica papaya* L. (papaya)	St	Ax	MS	MS	MS Myo(100) Gly(2.0)	NAA(0.09) BA(0.45)	S(30)	—	212—214
	St	R	MS	MS	MS Myo(100) Gly(2.0)	NAA(0.09)	S(30)	—	
	St	Ax (initial)	MS	MS	MS Myo(100) Gly(2.0)	NAA(0.19) Kn(10.7)	S(30)	—	
	St	Ax (subculture)	MS	MS	MS	NAA(0.18)	S(30)	—	

					Myo(100) Gly(2.0)	BA(0.45)			
	St	R	MS	MS	MS Myo(100) Gly(2.0)	IBA(0.1—3.0) *or* NAA((0.18—0.9)	S(30)	—	
	A	Ad	MS	MS	MS Myo(100) Gly(2.0)	BA(0.45) NAA(0.09)	S(30)	C(10,000)	
"Solo"	S, B, St	C	MS	MS	MS Myo(100) Gly(2.0)	Kn(1.0)	S(30)	—	226
	S, B, St	R	MS	MS	MS Myo(100) Gly(2.0)	IBA(5.0) Kn(1.0)	S(30)	—	
No name given	Fr	C	N-51	N-51	N-51	IAA(10)	S(20)	—	330
"Solo No. 1"	Se	C	(—) + $NaH_2PO_4 \cdot H_2O$(170)	(—)	Myo(100) Gly(4.0) Thia(0.4) Nia(1.0) Pyr(1.0)	IAA(1.0) Kn(0.1)	S(30)	Cc(15%) Ad(80)	397
	Se	Ax	(—) + $NaH_2PO_4 \cdot H_2O$(170)	(—)	Myo(100) Gly(4.0) Thia(0.4) Nia(1.0) Pyr(1.0)	IAA(0—0.05) Kn(1.0—2.0)	S(30)	—	
	Se	Ax (multiple shoots)	(—) + $NaH_2PO_4 \cdot H_2O$(170)	(—)	Myo(100) Gly(4.0) Thia(0.4) Nia(1.0) Pyr(1.0)	IAA(0.05) Kn(5) or BA(0.5—1.0)	S(30)	—	
	Se	R	(—) + $NaH_2PO_4 \cdot H_2O$(170)	(—)	Myo(100) Gly(4.0) Thia(0.4) Nia(1.0) Pyr(1.0)	IAA(5.0)	S(30)	Light	
"Solo No. 8"	St	Ax, C, Em	(—)	(—)	(—)	(—)	(—)	(—)	10
Others	—	—	—	—	—	—	—	—	82, 226a
Dilleniaceae (see Actinidiaceae)									
Ebenaceae									
Diospyros Kaki Thumb (Japanese persimmon) "Tsurunoko"	E(hypocotyl)	C	MS	MS	Myo(100) Nia(0.1)	NAA(3.0) Kn(0.1)	S(30)	Ye(10)	398
	E(hypocotyl)	C(subculture)	MS	MS	Myo(100) Nia(0.1)	NAA(1.0) Kn(0.1)	S(30)	Ye(10)	
	E(hypocotyl)	Ax, R	MS	MS	Myo(100) Nia(0.1)	NAA(1.0) Kn(0.1—1.0)	S(30)	Ye(10)	
"Fuyu"	S	C	FN	FN	Myo(0.1) Bio(0.01)	—	S(20)	Cys(10) Ad(5)	118

Table 1 (Continued)
FRUIT CROPS GROWN IN TISSUE CULTURE

Family (Genus and/or species)	Explant source[a]	Type of differentiation encountered[b]	Medium: Salts[c] Macro	Medium: Salts[c] Micro	Medium: Vitamins[d]	Medium: Hormones[e]	Medium: Sugar[f]	Medium: Other[g]	Ref.
					CaP(10) Thia(1.0) Nia(1.0) Pyr(1.0)				
Ericaceae									
Vaccinium angustifolium Ait (Lowbush blueberry)	Rh	G	W—43	W—43	W—43	—	S(20)	—	28
No name given	Fr	C	MS	MS	MS Myo(100) Gly(2.0)	2,4-D(0.1) BA(0.1) *or* Kn(0.1)	S(30)	—	264
Seedlings	St, E	Ax, Ad	A	A	A	IAA(4.0) 2iP(15)	S(30)	Ads(80)	265, 404
	St, E	Ax (subculture)	A	A	A	2iP(4.0) IAA(2.0)	S(30)	—	
No name given	S	C	MS	MS	MS Myo(100) Gly(2.0)	2,4-D(0.1 *or* 0.5)	S(30)	—	266
No name given	St, B	Ax	(—)	(—)	(—)	IBA(0.5) 2iP(15)	S(15)	—	127
	St, B	R	(—)	(—)	(—)	IBA(0.5)	S(15)	—	
V. ashei Reade (Rabbiteye blueberry)	A	C	—	—	—	—	—	—	217 a-c
V. darrowi Camp	St	Ad, Ax, C	L	L	L Myo(100) Gly(2.0)	2iP(15)	S(30)	Cs(300) pH 5.7	
V. elliottii Chapman *V. fuscatum* complex									
V. atrococcum *V. corymbosum* L. (highbush blueberry) *V. darrowi* *V. elliottii*		Ax, R	(—)	(—)	(—)	(—)	—	—	406
Fagaceae									
Castanea crenata × *C. sativa* "M15" (chestnut)	L	C	H-53	H-53	—	NAA(0.1)	—	Gte(20%)	38
C. sativa (European chestnut) "Saint-Maixent"	L	C	H-53	H-53	—	NAA(0.1)	—	Gte(20%)	

No name given	Fr	C	H-53	NN-69	Myo(500) Bio(0.1) CaP(0.5)	2, 4-D (1 or 10) Kn (0.5) *or* BA (0.5)	S (30)	Cys (10) Cc (12%)	381
	Fr	R	H-53	NN-69	Myo (500) Bio (0.1) Ca P (0.5)	IBA(10) *or* NAA(10)	S(30)	Cys(10) Cc(12%)	
C. vesca Gaertn. (a 100-year-old tree)	—	C	½ K	B	Thia(1.0) Myo(50 *or* 500) Bio(0.1) CaP(0.5)	IAA(0.01)	G(20)	Cys(10)	161—163
C. vulgaris	S	C	H-53	H-53	H-53	NAA(1.0) Kn(1.0)	(—)	Cc(10—12%)	374, 375
C. crenata "Ginyose"	S	C	FN	FN	Myo(100) Bio(0.01) CaP(10) Thia(1.0) Nia(1.0) Pyr(1.0)	—	S(20)	Cys(10) Ad(5)	118
C. dentata (Marsh) Borkh. (American chestnut)	E	G and/or radicle elongation	W-63'	W-63	W-63	—	S(20)	—	225b
	E	Ax	SC	SC	SC Myo(100) Gly(2.0)	BA(2.0) NAA(0.1)	S(30)	As(50)	
	St	Ax (limited)	SC	SC	SC Myo(100) Gly(2.0)	BA(2.0) NAA(0.1)	S(30)	As(50)	
C. mollissima Blume (Chinese chestnut)	S,L St	C	W-63' SC	W-63 SC	W-63 SC Myo(100) Gly(2.0)	— BA(2.0) NAA(0.1)	S(20) S(30)	— As(50)	
C. mollissima × *C. dentata* seedlings	Cotyledon pieces	Ad(—), Em(—)	SC	SC	SC Myo(100) Gly(2.0)	BA(2.0) NAA(0.1)	S(30)	As(50)	
Juglandaceae *Juglans regia* (walnut) "Lake"	St	C	CH	CH	CH	2,4-D (5.0) μm)	S(30)	Cc(10%)	71
J. nigra L. (black walnut) "Kwik-Krop"	St	Ax	CH	CH	CH	BA(5 μm) IBA (0.5 μm) *or* 2,4-D(0.25 μm)	S(30)	Cc(10%)	
Lauraceae *Persea americana* (Avocado)	Fr	C	N-51	N-51	N-51	—	S(20)	—	329
	Fr	C	N-51	N-51	N-51	±IAA(10)	S(20)	—	330—333

Table 1 (Continued)
FRUIT CROPS GROWN IN TISSUE CULTURE

Family (Genus and/or species)	Explant source[a]	Type of differentiation encountered[b]	Medium: Salts[c] Macro	Medium: Salts[c] Micro	Medium: Vitamins[d]	Medium: Hormones[e]	Medium: Sugar[f]	Medium: Other[g]	Ref.
Leguminosae (Fabaceae)									
Ceratonia siligua L. (carob)	Fr	C	N-51	N-51	N-51	±IAA(10)	S(20)	—	330
Moraceae									
Morus alba L. (mulberry)	(—)	C	MS NaFe-EDTA (5.0)	MS $ZnSO_4$ (10.6)	W-43 Gly(2.0) Bio(0.01) CaP(0.1) Myo(100)	NAA(1.0) Kn(1.0)	S(30)	—	132
		R	MS NaFe-EDTA (5.0)	MS $ZnSO_4$ (10.6)	W-43 Gly(2.0) Bio(0.01) CaP(0.1) Myo(100)	IPA(1.0) Kn(1.0)	S(30)	—	
Others	—	—	—	—	—	—	—	—	283,284
Musaceae									
Musa sapientum (banana)	Fr	C	W-43	W-43	W-43	2,4-D(5.0)	S(20)	Cc(5%)	236
No name given	Fr	C	N-51	N-51	N-51	±IAA(10)	S(20)	—	330
M. cavendishii Lamb. (plantain)	St	Ad, Ax (—)	Smith & Murashige (1970)			—	—	Ads(160)	218, 355
Myrtaceae									
Feijoa sellowiana O. Berg. (feioja or pineapple guava)	Fr	C	N-51	N-51	N-51	±IAA(10)	S(20)	—	330
Psidium guava L.	E + En, Fr	C	MS (½ MS Fe)	MS	—	2,4-D(2.0) Kn(5.0)	S(20)	Cc(15%) Cs(400)	27
P. guajava (guava)	Fr	C	N-51	N-51	N-51	±IAA(10)	S(20)	—	330
Oleaceae									
Olea europea (olive)	C	C	W-43 (2×)	W-43 (2×)	W-43	IAA(2.0) Kn(0.2)	S(20)	—	202
"Manzanillo"	Fr	C	W-43 (2×)	W-43 (2×)	W-43	IAA(2.0) Kn(20.0)	S(20)	—	203
No name given	Fr	C	N-51	N-51	N-51	±IAA(10)	S(20)	—	330
"Oleaster"	Em	G	—	—	—	—	—	—	87a
Palmae									
Phoenix dactylifera L.	En	C, Ax	MS	MS	MS	NAA(1.0)	S(50)	Cc(15%)	18

						Kn(0.1)			
(date palm)									
Seedlings	F	C	E	E	E Myo(100)	NAA(0.047) BA(1.1)	S(68.4)	Glu(100) Arg(100) Asp(100)	90, 91
	F	R (initiation)	E	E	E	NAA(0.047) BA(0.011)	S(68.4)	Glu(100) Arg(100) Asp(100)	
	F	R (growth)	E	E	E	NAA(0.0092) BA(0.011) GA_3(0.17)	S(34.2)	Glu(100) Arg(100) Asp(100)	
	F	*Or* R (growth)	¼ MS	¼ MS	¼ MS	GA_3(0.17) NAA(0.0092) BA(0.011)	S(68.4)	Glu(100) Arg(100) Asp(100)	
No name given	Fr	C	MS + NaH_2PO_4 (170)	MS	Myo(100) Thia(0.4)	2,4-D(100)	S(30)	Ad(40) C(3000)	315
	Fr	Ad	MS + NaH_2PO_4 (170)	MS	Myo(100) Thia(0.4)	—	S(30)	Ad(40) C(3000)	
Others	—	—	—	—	—	—	—	—	10, 314, 335
Cocos nucifera L. (coconut) "Makapuno"	E	G	NN-67	NN-67	NN-67	GA_3(1.0)	(—)	(—)	26, 85, 86
	E	R	MS	MS	MS	IAA(10)	G(80)	Cc(25%)	
No name given	E	G	Various media were used			—	—	—	78
"Malayan Dwarf"	F	C	E	E	E Myo(100)	NAA(0.047) BA(1.1)	S(68.4)	Glu(100) Arg(100) Asp(100)	90, 91
	F	R (initiation)	E	E	E Myo(100)	NAA(0.047) BA(0.011)	S(68.4)	Glu(100) Arg(100) Asp(100)	90, 91
	F	R (growth)	E	E	E Myo(100)	NAA(0.0092) BA(0.011) GA_3(0.17)	S(34.2)	Glu(100) Arg(100) Asp(100)	
	F	*Or* R (growth)	¼ MS	¼ MS	¼ MS	NAA(0.0092) BA(0.011) GA_3(0.17)	S(68.4)	Glu(100) Arg(100) Asp(100)	
"Jamaican tall"	E	C	SH	SH	SH	NAA(2.5)	S(30)	—	113
"Green Malayan Dwarf"	E	Ax, Ad (—)	MS	MS	MS	IAA(10) IBA(20) BA(2.5) and/or 2iP(5.0)	S(30)	Cc(25%) C(2500—3600)	
Others	E	—	—	—	—	—	—	—	5
Passifloraceae (passion fruit) *Passiflora edulis* Sims	—	—	—	—	—	—	—	—	317a-b

Table 1 (Continued)
FRUIT CROPS GROWN IN TISSUE CULTURE

Family (Genus and/or species)	Explant source[a]	Type of differentiation encountered[b]	Medium: Salts[c] Macro	Medium: Salts[c] Micro	Medium: Vitamins[d]	Medium: Hormones[e]	Medium: Sugar[f]	Medium: Other[g]	Ref.
P. mollissima Bailey									
P. caerulea L. *P. edulis* Sims *P. foetida* L. *P. trifasciata* Lem.	S, L	Ad	MS	MS	MS Myo(100) Gly(2.0)	BA(0.1)	S(30)	—	336, 337
P. suberosa	S, L	Ad, F	MS	MS	MS Myo(100) Gly(2.0)	BA(0.1)	S(30)	—	
Proteaceae									
Macademia tetraphylla (Macademia nut)	Fr	C	N-51	N-51	N-51	±IAA(10)	S(20)	—	330
Rosaceae									
Cydonia (Quince) "MA" "Adams" "C7/1"	M	Ax	MS	H-53 (100×)	Myo(50) Bio(0.1) CaP(0.5)	2,4-D (.02) BA(1.0) GA_3(0.1)	S(20)	—	49
"Van Deman" "Missouri Mammoth"	Fr	C	W-43	W-43	Myo(0.2) Bio(0.2) CaP(0.2)	2,4-D(2-0)	(—)	Cc(10%) Cys(10)	208
Amelanchier alnifolia Nutt (saskatoon)	St	C	MS	MS	MS Myo(100) Gly(2)	2,4-D (0.25) Kn(0.2)	S(30)	—	76
A. arborea (Michx. f.) Fern (downy service berry)									
Mespilus germanica L. (medlar)	Fr	C	W-43	W-43	Myo(0.2) Bio(0.2) CaP(0.2)	2,4-D (2.0)	(—)	Cc(10%) Cys(10)	208
Chaenomeles superba Rehd (Japanese Quince)	Fr	C	W-43	W-43	Myo(0.2) Bio(0.2) CaP(0.2)	2,4-D(2.0)	(—)	Cc(10%) Cys(10)	208
Fragaria X. *ananassa* Duch. (strawberry)	St, M	Ax	MS + NaFe-EDTA (40)	MS	WS	IBA(1.0) BA(0.1)	S(30)	—	6
	St, M	R	MS + NaFe-EDTA (40)	MS	WS	—	S(30)	—	
No name given	Fr	C (friable)	MS	AK	AK	2,4-D(5.0) BA(0 or 0.1)	S(40)	—	24

	Fr	Ad	MS	AK	AK	2,4-D(0.1) BA(10)	S(40)	—	
	Fr	R	MS	AK	AK	—	S(40)	—	
"Sparkle"	F	C	W	W	Gly(7.5) CaP(0.25) Thia(0.25) Nia(1.25) Pyr(0.25)	2,4-D (1.0) Kn(0.2)	S(30)	Cc(5%)	25
"Dom anil" "Fanil" "Gorella" "Senga precosana" "Surprise de Halles"	St	Ax	K	MS	MS Myo(100) Gly(2.0)	IBA(1.0) BA(1.0)	G(40)	—	44—47
"Surecrop"	A	C	NN-69 + FeNaEDT A (2.5)	NN-69	NN-69	IAA(1.0) Kn(1.0)	S(20)	—	115, 116
No name given	St	Ax (initiation)	MS NaFe- EDTA(20)	MS	MS Myo(100) Gly(2.0)	NAA(1.0)	S(20)	Cc(5%)	173
	St	Ax (moved to this medium)	K NaFe- EDTA(20)	MS	—	—	G(40)	—	
	St	Ax (subculture)	K NaFe- EDTA(20)	MS	WS	IBA(1.0) GA_3(1.0) BA (0.2)	G(40)	—	
	St	R [basal 2—3 mm dipped in IBA(10) for 10 min, then moved to 2nd medium (above)]							
"Hoh koh-wase"	Fr	G(fruit development in vitro)	Various media were used				—	—	178
"Kendall" "Red Gauntlet"	A	C, R, Ad[j]	—	—	—	—	—	—	205, 206
"Torrey"	B, St	Ax	LF	LF	LF	—	—	—	
No name given	Fr	G	—	—	—	—	—	—	211
"Suwannee"	St	G	(A modified Knop-Berthelot's medium)						225
[Over 50 cvs.] including *Fragaria vesca* L. (wood strawberry)	M	Ax (initial)			Thia(1.0)	IAA(1.0)	G(30)	$Na_2CO_3 \cdot H_2O$ (8.0)	247, 248
		Ax (storage)	2XK NaFeEDTA (12.35)	B	MS Myo(100) Gly(2.0) Thia(1.0) Nia(0.5) Pyr(0.5)	IAA(2.5) Kn(0.1)	G(30)	$Na_2CO_3 \cdot H_20$ (8.0)	
No name given	St	G	W-43	W-43	W-43	IAA(0.5)	S(20)	Cc(10%)	31, 234
"Alpine"	F	G	ST	ST	ST	NAA(0.01 or 0.001)	S(?)	—	77, 247, 248
		Ax (initial)	2XK NaFeEDTA (12.35)	B	Thia(1.0)	IAA(1.0)	G(30)	$NaCO_3 \cdot H_2O$ (8.0)	

Table 1 (Continued)
FRUIT CROPS GROWN IN TISSUE CULTURE

Family (Genus and/or species)	Explant source[a]	Type of differentiation encountered[b]	Medium: Salts[c] Macro	Medium: Salts[c] Micro	Medium: Vitamins[d]	Medium: Hormones[e]	Medium: Sugar[f]	Medium: Other[g]	Ref.
F. Virginiana	—	Ax (storage)	2XK NaFe-EDTA (12.35)	B	MS Myo(100) Gly(2.0) Thia(1.0) Nia(0.5) Pyr(0.5)	IAA(2.5) Kn(0.1)	G(30)	$Na_2CO_3 \cdot H_2O$ (8.0)	247
"Aliso" "Gorello" "Hummy Gento" "Pocahontas" "Rabunda" "Tioga" "MDUS 3293" "MDUS 3773" "MDUS 3816"	A	C	GD-1	GD-1	GD-1 Myo(100) Gly(0.4)	(—)	S(20)	—	320
"Pocahontas" "Gorella"	C (from anthers)	Ad	GD-1	GD-1	GD-1 Myo(100) Gly(0.4)	(—)	S(20)	Darkness for 70 days→16-hr days	320
"Tioga"	C (from anthers)	C	LS	LS	LS	Kn(0.2) IAA(2.0) 2,4-D(0.4)	S(30)	For 35 days	320
		Ad	GD-1	GD-1	GD-1 Myo(100) Gly(0.4)	(—)	S(20)	Darkness for 70 days→16-hr days	
"Rabunda"	C (from anthers)	Ad	GD-1	GD-1	GD-1 Myo(100) Gly(0.4)	(—)	S(20)	Darkness for 70 days→16-hr days	320
"Axbridge Early" "Cambridge Favourite" "Cambridge Premier" "Huxley" "J.I. 390" "Merton Herald" "Merton Princess" "Prizewinner" "Royal Sovreign"	M	Ax	W-63	W-63	W-63	IBA(1.0)	S(20)	Cc(10%)	382
"Mikrovermehrung"	St	Ax	K	MS	MS Myo(100)	IBA(1.0) BA(1.0)	G(40)	—	385

					Gly(2.0)				
"Gento"	M	Ax	MS	MS	MS Myo(100) Gly(2.0)	BA(0.25—2.5 μmol) IBA(0.25—1.0 μmol)	S(20)	—	165
"Gem"	St	Ax	W-63	W-63	W-63	IBA(1.0)	S(20)	Cc(10%)	348d, 352
"Ozark Beauty"	St	Ax	W-63	W-63	W-63	IBA(1.0)	S(20)	Cc(10%)	
Others	—	—	—	—	—	—	—	—	50, 83
Malus pumila Mill. (apple)	St	C	LS	LS	Thia(1.0) Myo(100)	2,4-D(0.5) Kn(1.0)	S(20)	—	2, 3
"Cox's Orange Pippin" seedlings	St	Ad, Ax	LS	LS	Thia(1.0) Myo(100)	Kn(0.1—1.0)	S(20)	—	
	St	R	LS	LS	Thia(1.0) Myo(100)	IBA(1.0) (15 min)	S(20)	—	
"MMIII" "Jonagold" "Golden Delicious" "Mutsu" "MM106"	M	Ax	MS	H-53(100×)	J Myo(50)	2,4-D(0.02) BA(1.0) GA_3(0.1)	S(20)	—	49, 301
MM106 MMIII M26	M	Ax are moved to this medium then to rooting medium	MS	MS	MS Gly(5) Myo(100) Thia(0.4)	IAA(10) BA(1.0) GA_3(0.1)	S(30)	—	
	M	R	MS	MS	Myo(100) Thia(0.5)	IAA(10) GA_3(0.1)	S(30)	—	
"Stark Jumbo" "EMLA-7" "MAC-9" "EMLA-27" "Antonovaka KA313" "Royalty" crabapple	St	Ax	CH	CH	Thia(2.5)	IBA(0.1 or 0.5) BA(1.1)	S(30)	—	71
"Antonovaka KA313"	St	R	CH	CH	Thia(2.5)	IBA(0—0.001)	S(30)	—	
"MAC 9", "EMLA-27"	St	R	CH	CH	Thia(2.5)	IBA(0.005)	S(30)	—	
"Reinette du Canada"	Se	G,Gr	K	H-53	2× MS Myo(100)	—	S(15—45)	—	13
"Em-9"	St	G,Gr	K	H-53	2× MS Myo(100)	—	S(15—45)	—	
"Cortland" seedlings	St	C	MS	MS	MS	NAA(2.0)	So(30)	—	73—75
"McIntosh"	St	C	MS	MS	MS	Kn(0.2)	So(30)	—	73—75
"Red Delicious" Niedzwetzkyana	St	C	MS	MS	MS Myo(100) Gly(2.0)	NAA(2.0) Kn(0.2)	So(30)	—	76
"Cortland" "Northern Spy" seedlings	B	C,G,Ax	DP	DP	DP	BA(0.1—10)	S(30)	—	89,298
"Granny Smith"	St	Ax	El	El	El	BA(0.23) *or* Z(0.022)		Cs(500) Ad(20)	96
"Jonathan"	M	Ax	KN + Na_2EDTA (37.3)	MS	MS Myo(100) Gly(2.0)	NAA(0.2) BA(2.0)	S(30)	—	159

Table 1 (Continued)
FRUIT CROPS GROWN IN TISSUE CULTURE

Family (Genus and/or species)	Explant source[a]	Type of differentiation encountered[b]	Medium: Salts[c] Macro	Salts[c] Micro	Vitamins[d]	Hormones[e]	Sugar[f]	Other[g]	Ref.
			$FeSO_4 \cdot 7H_2O$ (27.8)						
		Ax (subculture)	KN + Na_2EDTA (37.3) $FeSO_4 \cdot 7H_2O$ (27.8)	MS	MS Myo(100) Gly(2.0)	NAA(0.06) BA(2.0)	S(30)	—	
	M	R, Gr	KN + Na_2EDTA (37.3) $FeSO_4 \cdot 7H_2O$ (27.8)	MS	MS Myo(100) Gly(2.0)	±GA_3(0.1) NAA (10), *or* IBA (10)	S(30)	—	
"M.26"	M	Ax,R	K NaFe-EDTA(20)	MS	(—)	BA(1.0)	S(30)	—	168
"M.7"	M	Ax(initiation)	MS	MS	WS	IAA(10) BA(1.0) GA_3(0.1)	S(30)	—	169
"M.26"	M	Ax(proliferation)	MS	MS	WS	IBA(1.0) GA_3(0.1)	S(30)	Pg(162)	
	M	R	MS	MS	Myo(100) Thia(0.4)	BA(0.5) IBA(1.0) GA_3(0.1)	S(30)	—	
"M7" M.26 M.27 "Golden Delicious" "Malling kent" "Malling suntan" "James Grieve" "Cox's Orange Pippin"	St	R	K	MS	(—)	IBA (10)	S(20)	Pg(162)	170
	St	Ax	MS NaFe-EDTA(20)	MS	Myo(100) Thia(0.4)	IBA(1.0) BA(1.0) GA_3(0.1)	S(30)	Pg(162)	171, 171a
	St	R	MS NaFe-EDTA(20)	MS	Myo(100) Thia(0.4)	IBA(1.0) GA_3(0.1)	S(30)	Pg(162)	
"McIntosh" seedlings	M	Ax	MS	MS	MS Myo(100) Gly(2.0)	BA(1.14)	S(20)	—	199
	M	R	½ MS	½ MS	Myo(50) Gly(1.0) ½ MS	NAA(1.9)	S(20)	28°C	
"Sturmer Pippin" "Cox's Orange Pippin" "Granny Smith"	FR	C	W-43	W-43	Myo(0.2) Bio(0.2) CaP(0.2)	2,4-D(6.0)	—	Cc(10%) Cys(10)	207

"Jonathan" Seedlings	(—)	C	(—)	(—)	(—)	(—)	(—)	(—)	209
M.13 M.9	(—)	C	2×MiS	2×MiS	MiS	IAA(2.0) Kn(0.2)	(—)	(—)	228
MM.104	St	C	MS	MS	MS Myo(100) Gly(2.0)	IAA(0.35) Kn(0.04)	S(30)	—	239
"Golden Delicious" "Belle de Boskoop"	Fr	C	G	G	Myo(100) CaP(40) Thia(10)	—	S(30—50)	Cm(15%) 2,3-DPA (0.1) Glut(100)	269
No name given	Fr	C	Various media were used		—	—	—	—	274
"Golden Delicious"	Fr	C	MS	MS	Myo(100) Bio(0.01) CaP(1.0) Thia(1.0) Nia(1.0) Pyr(1.0)	2,4-D(1.0) BA(0.1)	(—)	Asp(180) As(50) Th(25)	287,288
"Antonovka" seedlings	St	G	MS	MS	MS Myo(100) Gly(2.0)	(—)	S(30)	—	293
"Golden Delicious" "Kokko" *M. prunifolia* Borkh. (rootstock)	S	C	FN	FN	Myo(0.1) Bio(0.01) CaP(10) Thia(1.0) Nia(1.0) Pyr(1.0)	—	S(20)	Cys(10) Ad(5)	118
"Jonathan"	A	C,Em(—)	MS	MS	—	IAA(1.0) Kn(1.0)	S(30)	—	197, 230
"M-XVI" "3-430"	C	C	W-63 (2×)	W-63 (2×)	W-63	NAA(2.0)[k] Kn(0.2)	S(20)	—	97,98,204
"Golden Delicious" "Red Delicious" "Jonathan"	Fr,N	C	MS	MS	MS Myo(100) Gly(2.0)	NAA(10) BA(10)	S(30)	Light	92
	Fr,N	Em,Ad	MS	MS	MS Myo(100) Gly(2.0)	NAA(0.1) BA(10)	S(30)	Dark	
"Golden Delicious"	N[l]	C	MS	MS	MS Myo(100) Gly(2.0)	—	S(30)	Light	92a
	N[l]	Em,Ad	MS	MS	MS Myo(100) Gly(2.0)	—	S(30)	Dark	
"Cortland" "McIntosh"	S,P,L,Fr	C	W-63	W-63	W-63	(—)	(—)	Cc(—) Cs(—) 28—32°C Cs(—)	322
M9 M2 M4 M11 "Bittenfelder" seedling	Cm	C	MS	MS	MS Myo(100) Gly(2.0)	NAA(10) BA(0.1)	S(30)	—	324

Table 1 (Continued)
FRUIT CROPS GROWN IN TISSUE CULTURE

Family (Genus and/or species)	Explant source[a]	Type of differentiation encountered[b]	Medium: Salts[c] Macro	Micro	Vitamins[d]	Hormones[e]	Sugar[f]	Other[g]	Ref.
No name given	S	C	F	F	F Myo(100) Gly(2.0)	(—)	S(30)	—	328
"McIntosh" seedlings	St	Ax	MS	MS	Myo(100) Gly(5.0) Thia(0.4) Nia(0.5) Pyr(0.5)	IAA(10) Kn(1.25)	S(30)	Pvp(150) Cc(15%)	383
	St	R	MS	MS	Myo(100) Thia	IAA(10)	S(30)	Pvp(150) Cc(15%)	
"Golden Delicious" "Lodi" "York Imperial"	Fr	C	MS	MS	Myo(100) Bio(0.01) CaP(1.0) Thia(1.0) Nia(1.0) Pyr(1.0)	2,4-D(1.0) BA(0.1)	S(30)	Asp(180) As(50) Th(25)	384
"Ozark Gold" "Spartan" MM106	St	Ax	MS	MS	WS Myo(1.0)	IAA(1.0) BA(1.0) GA_3(0.1)	S(30)	—	20, 404, 405
	St	R	½ MS	½ MS	Myo(100) Thia(0.4)	IBA(0.5 or 0.1)	S(30)		
M. robusta Rehd.	S	C	MS	MS	MS	NAA(2.0) Kn(0.2)	So(30)	—	73—75
M. silvestris L. (crabapple)	M	C,Ad(—)	MS	MS	(—)	IAA(1.0—2.0) 2,4-D(0.5) Kn(0.1—1.0) GA_3(1.0)	S(30—40)	—	222
Others	—	—	—	—	—	—	—	—	18a, 172, 209a, 216, 258, 261a, 311, 323, 382a.
No name given	St	Ax(—)	W-63	W-63 +H-53	W-63	(—)	S(20)	—	262
Prunus accolade	M	Ax	MS	H-53(100×)	Myo(50) Bio(0.1) CaP(0.5)	2,4-D(0.02) BA(1.0) GA_3(0.1)	S(20)	—	48
		R	MS	H-53(100×)	Myo(50) Bio(0.1) CaP(0.5)	IBA(1.0)	S(20)	—	

No name given	St	Ax(—),R	MS	H-53	Myo(50) Bio(0.1) CaP(0.5)	IBA(0.2) ±GA_3(0.03)	S(20)	—	303,304
P. amygdalus (Tourn.) L. (almond) seedlings	Se (w/o testa)	G	½ MS	½ MS	½ MS Myo(50) Gly(1.0)	—	S(30)	—	227
	Roots	C	MS	MS	MS Myo(100) Gly(2.0)	NAA(1—3)	S(30)	Cc(10%)	
	S,L,St	C	MS	MS	MS Myo(100) Gly(2.0)	NAA(4—8) or 2,4-D(4—8)	S(30)	—	
	Cotyledons, Em	C	MS	MS	MS Myo(100) Gly(2.0)	NAA(4—10)	S(30)	—	
	Callus from L, Em, cotyledons	Ad	MS	MS	MS Myo(100) Gly(2.0)	NAA(5.0) Kn(1.0)	S(30)	Cs(1000)	
		R	MS	MS	MS Myo(100) Gly(2.0)	NAA(5.0) Kn(0.5-1.0)	S(30)	Cs(1000)	
No name given	A	C	M	M	M	IAA (0.1) 2, 4-D(1.0) Kn (0.2)	S(30)	—	229
"Nonpareil" almond and an almond × peach hybrid	St	C	TK	TK	TK Myo(100) Gly(2.0)	2,4-D *or* NAA (>0.01)	S(30)	—	180,366
	St	Ax(small size)	TK	TK	TK Myo(100) Gly(2.0)	BA(1.0)	S(30)	—	
	St	Ax(large size)	TK	TK	TK Myo(100) Gly(2.0)	BA(0.1)	S(30)	—	
	St	R(light)	TK	TK	TK Myo(100) Gly(2.0)	NAA(1.0)	S(30)	—	
	St	R(dark)	TK	TK	TK Myo(100) Gly(2.0)	IBA(—)	S(30)	—	
P. armeniaca L. (apricot) "Morden"	St	C	MS	MS	MS Myo(100) Gly(2.0)	2,4-D(0.25) Kn(0.2)	S(30)	—	76
No name given	St	Ax	W-63	W-63 + H-53	W-63	(—)	S(20)	—	262
"Stella"	St	Ax	SC	SC	SC Myo(100) Gly(2.0)	BA(2.0) NAA(0.1)	S(30)	As(50)	348b
Others	A	—	—	—	—	—	—	—	139
P. austera	M	Ax	MS	H-53 (100×)	J Myo(50)	2,4-D(0.02) BA(1.0)	S(20)	—	49

Table 1 (Continued)
FRUIT CROPS GROWN IN TISSUE CULTURE

Family (Genus and/or species)	Explant source[a]	Type of differentiation encountered[b]	Medium: Salts[c] Macro	Micro	Vitamins[d]	Hormones[e]	Sugar[f]	Other[g]	Ref.
		R	MS	H-53 (100×)	J Myo(50)	GA_3(0.1) IBA(1.0)	S(20)	—	
P. avium L. (sweet cherry) "Bing"	St	C	MS	MS	MS Myo(100) Gly(2.0)	2,4-D(0.25) Kn(0.2)	S(30)	—	76
No name given	S, Cm	C	MS	MS	MS Myo(100) Gly(2.0)	BA(0.5) NAA(1.0)	S(20)	—	107
	S, Co	C	MS	MS	MS Myo(100) Gly(2.0)	NAA(10.0) BA(0.5)	S(20)	—	
"Limburger Vogelkirsche"	S, Cm, Co	C	MS	MS	MS Myo(100) Gly(2.0)	NAA(5.0) BA(1.0)	S(20)	—	108
No name given	C	R	MS	MS	MS Myo(100) Gly(2.0)	NAA(1) ABA(1)	S(20)	—	79,110
No name given	S	C	½ MS	½ MS	½ MS Myo(50) Gly(1.0)	BA(0.1)	S(20)	Cg(15) Try(5)	111,112
No name given	L	C,R	½ MS	MS	MS Myo(100) Gly(2.0)	NAA(4.0) BA(1.0)	S(20)	—	143
"Limburger Vogelkirsche"	A	C	N	N	—	—	S(20)	—	174
	A	C	NN-69	NN-69	NN-69	NAA(1.0) BA(1.0)	S(20)	—	175,176
	A	R	MS	MS	MS Myo(100) Gly(2.0)	NAA(1.0) BA(0.01)	S(20)	—	
"F 12/1" "Early Rivers" "Redelfinger" Schneider sp Knoro	M	Ax	MS	H-53(100×)	J Myo(50)	2,4-D(0.2) BA(1.0) GA_3(0.1)	S(20)	—	49
	M	R	MS	H-53(100×)	J Myo(50)	IBA(1.0)	S(20)	—	
"Early Rivers"	St	Ax	SC	SC	SC Myo(100) Gly(2.0)	BA(2.0) NAA(0.1)	S(30)	As(50)	348b

"F 12/1"	St	Ax,R(—)	MS NaFe-EDTA (20)	MS	Thia(0.4) Myo(100)	IBA(1.0) BA(1.0) GA_3(0.1)	S(30)	Pg(162)	171
"F 12/1"	St	Ax,R	MS	MS	MS Myo(100) Gly(2.0)	IAA(0.35) Kn(0.04)	S(30)	—	239
"V1629" "V340"	A	C	S	S	S Myo(250)	BA(5.0 *or* 10) 2,4-D(—)	S(30)	—	338
"1021" "811"	A	Ad	S	S	S Myo(100)	BA(10) IAA(1.0)	S(30)	—	
"978"	A	R	S	S	S Myo(250)	2,4-D(5.0)	S(30)	—	
No name given	Em	G	T	T	—	—	G(10)	Asp(2000)	377
Others	—	—	—	—	—	—	—	—	164
P. avium × *P. fruitcosa*	S, Cm, Co	C	MS	MS	MS	NAA(5.0) BA(1.0)	S(20)	—	108
Cherry root stocks "Colt"	St	Ax	CH	CH	CH	IBA(0.5 or 5.0 μM) BA(5.0 μM)	S(30)	—	71
"Maheleb × Mazzard 14"	St	R	CH	CH	CH	IBA(0 *or* 0.5 μM)	S(30)	—	
P. besseyi Bailey (Sand cherry)	St	C	MS	MS	MS CaP(0.5) Myo(100)	2,4-D (1.0) Kn(1.0)	S(30)	Cs(1000)	152
P. besseyi × *P. cerasus* "P.2037"	St	Ax,R	MS	MS	MS Myo(100) Gly(2.0)	IAA(0.35) Kn(0.04)	S(30)	—	239
P. cerasifera (ornamental plum) "Newport"	St	Ax	CH	CH	CH	IBA(0.5—5.0 μM) BA(5.0 μM)	S(30)	—	71
	St	R	CH	CH	CH	IBA(1—5 μM)	S(30)	—	
"P 2175"	A	C(very poor callus G)	S	S	S	(—)	S(30)	—	338
	A	Le,Ax	S-P7	S-P7	S-P7	IBA(1.0) BA(0.5)	S(20)	—	
	A	R	S-P7	S-P7	S-P7	IBA(1.0)	S(20)	—	
"Thundercloud"	St	Ax	MS	MS	MS Myo(100) Gly(2.0)	BA(1.0) NAA(0.1)	S(30)	—	407
	St	Ax (subculture)	MS	MS	MS Myo(100) Gly(2.0)	BA(1.0)	S(60)	—	
	St	R	MS	MS	MS Myo(100) Gly(2.0)	IBA(0.5)	—	—	
P. cerasus L. (sour cherry) "Montmorency"	St	C	MS	MS	MS Myo(100) Gly(2)	2,4-D(0.25) Kn(0.2)	S(30)	—	76,396a
	Fr	G	—	—	—	—	—	—	
	St	Ax	SC	SC	SC Myo(100) Gly(2.0)	BA(2.0) NAA(0.1)	S(30)	As(50)	348b-c

Table 1 (Continued)
FRUIT CROPS GROWN IN TISSUE CULTURE

Family (Genus and/or species)	Explant source[a]	Type of differentiation encountered[b]	Medium: Salts[c] Macro	Medium: Salts[c] Micro	Medium: Vitamins[d]	Medium: Hormones[e]	Medium: Sugar[f]	Medium: Other[g]	Ref.
		R, G	SC-H	SC-H	SC-H Myo(200) Gly(2.0)	IBA(3.0) NAA(1.0) Kn(0.04) BA(0.01) GA_3(0.1)	S(20)	Cs(1000)	
"Apelala" "Schaerbeck" "Visé"	M	Ax	MS	H-53(100×)	J Myo(50)	2,4-D(0.02) BA(1.0) GA_3(0.1)	S(20)	—	49
	M	R	MS	H-53(100×)	J Myo(50)	IBA(1.0)	S(20)	—	
"Besseyi"	St	Ax	W-63	W-63 + H-53	W-63	(—)	S(20)	—	262
	St	R	W-63	W-63 + H-53	W-63	IAA(1.0)	S(20)	—	
"Shubinka"	St	Ax	MS	MS	Thia(0.5) Nia(0.5) Pyr(0.5)	BA(0.1—0.5)	S(30)	As(1.0)	297
	St	R 18 hr in IBA (50 mg/ℓ) then moved							
	St	R	W-63	W-63	W-63	—	S(20)	—	
"V 603" "1280"	A	C	S	S	S Myo(250)	BA(5.0 *or* 10) 2,4-D(—)	S(30)	—	338
	A	Ad	S	S	S Myo(100)	IAA(1.0) BA(10.0)	S(30)	—	
	A	R	S	S	S Myo(250)	2,4-D(5.0)	S(30)	—	
"Schattenmorelle"	St, Cm, Co	C	MS	MS	MS Myo(100) Gly(2.0)	NAA(5.0) BA(1.0)	S(20)	—	107,108
P. dawkensis fastigiata	M	Ax	MS	H-53 (100 ×)	J Myo(50)	2,4-D(0.02) BA(1.0) GA_3(0.1)	S(20)	—	49
		R	MS	H-53 (100 ×)	J Myo(50)	IBA(1.0)	S(20)	—	
P. domestica L. (plum) "Fellenberg"	St	C	MS	MS	MS Myo(100) Gly(2.0)	2,4-D(0.25) Kn(0.2)	S(30)	—	76
"Brompton" "St. Julien"	M	Ax	MS	H-53 (100 ×)	J Myo(50)	2,4-D(0.02) BA(1.0) GA_3(0.1)	S(20)	—	49

	M	R	MS	H-53 (100 ×)	J Myo(50)	IBA(1.0)	S(20)	—	
"P107" "707"	A	C(very poor callus G)	S	S	S	(—)	S(30)	—	338
"248" "149"	A	Le,Ax	S-P7	S-P7	S-P7	IBA(1.0) BA(0.5)	S(20)	—	
"2444" "994"	A	R	S-P7	S-P7	S-P7	IBA(1.0)	S(20)	—	
"Stanley"	St	Ax	SC	SC	SC Myo(100) Gly(2.0)	BA(2.0) NAA(1.0)	S(30)	As(50)	348b-c
		R, G	SC-H	SC-H	SC-H Myo(200) Gly(2.0)	IBA(3.0) NAA(1.0) Kn(0.04) BA(0.01) GA_3(0.1)	S(20)	Cs(1000)	
Plum root stocks "Pixy" "St. Julien X"	St	Ax	CH	CH	CH	IBA(0.5—5.0 μM) BA(5.0μM) (+ others)	S(30)	—	71
	St	R	CH	CH	CH	BA(5.0 μM)	S(30)	—	
"Pixy"	St	Ax,R(—)	MS NaFe-EDTA (20)	MS	Thia(0.4) Myo(100)	IBA(1.0) BA(1.0) GA_3(0.1)	S(30)	Pg(162)	171
P. section *eucerasus*	St	C	½ MS	½ MS Myo(50) Gly(1.0)	½ MS	IAA(10) BA(0.1)	S(20) So(5)	—	112a
P. fruticosa (European Dwarf cherry) seedlings	S, Cm, Co	C	MS	MS Myo(100) Gly(2.0)	MS	NAA(5.0) BA(1.0)	S(20)	—	108
P. hally jolivette	M	Ax	MS	H-53 (100 ×)	J Myo(50)	2,4-D(0.02) BA(1.0) GA_3(0.1)	S(20)	—	49
P. incisa × *P. serrula*	M	R	MS	H-53 (100 ×)	J Myo(50)	IBA(1.0)	S(20)	—	
P. kursar *P. mahaleb* L. (mahaleb cherry)	St	C	MS	MS	MS Myo(100) Gly(2.0)	2,4-D(0.25) Kn(0.2)	S(30)	—	76
No name given	L	C,R	½ MS	MS	MS Myo(100) Gly(2.0)	NAA(4.0) BA(1.0)	S(20)	—	143
P. mariana "GF 8.1"	St	Ax,R	MS	MS	MS Myo(100) Gly(2.0)	IAA(0.35) Kn(0.04)	S(30)	—	239
P. nigra Ait (Canadian plum)	St	C	MS	MS	MS Myo(100) Gly(2.0)	2,4-D(0.25) Kn(0.2)	S(30)	—	76
P. nipponica *P. okame*	M	Ax	MS	H-53 (100 ×)	J Myo(50)	2,4-D(0.02) BA(1.0)	S(20)	—	49

Table 1 (Continued)
FRUIT CROPS GROWN IN TISSUE CULTURE

Family (Genus and/or species)	Explant source[a]	Type of differentiation encountered[b]	Medium: Salts[c] Macro	Medium: Salts[c] Micro	Medium: Vitamins[d]	Medium: Hormones[e]	Medium: Sugar[f]	Medium: Other[g]	Ref.
P. padus L. (European bird cherry)	S	C	G	G	G	GA_3(0.1) (—)	(—)	—	60
No name given	St	C	MS	MS	MS Myo(100) Gly(2.0)	2,4-D(0.25) Kn(0.2)	S(30)	—	76
P.X. pandora	L	C,R	½ MS	MS	MS Myo(100) Gly(2.0)	NAA(4.0) BA(1.0)	S(20)	—	143
No name given	A	C,R	NN-69	NN-69	NN-69	NAA(1.0) BA(1.0)	S(20)	—	177
No name given	M	Ax	MS	H-53 (100 ×)	J Myo(50)	2,4-D(0.02) BA(1.0) GA_3(0.1)	S(20)	—	43,49
	M	R	MS	H-53 (100 ×)	J Myo(50)	IBA(1.0)	S(20)	—	
Prunus persica Batsch (peach)	Se	G	K	H-53	—	—	S(50)	—	11,223
No name given	Se	Gr	K	H-53	Thia(0.2) Pyr(1.0) Myo(100)	—	S(50)	—	
"Frau Anneliese Rudolf"	C[m]	C	H-53	H-53	Thia(1.0)	2,4-D(5.0) Kn(0.5)	(—)	—	63
"South Haven" "Mme Roignat" "Elberta"		C (subculture)	MS	MS	Thia(0.2)	2,4-D(0.2—0.5) Kn(0.2)	(—)	—	
"Reliance"	St	C	MS	MS	MS Myo(100) Gly(2.0)	2,4-D(0.25) Kn(0.2)	S(30)	—	76
"Polly"	St	C	MS	MS	MS Myo(100) CaP(0.5)	2,4-D(1.0) Kn(1.0)	S(30)	Cs(1000)	152
No name given	M	C,Ad(—)	MS	MS	MS	IAA(1.0—2.0) 2,4-D(0.5) Kn(0.1—1.0) GA_3(1.0)	S(30)	—	222
"Harbrite"	St and Ax buds from tissue culture	R, G	SC-H	SC-H	SC-H Gly(2.0) Myo(200)	IBA(3.0) NAA(1.0) Kn(0.04) BA(0.01) GA_3(0.1)	S(30)	Cs(1000)	348c

"Dixired" "Nectared IV"	A	C	M	M	M	IAA(0.1) 2,4-D(1.0) Kn(0.2)	S(30)	—	229
No name given	St	C	W-63	W-63 + H-53	W-63	NAA(0.5)	S(20)	—	262
"Early White Giant"	Fr	C(endocarp)	K	N-51 $NaMoO_4 \cdot$ $2H_2O$(0.025)	Ms Myo(100)	NA(—) Kn(10)	S(50)	Glu(200) Asp(100) Tyrp(100)	272
		C (mesocarp)	K	N-51 $NaMoO_4 \cdot$ H_20(0.025)	MS Myo (100)	NA(—)	S(50)	Glu(200) Asp(100) Tryp(100)	272
"Springtime"	A	C,Em(—)	S	S	M Myo(100)	IAA(1.0) BA(1.0 *or* 5.0)	S(30)	—	338
"Redhaven" "Harbrite"	St	Ax	SC	SC	SC Myo(100) Gly(2.0)	NAA(0.1) BA(2.0)	S(30)	As(50)	346, 348, 348b
"Miller's Late"	Fr	C	ST	ST	ST	NAA(1.0 *or* 10) Kn(2.0)	S(20)	—	356,356a
Others	Em	—	—	—	—	—	—	—	402
"Okubo"	S	C	FN	FN	Myo(0.1) Bio(0.01) CaP(10) Thia(1.0) Nia(1.0) Pyr(1.0)	—	S(20)	Cys(10) Ad(5)	118
P. persica × *P. amygdalus* "GF 677"	St	Ax,R	MS	MS	MS Myo(100) Gly(2.0)	IAA(0.35) Kn(0.04)	S(30)	—	239
P. pseudocerasus not *P. pseudocerasus* Lindl. but belonging to section *pseudocerasus*	C	R	MS	MS	MS Myo(100) Gly(2.0)	IBA(10) ABA(0.1) BA(0.1)	S(20)	—	79,110
No name given	L	C,R	½ MS	MS	MS Myo(100)	NAA(4.0) BA(1.0)	S(20)	—	143
P. rhexii	M	Ax	MS	H-53 (100 ×)	J Myo(100)	2,4-D(.02) BA(1.0) GA_3(0.1)	S(20)	—	49
	M	R	MS	H-53 (100 ×)	J Myo(100)	IBA(1.0)	S(20)	—	
P. salicina (Japanese plum) "Calita"	St	Ax	MS NaFe-EDTA (20)	MS	—	IBA(0.1)	S(30)	—	321
	St	Ax (after 6 weeks)	MS NaFe-EDTA (20)	MS	MS Myo(100) Thia(0.4)	IBA(0.1)	S(30)	—	
	St	R	MS NaFe-EDTA (20)	MS	MS Myo(100) Thia(0.4)	GA_3(0.1) IBA(4.0)	S(30)	26°C	
P. Serotina Ehrh. (black cherry)	S,Cm	C	WR	WR	WR	2,4-D(1.0)	S(40)	Cs(1000) Cc(15%)	62

Table 1 (Continued)
FRUIT CROPS GROWN IN TISSUE CULTURE

Family (Genus and/or species)	Explant source[a]	Type of differentiation encountered[b]	Medium: Salts[c] Macro	Medium: Salts[c] Micro	Medium: Vitamins[d]	Medium: Hormones[e]	Medium: Sugar[f]	Medium: Other[g]	Ref.
P. serrulata (Japanese flowering cherry) "Kanzan" "Yedosakvra"	M	Ax	MS	H-53 (100 ×)	J Myo(50)	2,4-D(0.02) BA(1.0) GA_3(0.1)	S(20)	—	48
P. subhirtella autumnalis (Higan cherry)	M	R	MS	H-53 (100 ×)	J Myo(50)	IBA(1.0)	S(20)	—	
P. tomentosa Thunb. (Nanking cherry)	St	C	MS	MS	MS CaP(0.5) Myo(100)	2,4-D(1.0) Kn(1.0)	S(30)	Cs(1000)	152
P. yedoensis moerheimi (Japanese flowering cherry)	M	Ax	MS	H-53 (100 ×)	J Myo(50)	2,4-D(0.02) BA(1.0) GA_3(0.1)	S(20)	—	49
Shidare yoshimo	M	R	MS	H-53 (100 ×)	J Myo(50)	IBA(1.0)	S(20)	—	
Prunus sp. "Amur chokeberry"	E	G	—	—	—	—	—	—	4
"Common bird cherry"	S	C	G	G	G	—	—	—	60
No name given	S	Protoplasts	—	—	—	—	—	—	246,326a
Many species	S	Ax	MS	H-53 (+ 100 × Mn, − Al,I)	J Myo(100)	2,4-D(0.02) BA(1.0) GA_3(0.1)	S(20)	—	302
	S	Ax(proliferation)	NH_4NO_3(400) KNO_3(1800) $Ca(NO_3)\cdot 4H_2O$(200) $MgSO_4\cdot 7H_2O$(3600) KH_2PO_4 (2700)	H-53 + 100 × Mn, − Al,I)	J Myo(100)	BA(1.0)	S(20)	—	
Others	—	—	—	—	—	—	—	—	72, 106, 109,
Pyrus sp. (pear)	S	C	G	G	G	(—)	(—)	—	60
"Old Home" × "Farmingdale 51"	S	Ax	CH	CH	Thia(2.5)	2,4-D(0.05 or 0.005μM) BA(5.0μM)	S(30)	—	71
	S	R	CH	CH	Thia(2.5)	IBA(0.5, 2.5, or 5.0μM)	S(30)	—	

Pyrus fauri (Dwarf Korean Pear) Bradford pear	St	Ax	CH	CH	Thia(2.5)	IBA(0.5—5.0μM) BA(5.0μM)	S(30)	—	71
P. communis L. "Bartlett"	St	C	MS	MS	MS Myo(100) Gly(2.0)	2,4-D(0.25) Kn(0.2)	S(30)	—	76
"Grafin von Paris"	A	C	NN-69	NN-69	NN-69	NAA(1.0) BA(1.0)	S(20)	—	175,176
"Passe Crassne"	Fr	Studied ripening	—	—	—	—	—	—	201
"Packhams Late"	Fr	C	W-43	W-43	Myo(0.2) Bio(0.2) CaP(100)	2,4-D(2.0) Kn(0.1)	S(20)	Cc(10%) Cys(10) Asp(180)	208
"Conference"	(—)	C	Various media were used			—	—	—	242
"Bartlett" "Précoce de Trévoux"	Fr	C	—	—	Myo(50) Thia(10)	2,4-D(1.0) Kn(0.1)	S(50)	So(1000) Asp(1000) Glu(200) Ad(50) Tryp(100)	270
No name given	Fr	C	Various media were used			(—)	—	—	
No name given	—	C	SH	SH	SH Myo(1000)	—	S(30)	—	327
P. calleryana Deche (oriental pear)	Fr	C	W-43	W-43	Myo(0.2) Bio(0.2) CaP(0.2)	2,4-D(2.0)	S(20)	Cc(10%) Cys(10)	208
P. pyraster (wild pear)	M	C,Ad(—)	MS	MS	MS	IAA(1.0—2.0) 2,4-D(0.5) Kn(0.1—1.0) GA_3(1.0)	S(30—40)	—	222
P. serotina "Chojura"	S	C	FN	FN	Myo(0.1) Bio(0.01) CaP(10) Thia(1.0) Nia(1.0) Pyr(1.0)	—	S(20)	Cys(10) Ad(5)	118
Others	—	—	—	—	—	—	—	—	200
Rubus hispidus L. (Swamp dewberry)	C	C	H-53	H-53	H-53	(—)	(—)	(—)	30, 295, 319
No name given	C	C	K	(—)	(—)	NAA(0.05)	G(50)	Ye(2000)	32
Rubus sp. (blackberry) "Black Satin" "Senator Dirksen"	St	C,Ax	MS	MS	WS Myo(10)	IAA(10) BA(1.0) GA_3(0.1)	S(30)	Pg(162)	52
"Smoothstem" "US 64-39-2"	St	R	MS	MS	Myo(100) Thia(0.4)	IBA(1.0)	S(30)	Pg(162)	
"Darrow" "Senator Dirksen" "Thornless Evergreen"	St	Ax	SC	SC	SC Myo(100) Gly(2.0)	NAA(0.1) BA(2.0)	S(30)	As(50)	225, 347, 348a

Table 1 (Continued)
FRUIT CROPS GROWN IN TISSUE CULTURE

Family (Genus and/or species)	Explant source[a]	Type of differentiation encountered[b]	Medium: Salts[c] Macro	Medium: Salts[c] Micro	Medium: Vitamins[d]	Medium: Hormones[e]	Medium: Sugar[f]	Medium: Other[g]	Ref.
"Cherokee" "Comanche" "NY 95" "Thornfree" "Thornless Boysenberry" "Thornless Youngberry"	St	Ax,R	SC	SC	SC Myo(100) Gly(2.0)	NAA(0.1) BA(2.0)	S(30)	As(50)	225a, 347, 348a
9001/174 (blackberry × raspberry hybrid)	St	G,R	NN-69	NN-69	NN-69	IBA(1.0)	S(30)	—	181
"Bedford Giant"	St	Ax	LS	LS	LS	IBA(0.1) BA(1.0)	S(30)	Pg(160)	140
"Tayberry"	St	R	LS	LS	LS	IBA(1.0)	S(30)	Pg(160)	
"Novost" "Ku Z'mina"	M(from root suckers)	Ax,R	K	H-53	W-63	IBA(0.05)	G(20)	—	340
R. kinashii Lev. *R. reflexus* Ker	—	C	Various media were used				—	—	242
R. caesius L.	Ov	G	NN-69	NN-69	NN-69	IAA(2.0)	S(20)	—	400
R. fruticosus	C	Enz	—	—	—	—	—	—	94
No name given	C	Enz	H-53	H-53	Thia(1.0)	—	G(50)	—	114
No name given	C	Enz	—	—	—	—	—	—	146
No name given	—	C	Various media were used			—	—	—	242
R. idaeus (Red raspberry) "Schoenemann"	M	Ax	K	MS	MS Myo(100) Gly(2)	IBA(0.1) BA(0.05) GA_3(0.1)	S(20)	—	160
	M	R	K	MS	MS Myo(100) Gly(2)	IBA(0.1) GA_3(0.1)	S(20)	—	
No name given	St	G	(—)	(—)	(—)	(—)	(—)	(—)	296
"Malling Promise"	Ov	G	NN-69	NN-69	NN-69	IAA(2.0)	S(20)	—	400
"September"	Ov	G	MS	MS	SC Myo(100)	BA(3.0) NAA(3.0)	S(30)	Cs(1000) As(50)	150
"Bois blancs"	St	Ax	MS	MS	—	NAA(1.0)	S(30)	Cc(10%)	299
	St	R	MS	MS	—	NAA(1.0) Kn(1.0)	S(30)	Cc(10%)	
R. occidentalis L. (black raspberry) "Bristol"	St	Ax,R	SC	SC	SC Myo(100) Gly(2.0)	NAA(0.1) BA(2.0)	S(30)	AS(50)	347
Others	—	—	—	—	—	—	—	—	215, 312

Rubiaceae *Coffea arabica* L. (coffee)	L	C, Em	LS	LS	LS Myo(100)	Kn(0.1) 2,4-D(0.1)	S(20)	—	151
	L	R	GD-4	GD-4	GD-4 Myo(10) Gly(0.4)	NAA(0.1)	S(20)	—	
No name given	C	Caffeine synthesis	—	—	—	—	—		179
"Mundo Novo" "Bourbon Amarelo"	Se	G	LS	LS	Thia (4.0) Myo(100)	Kn(0.1) NAA(2.0) *or* 2,4-D(0.1)	S(40)	Cys(10) Cc(5%)	339
	Fr	C (slow, brown, friable)	LS	LS	Thia (4.0) Myo(100)	2,4-D(0.1)	S(40)	Cys(10) Cc(5%)	
	St (orthotropic)	C (fast, white or yellow, firm)	LS	LS	Thia (4.0) Myo(100)	2,4-D(0.1)	S(40)	Cys(10) Cc(5%)	
	St (plagiotropic)	C (no growth or slow, white, firm)	LS	LS	Thia (4.0) Myo(100)	2,4-D(0.1)	S(40)	Cys(10) Cc(5%)	
	P	C (friable)	LS	LS	Thia (4.0) Myo(100)	2,4-D(0.1) *or* NAA(2.0)	S(40)	Cys(10) Cc(5%)	
	L	C (slow, white, firm)	LS	LS	Thia (4.0) Myo(100)	2,4-D(0.1)	S(40)	Cys(10) Cc(5%)	
	L	R	LS	LS	Thia (4.0) Myo(100)	NAA(2.0)	S(40)	Cys(10) Cc(5%)	
	A	C	LS	LS	Thia (4.0) Myo(100)	2,4-D (0.1)	S(40)	Cys(10) Cc(5%)	
No name given	L	Conditioning medium for 7 weeks	MS	MS	Myo(103) Thia(10)	—	S(30)	Cys(25)	357
	L	C, Em	½ MS + KNO_3 (3800)	½ MS	Myo(103) Thia(10)	Kn(2.0) 2,4-D(1.0) *or* NAA(4.0—8.0)	S(30)	Cys(25)	
	L	G (to shoots)	½ MS + KNO_3 (3800)	½ MS	Myo(103) Thia(10)	NAA(0.05) Kn(0.5)	S(30)	Cys(25)	
C. arabica L. *C. canephora* Pierre ex Froehner "Robusta" *C. liberica* Bull. (Liberian coffee)	S (orthotropic)	C	MS	MS KI(0.8)	Myo(100) Thia(1.0)	Kn(0.1) 2,4-D(0.1) *or* NAA(1.0)	S(30)	Cys(10)	362
No name given	S	C (+ coffee)	GE-PRL-4	GE-PRL-4	GE-PRL-4	(—)	(—)	Cc(10%)	373, 378
Others	—	—	—	—	—	—	—	—	53,237
Rutaceae *Citrus aurantium* L. (sour orange) "Bigaradier"	St	C, Ax, Ad, R	MS	MS	Mo	NAA(1.0) Kn(1.0)	S(30)	—	39
No name given	Seedlings	C	½ K	—	—	NAA(1.0 or 0.1)	G(30)	Cc(15%)	42
C. aurantium ×	Ov (sexual	G (of zygote)	Ra-Z	Ra-Z	Ra-Z	IAA(1.0)	S(40)	Cs(400)	309

Table 1 (Continued)
FRUIT CROPS GROWN IN TISSUE CULTURE

Family (Genus and/or species)	Explant source[a]	Type of differentiation encountered[b]	Medium: Salts[c] Macro	Medium: Salts[c] Micro	Medium: Vitamins[d]	Medium: Hormones[e]	Medium: Sugar[f]	Medium: Other[g]	Ref.
Poncirus trifoliata	offspring product trifoliate leaves)				Gly(7.5)				
	Ov (After 6—7 weeks)	G (of zygote)	Ra-Z	Ra-Z	Ra-Z Gly(7.5)	IAA(1.0)	S(20)	Cs(400)	
	Ov (After 4 weeks)	G (of zygote)	Ra-Z	Ra-Z	Ra-Z Gly(7.5)	IAA(1.0)	—	—	
	Ov	Em	Ra-N	Ra-N	Ra-N Myo(100) Gly(2.0)	NAA(0.5)	S(50)	Oj(5%) Cs(400) Ad(25)	
C. aurantifolia Swingle (Lime)	E, N	C, Em (—)	(—)	(—)	(—)	(—)	(—)	(—)	325
C. aurantifolium	Seedlings	C	½ K	—	—	NAA(1.0 or 0.1)	G(30)	Cc(15%)	42
No name given	Ov (pollinated and unpollinated)	Em	MC	MC	MC	IAA(0.25) Kn(0.5) *or* IAA(0.25) Kn(0.5) GA_3(5.0)	S(30)	As(3.0)	235
C. grandis (L.) Osbeck. (Shaddock)	St	C	CM	CM	CM	NAA(2.5) 2,4-D(0.25) Kn(0.25)	S(50)	Me(500) As(5)	68
	St	Em	CM	CM	CM (with ½ thia, nia, and pyr)	NAA(0.1) BA(0.25)	S(50)	Me(500)	
No name given	St	R (low vigor shoots)	CM	CM	CM (with ½ thia, nia, and pyr)	NAA(0.1)	S(50)	Me(500)	69
	St	R (high vigor shoots)	CM	CM	CM (with ½ thia, nia, and pyr)	NAA(0.5) *or* NAA (0.25) + IBA (0.25)	S(50)	Me(500) As(5)	
No name given	Al, Jv	C	MT	MT	MT	2,4-D(3.0) Kn(1.0)	S(50)	Oj(3—30%)	256
C. limon Burm. f (lemon)	St	C, Ax, Ad, R	MS	MS	Mo	NAA(1.0) Kn(1.0)	S(30)	—	39
No name given	Seedlings	C	½ K	—	—	NAA(1.0 or 0.1)	G(30)	Cc(15%)	42
No name given	Al, Jv	C	MS	MS	Thia(5.0)	2,4-D(2.0)	S(30)	Oj(10%)	93
No name given	Al, Jv	C	MS	MS	MS Myo(100) Gly(2.0)	2,4-D(3.0) Kn(1.0)	S(30)	—	254,376

No name given	Al, Jv	C	MT	MT	MT	2,4-D(3.0) Kn(1.0)	S(50)	—	256
No name given	St	Ax (—)	W-43	W-43 + H-53	W-43	(—)	S(20)	—	262
"Eureka"	Jv	G	KO	KO	—	—	S(34.25)	pH 7.0—7.7	192—195
No name given	Jv	C	MS	MS	MS Myo(100)	—	S(30)	Glu(400)	271
"Ponderosa"	Fr, N	Em	MS	MS	MS	—	S(30)	Me(500)	308
No name given	Fr	C	N-51	N-51	N-51	±IAA(10)	S(20)	—	330
No name given	A	C (androgenic?)	MS	H	Thia(1.0) Nia(1.0)	IAA(1.0)	G(25)	—	88
C. madurensis = *C. microcarpa* bunge) seedlings	S	Ax, Ad	MS	MS	MS	NAA(0.1) BA(10)	S(50)	Me(500)	136
	S	R	MS	MS	MS	NAA(0.1)	S(50)	Me(500)	
No name given	Ov	Em (—)	RS	RS	RS Gly(7.5)	IAA(1.0)	S(20)	—	310
C. maxima Merr. (shaddock)	Fr	C	N-51	N-51	N-51	±IAA(10)	S(20)	—	330
"Pong yau" (pummelo)	Fr, N	Em	MS	MS	MS	—	S(30)	Me(500)	308
C. medica L. (Citron)	Al, Jv	C	MS	MS	Thia(5.0) Myo(100)	2,4-D(2.0) Kn(0.5)	S(50)	Oj(10%)	93
No name given	Al, Jv	C	MT	MT	MT	2,4-D(3.0) Kn(1.0)	S(50)	Oj(3—30%)	256
No name given	Fr	C	N-51	N-51	N-51	±IAA(10)	S(20)	—	330,334
No name given	A	C (androgenic?)	MS	H	Thia(1.0) Nia(1.0)	IAA(1.0)	G(25)	—	88
C. paradisi Macf. (grapefruit)	Al, Jv	C	MS	MS	Thia(5.0) Myo(100)	2,4-D(2.0) Kn(0.5)	S(50)	Oj(10%)	93
"Marsh Seedless"	N, O	G, Em(—)	MT	MT	MT	IAA(0.1) Kn(1.0)	S(50)	Me(500)	191
		R(—)	MT	MT	MT	IAA(10) Kn(0.1)	S(50)	Me(500)	
No name given	Al, Jv	C	MT	MT	MT	2,4-D(3.0) Kn(1.0)	S(50)	Oj(3—30%)	256
C. reticulata Blanco (Mandarin orange)	St	C, Ax, Ad, R	MS	MS	MO	NAA(1.0) Kn(1.0)	S(30)	—	39
No name given	Al, Jv	C	MS	MS	Thia(5.0) Myo(100)	2,4-D (2.0) Kn(0.5)	S(50)	Oj(10%)	93
"Ponkan"	N	C	MS	MS	Myo(100) Thia(0.2) Pyr(1.0) Nia(1.0) Gly(4.0)	Various	S(50)	Me(500)	370,371
C. reticulata × *C. sinensis* ("Temple" orange)	Fr, N	Em	MS	MS	MS	(—)	S(30)	Me(500)	308
C. sinensis Osbeck (sweet orange)	St	C, Ax, Ad, R	MS	MS	Mo	NAA(1.0) Kn(1.0)	S(30)	—	39

Table 1 (Continued)
FRUIT CROPS GROWN IN TISSUE CULTURE

Family (Genus and/or species)	Explant source[a]	Type of differentiation encountered[b]	Medium: Salts[c] Macro	Salts[c] Micro	Vitamins[d]	Hormones[e]	Sugar[f]	Other[g]	Ref.
"Maltaise"									
No name given	Jv	C	MS	MS	MS Myo(100)	2,4-D (0.3) Kn(1.0)	S(30)	Glu(400)	271
"Valencia"	N, O	G, Em	MT	MT	MT	IAA(0.1) Kn(1.0)	S(50)	Me(500)	191
	N, O	R	MT	MT	MT	IAA(1.0) Kn(0.1) GA_3(1.0)	S(50)	Me(500)	
No name given	Ov (pollinated and unpollinated)	Em	MC	MC	MC	IAA(0.25) Kn(0.5) *or* IAA(0.25) Kn(0.5) GA_3(5.0)	S(30)	As(3.0)	235
No name given	A, Jv	C	MT	MT	MT	2,4-D(3.0) Kn(1.0)	S(50)	Oj(3—30%)	256
"Washington Navel"	N	Em	MS	MS	BB Myo(100) Gly(5.0)	(—)	S(40)	Me(400) Ad(40)	56, 57, 184
	N	R	MS	MS	BB Myo(100) Gly(5.0)	GA_3(1.0)	S(40)	Me(400) Ad(40)	
	Jv	C	MS	MS	MS Myo(100)	2,4-D(0.3) Kn(1.0)	S(30)	Glu(400)	271
"Shamouti"	St	C	MT	MT	MT	ABA(2.6) *or* IAA(1.8) *or* BA(2.25)	S(50)	—	14—17, 133
	Peel	C	(—)	(—)	(—)	(—)	(—)	(—)	313
	Al, Jv	C	MS	MS	Thia(5) Myo(100)	2,4-D(2.0) Kn(0.5)	S(50)	Oj(10%)	93
	—	—	—	—	—	—	—	—	99
	C	C	MT	MT ($CoCl_2$)	MT	2,4-D (3×10^{-6} *M*) (3×10^{-6} *M*) Kn(10^{-6} *M*)	S(50)	—	100
	N	Em	MT	MT	MT	IAA(1.0) Kn(1.0)	S(50)	—	59, 186—189
	N, O	Em	MT	MT	MT	IAA(0.1) Kn(1.0)	S(50)	Me(500)	190, 191
	N	Em (subculture)	MT	MT	MT	Habituated	S(50)	Me(500) Ad(20)	183, 185, 190
	N	R	MT	MT	MT	GA_3(1.0)	S(50)	Ad(15)	184

	Effects of radiation on media and cells								358
	N	Protoplasts	—	—	—	—	—	—	379, 380
No name given	St	C	CM	CM	CM	NAA(2.5) 2,4-D(0.25) Kn(0.25)	S(50)	Me(500) As(5)	69
	St	Em	CM	CM	CM (with ½ thia, nia, pyr)	NAA(0.1) BA(0.25)	S(50)	Me(500) As(5.0)	
	St	R (low vigor shoots)	CM	CM	CM (with ½ thia, nia, and pyr)	NAA(0.1)	S(50)	Me(500) As(5.0)	
	St	R (high vigor shoots)	CM	CM	CM (with ½ thia, nia, and pyr)	NAA(0.5) *or* NAA(0.25) + IBA(0.25)	S(50)	Me(500) As(5.0)	
Citrus sp.	Se	R	K	H-53	MS(2 ×) Myo(100)	NAA(0.1)	S(40)	—	12
No name given		Grafting	—	—	—	—	—	—	253, 259—261, 318, 399
Others	—	—	—	—	—	—	—	—	35, 36, 40, 41, 54, 55, 58, 60a, 87, 181a-b, 190a, 326, 341, 359, 360
Sapotaceae									
Achras sapota (sapodilla)	E + En, Fr	C	MS (½ MS Fe)	MS	—	2,4-D(2.0) Kn(5.0)	S(20)	Cc(15%) Cs(400)	27
Saxifragaceae									
Ribes nigrum L. (black currant)	St	Ax (initiation)	MS NaFe-EDTA(20)	MS	MS Myo(1.0) Gly(2.0)	NAA(1.0)	S(20)	Cc(5%)	173
	St	Ax (moved to this medium)	K NaFe-EDTA(20)	MS	—	—	G(40)	—	
	St	Ax (subculture)	K NaFe-EDTA(20)	MS	WS	IBA(1.0) GA_3(1.0) BA(0.2)	G(40)	—	
	St	R [basal 2—3 mm of shoot in IBA(10) for 10 min, then moved to the 2nd medium (above)]							
"Wellington XXX" MPI-Klon 36 MPI-Klon 89	A	C	NN-69	NN-69	NN-69	NAA(1.0) BA(1.0)	S(20)	—	175,176
No name given	St	Ax	W-63	W-63 + H-53	W-63	(—)	S(20)	—	262

Table 1 (Continued)
FRUIT CROPS GROWN IN TISSUE CULTURE

Family (Genus and/or species)	Explant source[a]	Type of differentiation encountered[b]	Medium: Salts[c] Macro	Medium: Salts[c] Micro	Medium: Vitamins[d]	Medium: Hormones[e]	Medium: Sugar[f]	Medium: Other[g]	Ref.
R. rubrum L. (red currant)	E, Ov (7—9 mm berries *only*)	C	NN-69	NN-69	NN-69	—	S(20)	—	401
"Hosszúfurtű"		Em (within seed)	NN-69	NN-69	NN-69	GA_3(1.0)	S(20)	—	
R. grossularia = *R. uva crispa* L. (gooseberry) "Careless"	St	Ax (initiation)	MS NaFe-EDTA(20)	MS	MS Myo(100) Gly(2.0)	NAA(1.0)	S(20)	Cc(5%)	173
	St	Ax (moved to this medium)	K NaFe-EDTA(20)	MS	—	—	G(40)	—	
	St	Ax (subculture)	K NaFe-EDTA(20)	MS	WS	IBA(1.0) GA_3(1.0) BA(0.2)	G(40)	—	
	St	R [basal 2—3 mm of shoot in IBA(10) for 10 min, then moved to the 2nd media (above)]							
Others	—	—	—	—	—	—	—	—	141
Vitaceae									
Vitis sp. (grape) *V. berlandieri* Planch. (Spanish grape) *V. candicans* Engl *V. labrusca* L. (Fox grape) "Isabelle" *V. riparia* Michx. (River bank grape)	S	C	½ K	B	—	NAA(0.19)	G(30)	—	101,102
V. riparia × *V. rupestris*	St	Ax, R	K	B	(—)	(—)	S(17.5)	—	279
V. rupestris Sch. (Sand grape) "du Lot"	S	C	½ K	B	—	NAA(0.19)	G(30)	—	101,102
No name given	M	G	K	(—)	(—)	(—)	(—)	(—)	123,124
No name given	M	Gr	½ K	B	Myo(100) Bio(0.01) CaP(1.0) Thia(1.0) Nia(1.0) Pyr(1.0)	NAA(1.0)	S(15)	—	125
	M	R	½ K	B	Myo(100) Bio(0.01) CaP(1.0) Thia(1.0) Nia(1.0) Pyr(1.0)	NAA(0.01 *or* 0.05 *or* 0.1)	S(15)	—	

No name given	St	Ax, R	K	B	(—)	(—)	S(17.5)	—	277,278
V. thunbergii	F, A	C	N	N	—	NAA(1.9) BA(0.22)	S(20)	—	154
V. vinifera (European grape)	S	C	½ K	B	—	NAA(0.19)	G(30)	—	101,102
"Prodigiosa"	L	C, Ad	MS	H-53	Mo	BA(0.5) IBA(1.0) or IAA(1.0)	S(20)	—	103—105
"Rupestris du Lot" "Aramon × Rupestris Ganzin No. 9" "3309" "Riparia Gloire de Montpellier" "Chasselas" "Malaga" "Chardonnay" "Muscat de Frontignan" "Muscat d'Alexandria"	M	G, R	½ K	B	Myo(100) Bio(0.01) CaP(1.0) Thia(1.0) Nia(1.0) Pyr(1.0)	NAA(1.0)	S(15)	—	119,120
"Muscat de Hambourg" "Cardinal" "Araman" "161-49" "41 B"	M	G	½ K	B	Myo(100) Bio(0.01) CaP(1.0) Thia(1.0) Nia(1.0) Pyr(1.0)	NAA(1.0)	S(15)	—	119,120
"JS 23-46" "Black Corinth" "Grenache"	A	C (haploid)	GD-74	GD-74	GD-74	NAA(2.0) IAA(2.0) Kn(1.5)	S(20)	—	135
		C (continued growth)	GD-74	GD-74	GD-74	NAA(0.5) Kn(0.1)	S(20)	—	
"Carbernet Sauvignon" "Rhine Riesling" "Chenin blanc" "Sultana" "Waltham Cross" "Waltham Cross" "Bruce's Sport"	Fr	C	GE-B5	GE-B5	GE-B5	NAA(0.1) Kn(0.2)	S(30)	Cs(2000)	142
"Sylander Riesling"	S	C	SH	SH	SH	2,4-D(0.5) *P*CPA(2.0) Kn(0.1)	S(20)	—	70,167
	S	*Or* C	GE-B5	GE-B5	GE-B5	NAA(0.1) Kn(0.2)	S(20)	—	
	S	Ax, R	CD	CD	CH	BA(1.0)	S(30)	—	
	S	R (best rooting)	CD	CD	CH	BA(0 *or* 0.1)	S(30)	—	
"Cabernet Sauvignon"	O (unfertilized)	C (1st 3 weeks)	N-51	N-51	N-51	BA(1.1)	S(20)	—	249
	O	C (2nd 2 weeks)	N-51	N-51	N-51	BA(1.1) NOA(1.0)	S(20)	—	
	O	Em(3rd)	N-51	N-51	N-51	BA(0.55) NOA(1.0)	—		
	O	Em	N-51	N-51	N-51	2iP(5 μM)	S(20)	—	

Table 1 (Continued)
FRUIT CROPS GROWN IN TISSUE CULTURE

Family (Genus and/or species)	Explant source[a]	Type of differentiation encountered[b]	Medium: Salts[c] Macro	Medium: Salts[c] Micro	Medium: Vitamins[d]	Medium: Hormones[e]	Medium: Sugar[f]	Medium: Other[g]	Ref.
						GA_3(0.35)			
	O	R	W-63	W-63	W-63	—	S(20)	—	
No name given	O	C	SH	SH	SH	2,4-D(0.5) PCPA(2.0) Kn(0.1)	S(30)	—	327
"Sultana" ("Thompson seedless")	(—)	C	GE-B5	GE-B5	GE-B5	NAA(0.1) Kn(0.2)	S(20)	Cs(2000)	342
	C	Protoplasts	O	O	GE-B5 Gly(4.0)	2,4-D(0.1) BA(1.0)	S(10) R(0.25) G(0.25) X(0.15) A(0.15) So(44—58)	Cs(2000) Glu(2000)	343
	C	Protoplasts to callus	O	O	GE-B5 Gly(4.0) Myo(1000)	BA(1.0) 2,4-D(0.1) *or* NAA(1.0—4.0)	S(10) R(0.25) G(0.25) X(0.15) A(0.15) So(44—58)	Cs(2000) Glu(2000) Se(100)	
	C	Callus to colonies	GE-B5	GE-B5	GE-B5	NAA(0.1) Kn(0.2)	S(30)	Cs(2000)	
"Cabernet Sauvignon" "Muscat of Alexandria" "Shiraz" "Sultana" ("Thompson seedless") seedlings	Tendrils	F	N-68	N-68	N-68	BA(1.1 or 2.2)	S(20)	Cs(1000)	361
"Riesling" "Portugiesen"	S	C, R	MS	MS	MS Myo(100) Gly(2.0)	NAA(1.0) Kn(0.5)	G(30)	—	363
"Gros Colman"	S	C	FN	FN	Myo(0.1) Bio(0.01) CaP(10) Thia(1.0) Nia(1.0) Pyr(1.0)	—	S(20)	Cys(10) Ad(5.0)	118
"Seyval"	S, P, F, L	C	MS	MS	MS Myo(100) Gly(2.0)	2,4-D(1.0) BA(0.1)	S(30)	—	196
	C	Em	MS	MS	MS Myo(100)	NAA(2.0) BA(0.1)	S(30)	—	

	C	G (of embryoids)	MS	MS	Gly(2.0) —	—	S(20)	—	
"Healthy"	C (from Morel 1944ab)	C	(—)	(—)	(—)	(—)	(—)	(—)	240, 241, 354
Vitis sp. "F S.4-201-39" "Kober 5 BB"	S	C	H-53	H-53	H-53	? (1.0)	(—)	—	8,9
	S	R	H-53	H-53	H-53	? (0.1)	(—)	—	
	S	C	K	B	Myo(100) Thia(1.0)	IAA(0.18 or 1.8)	(—)	Cys(10)	240—242
	S	C, R	K	B	Myo(100) Thia(1.0)	IAA(0.018)	(—)	Cys(10)	
No name given	S, G	C	W-43	W-43	W-43	(—)	S(20)	—	289
No name given	Insect galls (+ insects)	C	Mo	Mo	Mo	(—)	(—)	(—)	316
Phylloxera vastatrix (incited galls)	C	C	W-43	W-43	W-43 CaP(2.5)	NAA (0.1)	Various	Cc(15%)	22,23
No name given	F	C	H-53	H-53	H-53	NAA(0.1)	G(10)	—	305
Others	—	—	—	—	—	—	—	—	28a, 29, 34, 290, 307

[a] Code for explant source — A = anther; Al = albedo; B = buds; C = callus; Cm = cambium; Co = cortex; E = embryo; En = endosperm; F = flowers; Fr = fruits; G = galls; Jv = juice vesicles; L = leaf; M = meristem; N = nucellus; O = ovules; Ov = ovaries; P = petiole; Rh = rhizome; S = stem; Se = seeds; and St = shoot tip.

[b] Code for differentiation — Ad = adventitious buds; Ax = axillary buds; C = callus; Em = embryoids; Enz = enzymes; F = flowers; G = growth (in a normal manner); Gr = grafting; Le = leaf expansion; and R = roots.

[c] Code for salts and vitamin mixtures (see Table 2).

[d] Code for vitamins — Bio = biotin; CaP = Ca pantothenate; Ch = Choline chloride; Fol = folic acid; Gly = glycine (this compound is not a vitamin, but is listed under the vitamin column for convenience); Myo = (*iso*-) or *myo*-inositol (this compound is not a vitamin, but is listed under the vitamin column for convnience); Nia = nicotinamide (niacin); Paba = para-amino benzoic acid; Pyr = pyridoxide · HCl; Rib = riboflavin; and Thia = thiamine · HCl.

[e] Hormones that are commonly used in fruit tissue culture (see Table 3).

[f] Code for sugars — A = arabinose; G = glucose; Ma = mannose; R = ribose; S = sucrose; So = sorbitol; and X = xylose.

[g] Code for other additions to media — Ad = adenine; AdS = adenine sulfate; Arg = arginine; As = ascorbic acid; Asp = asparagine; C = charcoal (activated); Ca = DL-catechin; Cc = coconut milk; Cg = chlorogenic acid; Cm = corn milk; Cs = casein hydrolysate (edamin); Cys = cysteine; Glu = glutamine; Glut = glutathion; Gte = green tomato extract; Me = malt extract; Oj = orange juice; Pg = phloroglucinol; Pvp = polyvinylpyrrolidone; Se = serine; Th = thiourea; Try = tryptamine; Tryp = tryptophan; and Ye = yeast extract.

[h] (—) = information was not specified in the paper and — = dash indicates that the information was not relevant.

[i] With $Fe_2(SO_4)_3$ replaced by Na_2EDTA (37.3) + $FeSO_4 \cdot 7H_2O$ (27.8).

[j] The media are *very* complex and the author suggests that the reader study this article in its original form.

[k] Article lists 20 mg/ℓ, but this is probably an error.

[l] From fruits that were 50 days post-anthesis.

[m] Grown with *Taphrina deformans* (Berk) Tul.

Table 2
CODE FOR SALTS AND VITAMIN MIXTURES

	Anderson[19] [A]	Asahira & Kano[24] [AK]	Berthelot's solution (from Mullin & Schlegel[247]) [B]	Button & Bornman[56] [BB]
	ELEMENTS (mg/*l*)			
Macro				
a. $(NH_4)_2SO_4$	—	—	—	—
b. NH_4NO_3	2,000	1,650	—	1,650
c. KNO_3	950	1,900	—	1,900
d. $Ca(NO_3)_2 \cdot 4H_2O$	—	—	—	—
e. $CaCl_2 \cdot 2H_2O$	440	440	—	440
f. $MgSO_4 \cdot 7H_2O$	370	370	—	370
g. KH_2PO_4	170	170	—	170
h. $NaH_2PO_4 \cdot H_2O$	170	—	—	—
i. Na_2EDTA	37.3	37.3	—	37.3
j. Fe-EDTA	—	—	—	—
k. NaFe-EDTA	—	—	—	—
l. $FeSO_4 \cdot 7H_2O$	27.8	27.8	—	27.8
m. KCl	—	—	—	—
n. Na_2SO_4	—	—	—	—
o. Geigy® sequestrene 330 Fe	—	—	—	—
p. Fe citrate	—	—	—	—
q. $Fe_3(PO_4)_2$	—	—	—	—
r. $FeCl_3$	—	—	—	—
s. $Fe_2(SO_4)_3$	—	—	25.0	—
t. $FePO_4 \cdot 4H_2O$	—	—	—	—
u. $NaH_2PO_4 \cdot 2H_2O$	—	—	—	—
v. Na_2HPO_4	—	—	—	—
w. NH_4Cl	—	—	—	—
x. $CaSO_4$	—	—	—	—
y. $Ca_3(PO_4)_2$	—	—	—	—
z. $NH_4H_2PO_4$	—	—	—	—
zi. $NaNO_3$	—	—	—	—
Micro				
$MnSO_4 \cdot H_2O$	16.9	—	—	—
$MnSO_4 \cdot 2H_2O$	—	35.0	—	—
$MnSO_4 \cdot 4H_2O$	—	—	—	22.3
$MnSO_4 \cdot 7H_2O$	—	—	1.0	—
$ZnSO_4 \cdot 7H_2O$	8.6	10.0	0.05	8.6
H_3BO_3	6.2	10.0	0.025	6.2
KI	0.83	1.0	0.25	0.83
$Na_2MoO_4 \cdot 2H_2O$	0.25	25.0	—	0.25
MoO_3	—	—	—	—
$CuSO_4 \cdot 5H_2O$	0.025	0.035	0.025	0.025
$CoCl_2 \cdot 6H_2O$	0.025	0.035	0.025	0.025
$NiCl_2 \cdot 6H_2O$	—	—	0.025	—
$Cu(NO_3)_2 \cdot 3H_2O$	—	—	—	—
$(NH_4)_6Mo_7O_{24} \cdot 4H_2O$	—	—	—	—
$AlCl_3$	—	—	—	—
Vitamins				
Cyanocobalmin	—	—	—	—
p-Amino benzoic acid	—	—	—	—
Folic acid	—	0.1	—	—
Riboflavin	—	—	—	—
Biotin	—	0.05	—	—

Table 2 (continued)
CODE FOR SALTS AND VITAMIN MIXTURES

	Anderson[19] [A]	Asahira & Kano[24] [AK]	Berthelot's solution (from Mullin & Schlegel[247]) [B]	Button & Bornman[56] [BB]
	ELEMENTS (mg/ℓ)			
Choline chloride	—	—	—	—
Ca panotothenate	—	—	—	0.03
Thiamine · HCl	0.4	0.5	—	0.3
Nicotinamide (niacin)	—	5.0	—	1.0
Pyridoxine · HCl	—	0.5	—	0.05
Other				
(*iso*-) or *myo*-inositol	100	100	—	100
Glycine	—	2.0	—	5.0
Ascorbic acid	—	—	—	—
Sucrose	30,000	40,000	—	40,000

	Campbell & Durzan[61] [CD]	Chaturvedi & Mitra[69] [CM]	Cheng[71] [CH]	Dutcher & Powell[89] [DP]
	ELEMENTS (mg/ℓ)			
Micro				
a. $(NH_4)_2SO_4$	—	—	—	—
b. NH_4NO_3	800	1,500	825	1,650
c. KNO_3	340	1,500	950	1,900
d. $Ca(NO_3)_2 \cdot 4H_2O$	980	—	—	—
e. $CaCl_2 \cdot 2H_2O$	—	400	220	440
f. $MgSO_4 \cdot 7H_2O$	370	360	185	370
g. KH_2PO_4	170	150	85	170
h. $NaH_2PO_4 \cdot H_2O$	—	—	—	—
i. Na_2EDTA	37.3	37.3	7.2	37.3
j. Fe-EDTA	—	—	—	—
k. NaFe-EDTA	—	—	—	—
l. $FeSO_4 \cdot 7H_2O$	27.8	27.8	6.0	27.9
m. KCl	65	—	—	—
n. Na_2SO_4	—	—	—	—
o. Geigy® sequestrene 330 Fe	—	—	—	—
p. Fe citrate	—	—	—	—
q. $Fe_3(PO_4)_2$	—	—	—	—
r. $FeCl_3$	—	—	—	—
s. $Fe_2(SO_4)_3$	—	—	—	—
t. $FePO_4 \cdot 4H_2O$	—	—	—	—
u. $NaH_2PO_4 \cdot 2H_2O$	—	—	—	—
v. Na_2HPO_4	—	—	—	—
w. NH_4Cl	—	—	—	—
x. $CaSO_4$	—	—	—	—
y. $Ca_3(PO_4)_2$	—	—	—	—
z. $NH_4H_2PO_4$	—	—	—	—
zi. $NaNO_3$	—	—	—	—
Micro				
$MnSO_4 \cdot H_2O$	16.9	—	—	—
$MnSO_4 \cdot 2H_2O$	—	—	—	—

Table 2 (continued)
CODE FOR SALTS AND VITAMIN MIXTURES

	Campbell & Durzan[61] [CD]	Chaturvedi & Mitra[69] [CM]	Cheng[71] [CH]	Dutcher & Powell[89] [DP]
	ELEMENTS (mg/ℓ)			
$MnSO_4 \cdot 4H_2O$	—	22.3	11.4	44
$MnSO_4 \cdot 7H_2O$	—	—	—	—
$ZnSO_4 \cdot 7H_2O$	8.6	8.6	10.6	17
H_3BO_3	6.2	6.2	6.2	12
KI	0.83	0.83	0.8	—
$Na_2MoO_4 \cdot 2H_2O$	0.25	0.25	0.4	0.5
MoO_3	—	—	—	—
$CuSO_4 \cdot 5H_2O$	0.025	0.025	0.02	0.05
$CoCl_2 \cdot 6H_2O$	0.025	0.025	0.02	—
$NiCl_2 \cdot 6H_2O$	—	—	—	—
$Cu(NO_3)_2 \cdot 3H_2O$	—	—	—	—
$(NH_4)_6MO_7O_{24} \cdot 4H_2O$	—	—	—	—
$AlCl_3$	—	—	—	—
Vitamins				
Cyanocobalmin	—	—	—	—
p-Amino benzoic acid	—	—	—	—
Folic acid	—	0.1	—	—
Riboflavin	—	0.1	—	—
Biotin	—	0.1	—	—
Choline chloride	—	—	—	—
Ca pantothenate	—	—	—	—
Thiamine · HCl	—	5.0	2.5	—
Nicotinamide (niacin)	—	0.5	—	—
Pyridoxine · HCl	—	1.25	—	—
Other				
(*iso*-) or *myo*-Inositol	—	100	250	—
Glycine	—	2.0	—	—
Ascorbic acid	—	5.0	—	—
Sucrose	30,000	50,000	30,000	30,000

	Eeuwens[90] [E]	Elliott[95] [El]	Fox[117] [F]	Fujii and Nito[118] [FN]	Gamborg & Eveleigh[126] [GE-B5]
	ELEMENTS (mg/ℓ)				
a. $(NH_4)_2SO_4$	—	—	—	—	134
b. NH_4NO_3	—	1,650	1,000	60	—
c. KNO_3	1,600	1,900	1,000	—	2,500
d. $Ca(NO_3)_2 \cdot 4H_2O$	—	—	500	170	—
e. $CaCl_2 \cdot 2H_2O$	294	440	—	—	150
f. $MgSO_4 \cdot 7H_2O$	246.5	370	300	240	250
g. KH_2PO_4	—	170	250	40	—
h. $NaH_2PO_4 \cdot H_2O$	—	—	—	—	150
i. Na_2EDTA	33.6	—	—	—	—
j. Fe-EDTA	—	—	—	—	—
k. NaFe-EDTA	—	20	35	—	—
l. $FeSO_4 \cdot 7H_2O$	13.9	—	—	—	—
m. KCl	149.12	—	50	80	—

Table 2 (continued)
CODE FOR SALTS AND VITAMIN MIXTURES

	Eeuwens[90] [E]	Elliott[95] [El]	Fox[117] [F]	Fujii and Nito[118] [FN]	Gamborg & Eveleigh[126] [GE-B5]
	ELEMENTS (mg/ℓ)				
n. Na_2SO_4	—	—	—	—	—
o. Geigy® sequestrene 330 Fe	—	—	—	—	28
p. Fe citrate	—	—	—	10	—
q. $Fe_3(PO_4)_2$	—	—	—	—	—
r. $FeCl_3$	—	—	—	—	—
s. $Fe_2(SO_4)_3$	—	—	—	—	—
t. $FePO_4 \cdot 4H_2O$	—	—	—	—	—
u. $NaH_2PO_4 \cdot 2H_2O$	475.4	—	—	—	—
v. Na_2HPO_4	—	—	—	—	—
w. NH_4Cl	535	—	—	—	—
x. $CaSO_4$	—	—	—	—	—
y. $Ca_3(PO_4)_2$	—	—	—	—	—
z. $NH_4H_2PO_4$	—	—	—	—	—
zi. $NaNO_3$	—	—	—	—	—
Micro					
$MnSO_4 \cdot H_2O$	—	—	5.0	0.4	10
$MnSO_4 \cdot 2H_2O$	—	—	—	—	—
$MnSO_4 \cdot 4H_2O$	11.2	22.3	—	—	—
$MnSO_4 \cdot 7H_2O$	—	—	—	—	—
$ZnSO_4 \cdot 7H_2O$	7.2	8.6	7.5	0.05	3.0
H_3BO_3	3.1	6.4	5.0	0.6	3.0
KI	0.83	0.83	0.8	—	0.75
$Na_2MoO_4 \cdot 2H_2O$	0.24	—	—	—	0.25
MoO_3	—	—	—	—	—
$CuSO_4 \cdot 5H_2O$	0.25	0.025	—	0.05	0.25
$CoCl_2 \cdot 6H_2O$	0.24	0.025	—	—	0.25
$NiCl_2 \cdot 6H_2O$	0.016	—	—	—	—
$Cu(NO_3)_2 \cdot 3H_2O$	—	—	—	—	—
$(NH_4)_6Mo_7O_{24} \cdot 4H_2O$	—	0.18	—	—	—
$AlCl_3$	—	—	—	—	—
Vitamins					
Cyanocobalmin	—	—	—	—	—
p-Amino benzoic acid	—	—	—	—	—
Folic acid	—	—	—	—	—
Riboflavin	—	—	—	—	—
Biotin	—	0.02	—	0.01	—
Choline chloride	—	—	—	—	—
Ca pantothenate	—	0.02	—	10.0	—
Thiamine · HCl	1.0	4.0	0.1	1.0	10.0
Nicotinamide (niacin)	1.0	0.5	0.5	1.0	1.0
Pyridoxine · HCl	1.0	0.05	0.5	1.0	1.0
Other					
(*iso*-) or *myo*-Inositol	102	100	100	0.1	100
Glycine	—	2.0	2.0	—	—
Ascorbic acid	—	—	—	—	—
Sucrose	68,400	?	30,000	20,000	20,000
$H_2MoO_4 \cdot H_2O$	—	—	—	0.02	—

Table 2 (continued)
CODE FOR SALTS AND VITAMIN MIXTURES

	Gamborg & Eveleigh[126]	Gautheret[128]	Gresshoff & Doy[134,135]		
				Med 1	Med 4
	[GE-PRL-4]	[G]	[GD74]	[GD-1]	[GD-4]
	ELEMENTS (mg/l)				
Macro					
a. $(NH_4)_2SO_4$	200	—	200	200	—
b. NH_4NO_3	—	—	—	—	1,000
c. KNO_3	1,000	125	1,000	1,000	1,000
d. $Ca(NO_3)_2 \cdot 4H_2O$	—	500	—	—	367
e. $CaCl_2 \cdot 2H_2O$	150	—	150	150	—
f. $MgSO_4 \cdot 7H_2O$	250	125	250	250	35
g. KH_2PO_4	—	125	—	—	300
h. $NaH_2PO_4 \cdot H_2O$	90	—	90	90	—
i. Na_2EDTA	—	—	37.3	37.3	37.3
j. Fe-EDTA	—	—	—	—	—
k. NaFe-EDTA	—	—	—	—	—
l. $FeSO_4 \cdot 7H_2O$	—	—	27.8	27.8	27.8
m. KCl	300	—	300	—	65
n. Na_2SO_4	—	—	—	—	—
o. Geigy® sequestrene 330 Fe	28	—	—	—	—
p. Fe citrate	—	—	—	—	—
q. $Fe_3(PO_4)_2$	—	—	—	—	—
r. $FeCl_3$	—	0.001	—	—	—
s. $Fe_2(SO_4)_3$	—	—	—	—	—
t. $FePO_4 \cdot 4H_2O$	—	—	—	—	—
u. $NaH_2PO_4 \cdot 2H_2O$	—	—	—	30	—
v. Na_2HPO_4	30	—	30	—	—
w. NH_4Cl	—	—	—	—	—
x. $CaSO_4$	—	—	—	—	—
y. $Ca_3(PO_4)_2$	—	—	—	—	—
z. $NH_4H_2PO_4$	—	—	—	—	—
zi. $NaNO_3$	—	—	—	—	—
Micro					
$MnSO_4 \cdot H_2O$	10	—	1.0	10	10
$MnSO_4 \cdot 2H_2O$	—	—	—	—	—
$MnSO_4 \cdot 4H_2O$	—	—	—	—	—
$MnSO_4 \cdot 7H_2O$	—	1.0	—	—	—
$ZnSO_4 \cdot 7H_2O$	3.0	0.5	0.3	3.0	3.0
H_3BO_3	3.0	0.025	0.3	3.0	3.0
KI	0.75	0.25	0.75	0.35	0.8
$Na_2MoO_4 \cdot 2H_2O$	0.25	—	0.025	0.25	0.25
MoO_3	—	—	—	—	—
$CuSO_4 \cdot 5H_2O$	0.25	0.025	0.025	0.25	0.25
$CoCl_2 \cdot 6H_2O$	0.25	0.025	0.025	0.25	0.25
$NiCl_2 \cdot 6H_2O$	—	0.025	—	—	—
$Cu(NO_3)_2 \cdot 3H_2O$	—	—	—	—	—
$(NH_4)_6Mo_7O_{24} \cdot 4H_2O$	—	—	—	—	—
$AlCl_3$	—	—	—	—	—
Vitamins					
Cyanocobalmin					
p-Amino benzoic acid	—	—	—	—	—
Folic acid	—	—	—	—	—
Riboflavin	—	—	—	—	—
Biotin	—	—	2.0	—	—
Choline chloride	—	—	—	—	—

Table 2 (continued)
CODE FOR SALTS AND VITAMIN MIXTURES

	Gamborg & Eveleigh[126] [GE-PRL-4]	Gautheret[128] [G]	Gresshoff & Doy[134,135] [GD74]	Gresshoff & Doy[134,135] Med 1 [GD-1]	Gresshoff & Doy[134,135] Med 4 [GD-4]
	ELEMENTS (mg/ℓ)				
Ca pantothenate	—	—	—	—	—
Thiamine · HCl	—	—	10	1.0	1.0
Nicotinamide (niacin)	—	—	1.0	0.1	0.1
Pyridoxine · HCl	—	—	1.0	0.1	0.1
Other					
(*iso*-) or *myo*-Inositol	—	—	100	100	10
Glycine	—	—	—	0.4	0.4
Ascorbic acid	—	—	—	—	—
Sucrose	—	—	20,000	20,000	20,000

	Harada[138] [H]	Heller[149] [H-53]	Jacquoit (Gautheret[131]) [J]	Knop's (from Morel & Muller[245]) [K]	Kordan[192] [KO]	Knudsen's[182] [KN]
	ELEMENTS (mg/ℓ)					
Macro						
a. $(NH_4)_2SO_4$	—	—	—	—	—	500
b. NH_4NO_3	1,650	—	—	—	—	—
c. KNO_3	1,900	—	—	125	509	—
d. $Ca(NO_3)_2 \cdot 4H_2O$	—	—	—	500	—	1,000
e. $CaCl_2 \cdot 2H_2O$	440	75	—	—	147	—
f. $MgSO_4 \cdot 7H_2O$	370	250	—	125	108	250
g. KH_2PO_4	170	—	—	125	—	250
h. $NaH_2PO_4 \cdot H_2O$	—	125	—	—	—	—
i. Na_2EDTA	37.3	—	—	—	—	—
j. Fe-EDTA	—	—	—	—	—	—
k. NaFe-EDTA	—	—	—	—	—	—
l. $FeSO_4 \cdot 7H_2O$	27.8	—	—	—	—	—
m. KCl	—	750	—	—	—	—
n. Na_2SO_4	—	—	—	—	—	—
o. Geigy® sequestrene 330 Fe	—	—	—	—	—	—
p. Fe citrate	—	—	—	—	27.3	—
q. $Fe_3(PO_4)_2$	—	—	—	—	—	—
r. $FeCl_3$	—	1.0	—	—	—	—
s. $Fe_2(SO_4)_3$	—	—	—	—	—	—
t. $FePO_4 \cdot 4H_2O$	—	—	—	—	—	50
u. $NaH_2PO_4 \cdot 2H_2O$	—	—	—	—	—	—
v. Na_2HPO_4	—	—	—	—	—	—
w. NH_4Cl	—	—	—	—	—	—
x. $CaSO_4$	—	—	—	—	—	—
y. $Ca_3(PO_4)_2$	—	—	—	—	—	—
z. $NH_4H_2PO_4$	—	—	—	—	—	—
zi. $NaNO_3$	—	600	—	—	—	—
Micro						
$MnSO_4 \cdot H_2O$	—	—	—	—	—	—
$MnSO_4 \cdot 2H_2O$	—	—	—	—	—	—
$MnSO_4 \cdot 4H_2O$	25	0.1	—	—	5.0	—

Table 2 (continued)
CODE FOR SALTS AND VITAMIN MIXTURES

	Harada[138] [H]	Heller[149] [H-53]	Jacquoit (Gautheret[131]) [J]	Knop's (from Morel & Muller[245]) [K]	Kordan[192] [KO]	Knudsen's[182] [KN]
			ELEMENTS (mg/ℓ)			
$MnSO_4 \cdot 7H_2O$	—	—	—	—	—	—
$ZnSO_4 \cdot 7H_2O$	10	1.0	—	—	2.2	—
H_3BO_3	10	1.0	—	—	2.8	—
KI	—	0.01	—	—	—	—
$Na_2MoO_4 \cdot 2H_2O$	0.25	—	—	—	—	—
MoO_3	—	—	—	—	—	—
$CuSO_4 \cdot 5H_2O$	0.025	0.03	—	—	1.2	—
$CoCl \cdot 6H_2O$	—	—	—	—	—	—
$NiCl_2 \cdot 6H_2O$	—	0.03	—	—	—	—
$Cu(NO_3)_2 \cdot 3H_2O$	—	—	—	—	—	—
$(NH_4)_6Mo_7O_{24} \cdot 4H_2O$	—	—	—	—	3.7	—
$AlCl_3$	—	0.03	—	—	—	—
Vitamins						
Cyanocobalmin	—	—	—	—	—	—
p-Amino benzoic acid	—	—	1.0	—	—	—
Folic acid	0.5	—	0.01	—	—	—
Riboflavin	—	—	0.1	—	—	—
Biotin	0.05	—	0.1	—	—	—
Choline chloride	—	—	—	—	—	—
Ca pantothenate	—	—	0.5	—	—	—
Thiamine · HCl	0.5	—	1.0	—	—	—
Nicotinamide (niacin)	5.0	—	1.0	—	—	—
Pyridoxine · HCl	0.5	—	—	—	—	—
Other						
(*iso*-) or *myo*-Inositol	500	—	50	—	—	—
Glycine	2.0	—	—	—	—	—
Ascorbic acid	—	—	—	—	—	—
Sucrose	20,000	—	—	—	34,250	20,000

	Lee & de Fossard[205] [LF]	Linsmaier & Skoog[210] [LS]	Lyrene[217c] [L]	Miller[231,232] [M]	Miller & Skoog[233] [MiS]
		ELEMENTS (mg/ℓ)			
Macro					
a. $(NH_4)_2SO_4$	—	—	—	—	—
b. NH_4NO_3	800	1,650	—	1,000	—
c. KNO_3	1,010	1,900	190	1,000	80
d. $Ca(NO_3)_2 \cdot 4H_2O$	—	—	1,140	500	100
e. $CaCl_2 \cdot 2H_2O$	294	440	—	—	—
f. $MgSO_4 \cdot 7H_2O$	370	370	370	71.5	35
g. KH_2PO_4	—	170	170	300	37.5
h. $NaH_2PO_4 \cdot H_2O$	138	—	—	—	—
i. Na_2EDTA	16.8	37.3	74.6	—	—
j. Fe-EDTA	—	—	—	—	—
k. NaFe-EDTA	—	—	—	13.2	—
l. $FeSO_4 \cdot 7H_2O$	13.9	27.8	55.6	—	—

Table 2 (continued)
CODE FOR SALTS AND VITAMIN MIXTURES

	Lee & de Fossard[205] [LF]	Linsmaier & Skoog[210] [LS]	Lyrene[217c] [L]	Miller[231,232] [M]	Miller & Skoog[233] [MiS]
	ELEMENTS (mg/*l*)				
m. KCl	—	—	—	65	65
n. Na_2SO_4	63.9	—	—	—	—
o. Geigy® sequestrene 330 Fe	—	—	—	—	—
p. Fe citrate	—	—	—	—	—
q. $Fe_3(PO_4)_2$	—	—	—	—	—
r. $FeCl_3$	—	—	—	—	—
s. $Fe_2(SO_4)_3$	—	—	—	—	2.5
t. $FePO_4 \cdot 4H_2O$	—	—	—	—	—
u. $NaH_2PO_4 \cdot 2H_2O$	—	—	—	—	—
v. Na_2HPO_4	—	—	—	—	—
w. NH_4Cl	—	—	—	—	—
x. $CaSO_4$	—	—	—	—	—
y. $Ca_3(PO_4)_2$	—	—	—	—	—
z. $NH_4H_2PO_4$	—	—	—	—	—
zi. $NaNO_3$	—	—	—	—	—
Micro					
$MnSO_4 \cdot H_2O$	—	—	—	—	—
$MnSO_4 \cdot 2H_2O$	—	—	—	—	—
$MnSO_4 \cdot 4H_2O$	11.1	22.3	22.3	14	4.4
$MnSO_4 \cdot 7H_2O$	—	—	—	—	—
$ZnSO_4 \cdot 7H_2O$	5.8	8.6	8.6	3.8	1.5
H_3BO_3	3.1	6.2	6.2	1.6	1.6
KI	0.4	0.83	0.83	—	0.75
$Na_2MoO_4 \cdot 2H_2O$	0.024	0.25	0.25	—	—
MoO_3	—	—	—	—	—
$CuSO_4 \cdot 5H_2O$	0.025	0.025	0.02	—	—
$CoCl_2 \cdot 6H_2O$	0.118	0.025	0.02	—	—
$NiCl_2 \cdot 6H_2O$	—	—	—	—	—
$Cu(NO_3)_2 \cdot 3H_2O$	—	—	—	0.35	—
$(NH_4)_6Mo_7O_{24} \cdot 4H_2O$	—	—	—	0.1	—
$AlCl_3$	—	—	—	—	—
Vitamins					
Cyanocobalmin	—	—	—	—	—
p-Amino benzoic acid	—	—	—	—	—
Folic acid	0.44	—	—	—	—
Riboflavin	0.38	—	—	—	—
Biotin	0.49	—	—	—	—
Choline chloride	0.14	—	—	—	—
Ca pantothenate	0.48	—	—	—	—
Thiamine · HCl	0.67	0.4	0.1	0.1	0.1
Nicotinamide (Niacin)	1.46	—	0.5	0.5	0.5
Pyridoxine · HCl	0.61	—	0.5	0.1	0.5
Other					
(*iso*-) or *myo*-Inositol	54	100	100	100	—
Glycine	0.38	—	2.0	—	2.0
Ascorbic acid	1.8	—	—	—	—
Sucrose	60,000	30,000	30,000	30,000	20,000

Table 2 (continued)
CODE FOR SALTS AND VITAMIN MIXTURES

	Mitra & Chaturvedi[235] [MC]	Morel[242] [Mo]	Murashige & Skoog[255] [MS]	Murashige & Tucker[256] [MT]	Nitsch[268] [N-51]
Macro					
a. $(NH_4)_2SO_4$	—	—	—	—	—
b. NH_4NO_3	1,000	—	1,650	1,650	—
c. KNO_3	475	125	1,900	1,900	125
d. $Ca(NO_3)_2 \cdot 4H_2O$	—	500	—	—	500
e. $CaCl_2 \cdot 2H_2O$	110	—	440	440	—
f. $MgSO_4 \cdot 7H_2O$	92.5	125	370	370	125
g. KH_2PO_4	—	125	170	170	125
h. $NaH_2PO_4 \cdot H_2O$	300	—	—	—	—
i. Na_2EDTA	18.6	—	37.3	37.3	—
j. Fe-EDTA	—	—	—	—	—
k. NaFe-EDTA	—	—	—	—	—
l. $FeSO_4 \cdot 7H_2O$	13.9	—	27.8	27.8	—
m. KCl	—	—	—	—	—
n. Na_2SO_4	150	—	—	—	—
o. Geigy® sequestrene 330 Fe	—	—	—	—	—
p. Fe citrate	—	—	—	—	10
q. $Fe_3(PO_4)_2$	—	—	—	—	—
r. $FeCl_3$	—	—	—	—	—
s. $Fe_2(SO_4)_3$	—	25.0	—	—	—
t. $FePO_4 \cdot 4H_2O$	—	—	—	—	—
u. $NaH_2PO_4 \cdot 2H_2O$	—	—	—	—	—
v. Na_2HPO_4	—	—	—	—	—
w. NH_4Cl	—	—	—	—	—
x. $CaSO_4$	—	—	—	—	—
y. $Ca_3(PO_4)_2$	—	—	—	—	—
z. $NH_4H_2PO_4$	—	—	—	—	—
zi. $NaNO_3$	—	—	—	—	—
Micro					
$MnSO_4 \cdot H_2O$	—	—	—	—	—
$MnSO_4 \cdot 2H_2O$	—	—	—	—	—
$MnSO_4 \cdot 4H_2O$	11.2	—	22.3	22.3	3.0
$MnSO_4 \cdot 7H_2O$	—	1.0	—	—	—
$ZnSO_4 \cdot 7H_2O$	4.3	0.05	8.6	8.6	0.5
H_3BO_3	3.1	0.025	6.2	6.2	0.5
KI	0.42	0.25	0.83	0.83	—
$Na_2MoO_4 \cdot 2H_2O$	0.13	—	0.25	0.25	—
MoO_3	—	—	—	—	—
$CuSO_4 \cdot 5H_2O$	0.13	0.025	0.025	0.025	0.025
$CoCl_2 \cdot 6H_2O$	0.13	0.025	0.025	0.025	0.025
$NiCl_2 \cdot 6H_2O$	—	0.025	—	—	—
$Cu(NO_3)_2 \cdot 3H_2O$	—	—	—	—	—
$(NH_4)_6Mo_7O_{24} \cdot 4H_2O$	—	—	—	—	—
$AlCl_3$	—	—	—	—	—
Vitamins					
Cyanocobalmin	—	—	—	—	—
p-Amino benzoic acid	—	—	—	—	—
Folic acid	0.1	—	—	—	—
Riboflavin	0.05 —	—	—	—	—
Biotin	0.05	?[a]	—	—	—
Choline chloride	—	—	—	—	—
Ca pantothenate	—	?[a]	—	—	—
Thiamine · HCl	0.02	1.0	0.1	10.0	—

Table 2 (continued)
CODE FOR SALTS AND VITAMIN MIXTURES

	Mitra & Chaturvedi[235] [MC]	Morel[242] [Mo]	Murashige & Skoog[255] [MS]	Murashige & Tucker[256] [MT]	Nitsch[268] [N-51]
	ELEMENTS (mg/ℓ)				
Nicotinamide (niacin)	0.1	—	0.5	5.0	—
Pyridoxine · HCl	0.1	—	0.5	10.0	—
Other					
(*iso*-) or *myo*-Inositol	200	?[a]	100	100	—
Glycine	—	—	2.0	2.0	—
Ascorbic acid	3.0	—	—	—	—
Sucrose	30,000	—	30,000	50,000	50,000

	Nitsch[273] [N]	Nitsch et al.[276] [N-68]	Nitsch & Nitsch[267] [NN-67]	Nitsch & Nitsch[275] [NN-69]	Ohyama & Nitsch[282] [O]	Ranga Swamy[310] [RS]
	ELEMENTS (mg/ℓ)					
Macro						
a. $(NH_4)_2SO_4$	—	—	—	—	—	—
b. NH_4NO_3	725	720	—	720	800	—
c. KNO_3	950	950	—	950	2,000	80
d. $Ca(NO_3)_2 \cdot 4H_2O$	—	—	—	—	—	260
e. $CaCl_2 \cdot 2H_2O$	—	166	—	166	1,300	—
f. $MgSO_4 \cdot 7H_2O$	185	185	—	185	400	360
g. KH_2PO_4	88	68	—	68	200	—
h. $NaH_2PO_4 \cdot H_2O$	—	—	—	—	—	165
i. Na_2EDTA	37.3	37.3	—	37.3	37.3	—
j. Fe-EDTA	—	—	—	—	—	—
k. NaFe-EDTA	—	—	—	—	—	—
l. $FeSO_4 \cdot 7H_2O$	27.8	27.8	—	27.8	27.8	—
m. KCl	—	—	—	—	1,500	65
n. Na_2SO_4	—	—	—	—	—	200
o. Geigy® sequestrene 330 Fe	—	—	—	—	—	—
p. Fe citrate	—	—	—	—	—	10
q. $Fe_3(PO_4)_2$	—	—	—	—	—	—
r. $FeCl_3$	—	—	—	—	—	—
s. $Fe_2(SO_4)_3$	—	—	—	—	—	—
t. $FePO_4 \cdot 4H_2O$	—	—	—	—	—	—
u. $NaH_2PO_4 \cdot 2H_2O$	—	—	—	—	—	—
v. Na_2HPO_4	—	—	—	—	—	—
w. NH_4Cl	—	—	—	—	—	—
x. $CaSO_4$	—	—	—	—	—	—
y. $Ca_3(PO_4)_2$	—	—	—	—	—	—
z. $NH_4H_2PO_4$	—	—	—	—	—	—
zi. $NaNO_3$	—	—	—	—	—	—
Micro						
$MnSO_4 \cdot H_2O$	—	—	—	—	—	—
$MnSO_4 \cdot 2H_2O$	—	—	—	—	—	—
$MnSO_4 \cdot 4H_2O$	25	25	—	25	50	3.0
$MnSO_4 \cdot 7H_2O$	—	—	—	—	—	—
$ZnSO_4 \cdot 7H_2O$	10	10	—	10	20	0.5
H_3BO_3	10	10	—	3	20	0.5
KI	—	—	—	—	—	—
$Na_2MoO_4 \cdot 2H_2O$	0.25	0.25	—	0.25	0.5	0.25
MoO_3	—	—	—	—	—	—

Table 2 (continued)
CODE FOR SALTS AND VITAMIN MIXTURES

	Nitsch[273] [N]	Nitsch et al.[276] [N-68]	Nitsch & Nitsch[267] [NN-67]	Nitsch & Nitsch[275] [NN-69]	Ohyama & Nitsch[282] [O]	Ranga Swamy[310] [RS]
	ELEMENTS (mg/*l*)					
$CuSO_4 \cdot 5H_2O$	0.025	0.025	—	0.08	—	0.025
$CoCl_2 \cdot 6H_2O$	—	—	—	—	—	0.025
$NiCl_2 \cdot 6H_2O$	—	—	—	—	—	—
$Cu(NO_3)_2 \cdot 3H_2O$	—	—	—	—	—	—
$(NH_4)_6Mo_7O_{24} \cdot 4H_2O$	—	—	—	—	—	
$AlCl_3$	—	—	—	—	—	—
Vitamins						
Cyanocobalmin	—	—	—	—	—	—
p-Amino benzoic acid	—	—	—	—	—	—
Folic acid	—	0.5	0.5	0.5	0.5	—
Riboflavin	—	—	—	—	—	—
Biotin	—	0.05	0.05	0.05	0.5	—
Choline chloride	—	—	—	—	—	—
Ca pantothenate	—	—	—	—	—	0.025
Thiamine · HCl	—	0.5	0.5	0.5	0.5	0.25
Nicotinamide (niacin)	—	5.0	5.0	5.0	5.0	1.25
Pyridoxine · HCl	—	0.5	0.5	0.5	0.5	0.025
Other						
(*iso*-) or *myo*-Inositol	—	100	100	100	100	—
Glycine	—	2.0	2.0	2.0	—	7.5
Ascorbic acid	—	—	—	—	—	—
Sucrose	20,000	20,000	—	20,000	—	20,000

	Rangan et al.[309]			Seirlis et al.[338]	
	Nucellar [Ra-N]	Zygotic [Ra-Z]	Ringe & Nitsch[317] [RN]	[S]	[S-P7]
	ELEMENTS (mg/*l*)				
Macro					
a. $(NH_4)_2SO_4$	—	—	—	—	—
b. NH_4NO_3	1,650	—	—	1,000	1,650
c. KNO_3	1,900	80	125	1,000	1,900
d. $Ca(NO_3)_2 \cdot 4H_2O$	—	260	500	500	—
e. $CaCl_2 \cdot 2H_2O$	440	—	—	—	330
f. $MgSO_4 \cdot 7H_2O$	370	360	125	71.5	370
g. KH_2PO_4	170	—	125	300	170
h. $NaH_2PO_4 \cdot H_2O$	—	165	—	—	—
i. Na_2EDTA	37.3	37.3	37.3	—	—
j. Fe-EDTA	—	—	—	—	—
k. NaFe-EDTA	—	—	—	13.2	—
l. $FeSO_4 \cdot 7H_2O$	27.8	27.8	27.8	—	—
m. KCl	—	65	—	65	—
n. Na_2SO_4	—	200	—	—	—
o. Geigy® sequestrene 330 Fe	—	—	—	—	—
p. Fe citrate	—	—	—	—	—
q. $Fe_3(PO_4)_2$	—	—	—	—	—
r. $FeCl_3$	—	—	—	—	1.0

Table 2 (continued)
CODE FOR SALTS AND VITAMIN MIXTURES

		Rangan et al.[309]			Seirlis et al.[338]	
		Nucellar [Ra-N]	Zygotic [Ra-Z]	Ringe & Nitsch[317] [RN]	[S]	[S-P7]
		ELEMENTS (mg/*l*)				
s.	$Fe_2(SO_4)_3$	—	—	—	—	—
t.	$FePO_4 \cdot 4H_2O$	—	—	—	—	—
u.	$NaH_2PO_4 \cdot 2H_2O$	—	—	—	—	—
v.	Na_2HPO_4	—	—	—	—	—
w.	NH_4Cl	—	—	—	—	—
x.	$CaSO_4$	—	—	—	—	—
y.	$Ca_3(PO_4)_2$	—	—	—	—	—
z.	$NH_4H_2PO_4$	—	—	—	—	—
zi.	$NaNO_3$	—	—	—	—	—
Micro						
	$MnSO_4 \cdot H_2O$	—	—	—	—	—
	$MnSO_4 \cdot 2H_2O$	—	—	—	—	—
	$MnSO_4 \cdot 4H_2O$	16.9	3.0	25	14.0	0.1
	$MnSO_4 \cdot 7H_2O$	—	—	—	—	—
	$ZnSO_4 \cdot 7H_2O$	8.6	0.5	10	3.8	1.0
	H_3BO_3	6.2	0.5	10	1.6	1.0
	KI	0.83	—	1.0	—	0.01
	$Na_2MoO_4 \cdot 2H_2O$	0.25	0.025	0.25	—	—
	MoO_3	—	—	—	—	—
	$CuSO_4 \cdot 4H_2O$	0.025	0.025	0.025	—	0.03
	$CoCl_2 \cdot 6H_2O$	0.025	0.025	0.025	—	—
	$NiCl_2 \cdot 6H_2O$	—	—	—	—	0.03
	$Cu(NO_3)_3 \cdot 3H_2O$	—	—	—	0.35	—
	$(NH_4)_6Mo_7O_{24} \cdot 4H_2O$	—	—	—	0.10	—
	$AlCl_3$	—	—	—	—	0.03
Vitamins						
	Cyanocobalmin	—	—	—	—	—
	p-Amino benzoic acid	—	—	—	—	—
	Folic acid	—	—	0.5	—	—
	Riboflavin	—	—	—	—	—
	Biotin	—	—	0.05	—	0.01
	Choline chloride	—	—	—	—	—
	Ca pantothenate	—	0.025	—	—	1.0
	Thiamine · HCl	0.1	0.25	0.5	0.1	1.0
	Nicotinamide (niacin)	0.5	1.25	5.0	0.5	—
	Pyridoxine · HCl	0.5	0.025	0.5	0.1	—
Other						
	(*iso*-) or *myo*-Inositol	100	—	100	100	1.0
	Glycine	2.0	7.5	2.0	—	—
	Ascorbic acid	—	—	—	—	—
	Sucrose	50,000	40,000	40,000	30,000	20,000

		Schenk & Hildebrandt[327] [SH]	Skirvin & Chu[348b] [SC]	Skirvin, Chu & Rukan[348c] (1980) [SC-H]	Skoog & Tsui[353,b] [ST]	Tabachnik & Kester[366] [TK]
Macro						
a.	$(NH_4)_2SO_4$	—	—	—	—	—
b.	NH_4NO_3	—	1,650	1,650	—	—

Table 2 (continued)
CODE FOR SALTS AND VITAMIN MIXTURES

	Schenk & Hildebrandt[327] [SH]	Skirvin & Chu[348b] [SC]	Skirvin, Chu & Rukan[348c] (1980) [SC-H]	Skoog & Tsui[353,b] [ST]	Tabachnik & Kester[366] [TK]
	ELEMENTS (mg/ℓ)				
c. KNO_3	2,500	1,900	1,900	80	200
d. $Ca(NO_3)_2 \cdot 4H_2O$	—	—	—	100	1,140
e. $CaCl_2 \cdot 2H_2O$	200	440	440	—	—
f. $MgSO_4 \cdot 7H_2O$	400	370	370	35	410
g. KH_2PO_4	—	170	170	12.5	200
h. $NaH_2PO_4 \cdot H_2O$	—	—	—	—	—
i Na₂EDTA	15	37.3	37.3	—	37.3
j. Fe-EDTA	—	—	—	—	—
k. NaFe-EDTA	—	—	—	—	—
l. $FeSO_4 \cdot 7H_2O$	20	27.8	27.8	—	27.8
m. KCl	—	—	—	65	—
n. Na_2SO_4	—	—	—	—	—
o. Geigy® sequestrene 330 Fe	—	—	—	—	—
p. Fe citrate	—	—	—	—	—
q. $Fe_3(PO_4)_2$	—	—	—	—	—
r. $FeCl_3$	—	—	—	—	—
s. $Fe_2(SO_4)_3$	—	—	—	2.5	—
t. $FePO_4 \cdot 4H_2O$	—	—	—	—	—
u. $NaH_2PO_4 \cdot 2H_2O$	—	—	—	—	—
v. Na_2HPO_4	—	—	—	—	—
w. NH_4Cl	—	—	—	—	—
x. $CaSO_4$	—	—	—	—	—
y. $Ca_3(PO_4)_2$	—	—	—	—	—
z. $NH_4H_2PO_4$	300	—	—	—	—
zi. $NaNO_3$	—	—	—	—	—
Micro					
$MnSO_4 \cdot H_2O$	10	—	—	—	16.9
$MnSO_4 \cdot 2H_2O$	—	—	—	—	—
$MnSO_4 \cdot 4H_2O$	—	22.3	22.3	4.4	—
$MnSO_4 \cdot 7H_2O$	—	—	—	—	—
$ZnSO_4 \cdot 7H_2O$	1.0	8.6	8.6	1.5	8.6
H_3BO_3	5.0	6.2	6.2	1.6	6.2
KI	1.0	0.83	0.83	0.75	0.83
$Na_2MoO_4 \cdot 2H_2O$	0.1	0.25	0.25	—	0.25
MoO_3	—	—	—	—	—
$CuSO_4 \cdot 5H_2O$	0.2	0.025	0.025	—	0.025
$CoCl_2 \cdot 6H_2O$	0.1	0.025	0.025	—	0.025
$NiCl_2 \cdot 6H_2O$	—	—	—	—	—
$Cu(NO_3)_2 \cdot 3H_2O$	—	—	—	—	—
$(NH_4)_6Mo_7O_{24} \cdot 4H_2O$	—	—	—	—	—
$AlCl_3$	—	—	—	—	—
Vitamins					
Cyanocobalmin	—	0.0015	—	—	—
p-Amino benzoic acid	—	0.5	1.0	—	—
Folic acid	—	0.5	0.25	—	—
Riboflavin	—	0.5	—	—	—
Biotin	—	1.0	0.05	—	—
Choline chloride	—	1.0	1.0	—	—
Ca pantothenate	—	1.0	0.5	—	—

Table 2 (continued)
CODE FOR SALTS AND VITAMIN MIXTURES

	Schenk & Hildebrandt[327] [SH]	Skirvin & Chu[348b] [SC]	Skirvin, Chu & Rukan[348c] (1980) [SC-H]	Skoog & Tsui[353,b] [ST]	Tabachnik & Kester[366] [TK]
ELEMENTS (mg/ℓ)					
Thiamine · HCl	5.0	1.0	2.0	0.1	0.1
Nicotinamide (niacin)	5.0	2.0	2.5	0.5	0.5
Pyridoxine · HCl	0.5	2.0	0.25	0.5	0.5
Other					
(*iso*-) or *myo*-Inositol	1000	100	200	—	100
Glycine	—	2.0	2.0	2.0	2.0
Ascorbic acid	—	50.0	—	—	—
Sucrose	30,000	30,000	20,000	20,000	20,000

	Tukey[377] [T]	Wetmore & Sorokin[387] [WS]	Wetmore & Rier[386] [WR]	White's[388] [W-43]
ELEMENTS (mg/ℓ)				
Macro				
a. $(NH_4)_2SO_4$	—	—	—	—
b. NH_4NO_3	—	—	—	—
c. KNO_3	2,000	—	125	80
d. $Ca(NO_3)_2 \cdot 4H_2O$	—	—	500	200
e. $CaCl_2 \cdot 2H_2O$	—	—	—	—
f. $MgSO_4 \cdot 7H_2O$	187.5	—	125	360
g. KH_2PO_4	—	—	125	—
h. $NaH_2PO_4 \cdot H_2O$	—	—	—	16.5
i. Na_2EDTA	—	—	—	—
j. Fe-EDTA	—	—	—	—
k. NaFe-EDTA	—	—	—	—
l. $FeSO_4 \cdot 7H_2O$	—	—	—	—
m. KCl	750.2	—	—	65
n. Na_2SO_4	—	—	—	200
o. Geigy® sequestrene 330 Fe	—	—	—	—
p. Fe citrate	—	—	10	—
q. $Fe_3(PO_4)_2$	187.5	—	—	—
r. $FeCl_3$	—	—	—	—
s. $Fe_2(SO_4)_3$	—	—	—	2.5
t. $FePO_4 \cdot 4H_2O$	—	—	—	—
u. $NaH_2PO_4 \cdot 4H_2O$	—	—	—	—
v. Na_2HPO_4	—	—	—	—
w. NH_4Cl	—	—	—	—
x. $CaSO_4$	187.5	—	—	—
y. $Ca_3(PO_4)_2$	187.5	—	—	—
z. $NH_4H_2PO_4$	—	—	—	—
zi. $NaNO_3$	—	—	—	—
Micro				
$MnSO_4 \cdot H_2O$	—	—	—	—
$MnSO_4 \cdot 2H_2O$	—	—	—	—
$MnSO_4 \cdot 4H_2O$	—	—	15	—
$MnSO_4 \cdot 7H_2O$	—	—	—	8.25
$ZnSO_4 \cdot 7H_2O$	—	—	2.5	2.67
H_3BO_3	—	—	2.5	1.5
KI	—	—	—	0.75

Table 2 (continued)
CODE FOR SALTS AND VITAMIN MIXTURES

	Tukey[377] [T]	Wetmore & Sorokin[387] [WS]	Wetmore & Rier[386] [WR]	White's[388] [W-43]
ELEMENTS (mg/ℓ)				
$Na_2MoO_4 \cdot 2H_2O$	—	—	0.125	—
MoO_3	—	—	—	—
$CuSO_4 \cdot 5H_2O$	—	—	0.125	—
$CoCl_2 \cdot 6H_2O$	—	—	0.125	—
$NiCl_2 \cdot 6H_2O$	—	—	—	—
$Cu(NO_3)_2 \cdot 3H_2O$	—	—	—	—
$(NH_4)_6Mo_7O_{24} \cdot 4H_2O$	—	—	—	—
$AlCl_3$	—	—	—	—
Vitamins				
Cyanocobalmin	—	—	—	—
p-Amino benzoic acid	—	0.5	0.5	—
Folic acid	—	0.1	0.1	—
Riboflavin	—	0.5	0.5	—
Biotin	—	0.01	0.01	—
Choline chloride	—	—	1.0	—
Ca pantothenate	—	5.0	5.0	—
Thiamine · HCl	—	1.0	1.0	0.1
Nicotinamide (niacin)	—	5.0	5.0	0.5
Pyridoxine · HCl	—	1.0	1.0	0.1
Other				
(*iso*-) or *myo*-Inositol	—	10.0	100	—
Glycine	—	—	—	3.0
Ascorbic acid	—	—	—	—
Sucrose	—	—	40,000	20,000

	White's[389] [W-63]
ELEMENTS (mg/ℓ)	
Macro	
a. $(NH_4)_2SO_4$	—
b. NH_4NO_3	—
c. KNO_3	80
d. $Ca(NO_3)_2 \cdot 4H_2O$	200
e. $CaCl_2 \cdot 2H_2O$	—
f. $MgSO_4 \cdot 7H_2O$	720
g. KH_2PO_4	—
h. $NaH_2PO_4 \cdot H_2O$	17
i. Na_2EDTA	—
j. Fe-EDTA	—
k. NaFe-EDTA	—
l. $FeSO_4 \cdot 7H_2O$	—
m. KCl	65
n. Na_2SO_4	200
o. Geigy® sequestrene 330 Fe	—
p. Fe citrate	—
q. $Fe_3(PO_4)_2$	—
r. $FeCl_3$	—

Table 2 (continued)
CODE FOR SALTS AND VITAMIN MIXTURES

		White's[389] [W-63]
ELEMENTS (mg/l)		
s.	$Fe_2(SO_4)_3$	2.5
t.	$FePO_4 \cdot 4H_2O$	—
u.	$NaH_2PO_4 \cdot 2H_2O$	—
v.	Na_2HPO_4	—
w.	NH_4Cl	—
x.	$CaSO_4$	—
y.	$Ca_3(PO_4)_2$	—
z.	$NH_4H_2PO_4$	—
zi.	$NaNO_3$	—
Micro		
	$MnSO_4 \cdot H_2O$	—
	$MnSO_4 \cdot 2H_2O$	—
	$MnSO_4 \cdot 4H_2O$	5.0
	$MnSO_4 \cdot 7H_2O$	—
	$ZnSO_4 \cdot 7H_2O$	3.0
	H_3BO_3	1.5
	KI	0.75
	$Na_2MoO_4 \cdot 2H_2O$	—
	MoO_3	0.001
	$CuSO_4 \cdot 5H_2O$	0.01
	$CoCl_2 \cdot 6H_2O$	—
	$NiCl_2 \cdot 6H_2O$	—
	$Cu(NO_3)_2 \cdot 3H_2O$	—
	$(NH_4)_6Mo_7O_{24} \cdot 4H_2O$	—
	$AlCl_3$	—
Vitamins		
	Cyanocobalmin	—
	p-Amino benxoic acid	—
	Folic acid	—
	Riboflavin	—
	Biotin	—
	Choline chloride	—
	Ca pantothenate	—
	Thiamine · HCl	0.1
	Nicotinamide (niacin)	0.5
	Pyridoxine · HCl	0.1
Other		
	(*iso*-) or *myo*-Inositol	—
	Glycine	3.0
	Ascorbic acid	—
	Sucrose	20,000

[a] The amount was not specified in the paper.

[b] Sometimes mistakingly listed in the literature as Skoog and Cheng medium.

Table 3
HORMONES THAT ARE COMMONLY USED IN FRUIT TISSUE CULTURES

Hormone type code	Correct name	Molecular weight
Auxins		
2,4-D	2,4-Dichlorophenoxyacetic acid	221.04
2,3-DPA	2,3-Dichlorophenylacetic acid	205.04
IAA	Indole-3-acetic acid	175.18
IBA	Indole-3-butyric acid	203.23
NAA	α-Naphthalene-acetic acid	186.20
NOA	2-Naphthoxyacetic acid	202.20
p CPA	*para* Chlorophenoxy acetic acid	186.53
NA	1-Naphthaleneacetamide	185.14
Cytokinins		
BA	6-Benzylaminopurine	225.20
2-iP	6-(γγ-Dimethylallylamino) purine [2-isopentyladenine]	203.30
Kn	6-Furfurylaminopurine [kinetin]	215.21
Z	6-(4-Hydroxy-3-methyl-but-2-enylamino)-purine [zeatin]	
Dormins		
ABA	5-(Hydroxy-2,6,6-trimethyl-4-oxo-2-cyclohexen-1-yl)-3-methyl-2,4-pentadienoic acid [abscisic acid]	264.31
Gibberellins		
GA_3	2,4a,7-Trihydroxy-1-methyl-8-methylene-gibb-3-ene-1,10-carboxylic acid 1,4-lactone [gibberellic acid]	346.37

REFERENCES

1. **Abbott, A. J.**, Propagating temperate woody species in tissue culture, *Sci. Hortic.*, 28, 155, 1977.
2. **Abbott, A. J. and Whiteley, E.**, *In vitro* Regeneration and Multiplication of Apple Tissues, Abstr. 263, in 3rd Int. Congr. Plant Tissue and Cell Culture, Leicester, 1974.
3. **Abbott, A. J. and Whiteley, E.**, Culture of *Malus* tissues *in vitro*. I. Multiplication of apple plants from isolated shoot apices, *Sci. Hortic.*, 4, 183, 1976.
4. **Abou-Zeid, A.**, Many sided influences on the growth and morphogenesis in the *in vitro* culture of Prunus species embryo axes. I. Myo-inositol as a critical growth factor in the radicular meristem of the embryo axes of dormant cherry, peach and plum embryos, *Angew. Bot.*, 47, 227, 1973.
5. **Abraham, A. and Thomas, K. J.**, A note on the *in vitro* culture of excised coconut embryos, *Indian Coconut J.*, 15, 84, 1962.
6. **Adams, A. N.**, An improved medium for strawberry meristem culture, *J. Hortic. Sci.*, 47, 263, 1972.
7. **Aghion, D. and Beauchesne, G.**, Utilisation de la technique de culture sterile d'organes pour obtenir des clones d'ananas, *Fruits*, 15, 464, 1960.
8. **Alleweldt, G. and Radler, F.**, Das wachstum der gewebekulturen photoperiodisch vorbehandelter rebenpflanzen, *Naturwissenschaften*, 48, 109, 1961.
9. **Alleweldt, G. and Radler, F.**, Interrelationship between photoperiodic behavior of grapes and the growth of plant tissue cultures, *Plant Physiol.*, 37, 376, 1962.

10. **Al-Mehdi, A. A. and Hogan, L.**, *In vitro* growth and development of papaya (*Carica papaya* L.) and date-palm (*Phoenix dactylifera* L.), *HortScience,* 14 (Abstr.), 422, 1979.
11. **Alskief, J. and Gautheret, M. R.**, *In vitro* grafting of apex onto decapitated peach-tree seedlings, *C. R. Acad. Sci. Ser. D,* 284, 2499, 1977.
12. **Alskief, J. and Riedel, M.**, Oganogenesis phenomena induced *in vitro* in Citrus fruit, *C. R. Acac. Sci. Ser. D,* 287, 249, 1978.
13. **Alskief, J. and Villemur, P.**, Technique of shoot apex grafting, *in vitro* onto decapitated rootstock seedlings of apple-tree (*Malus pumila* Mill), *C. R. Acad. Sci. Ser. D,* 287, 1115, 1978.
14. **Altman, A. and Goren, R.**, Promotion of callus formation by abscisic acid in citrus bud cultures, *Plant Physiol.,* 47, 844, 1971.
15. **Altman, A. and Goren, R.**, Growth and dormancy cycles in citrus bud cultures and their hormonal control, *Physiol. Plant.,* 30, 240, 1974a.
16. **Altman, A. and Goren, R.**, Interrelationship of abscisic acid and gibberellic acid in the promotion of callus formation in the abscission zone of Citrus bud cultures, *Physiol. Plant.,* 32, 55, 1974b.
17. **Altman, A. and Goren, R.**, Horticultural and physiological aspects of Citrus bud cultures, *Acta Hortic.,* 78, 51, 1977.
18. **Ammar, S. and Benbadis, A.**, Vegetative propagation of the date palm (*Phoenix dactylifera* L.) by tissue cultures obtained from seedlings, *C. R. Acad. Sci. Ser. D,* 284, 1789, 1977.

18a. **Anderson, J. D., Lieberman, M., and Stewart, R. N.**, Ethylene production by apple protoplasts, *Plant Physiol.,* 63, 931, 1979.

19. **Anderson, W. C.**, Propagation of rhododendrons by tissue culture. I. Development of a culture medium for multiplication of shoots, *Proc. Int. Plant Prop. Soc.,* 25, 129, 1975.
20. **Anon.**, Tissue culture quickens plant propagation, *Agric. Res.,* 27(10), 3, 1979.
21. **Apte, P. V., Kaklij, G. S., and Heble, M. R.**, Proteolytic enzymes (*Bromelains*) in tissue cultures of *Ananas-sativus* (pineapple), *Plant Sci. Lett.,* 14, 57, 1979.

21a. **Arya, H. C. and Hildebrandt, A. C.**, Differential sensitivities to gamma radiation of *Phylloxera* gall and normal grape stem cells in tissue culture, *Can. J. Bot.,* 47, 1623, 1969a.

21b. **Arya, H. C. and Hildebrandt, A. C.**, Effect of gamma-radiation on callus growth of *Phylloxera* gall and normal grape stem tissues in culture, *Indian J. Exp. Biol.,* 7, 158, 1969b.

22. **Arya, H. C., Hildebrandt, A. C., and Riker, A. J.**, Growth in tissue culture of single-cell clones from grape stem and *Phylloxera* gall., *Plant Physiol.,* 37, 387, 1962a.
23. **Arya, H. C., Hildebrandt, A. C., and Riker, A. J.**, Clonal variation of grape-stem and phylloxera-gall callus growing *in vitro* in different concentrations of sugars, *Am. J. Bot.,* 49, 368, 1962b.
24. **Asahira, T. and Kano, Y.**, Shoot formation from cultured tissue of strawberry fruits, *J. Jpn. Soc. Hortic. Sci.,* 46, 317, 1977.
25. **Bajaj, Y. P. S. and Collins, W. B.**, Some factors affecting the in vitro development of strawberry fruits, *Proc. Am. Soc. Hortic. Sci.,* 93, 326, 1968.
26. **Balaga, H. Y. and De Guzman, E. V.**, The growth and development of coconut "Makapuno" embryos *in vitro.* II. Increased root incidence and growth in response to media composition and to sequential culture from liquid medium, *Philipp. Agric.,* 53, 551, 1970.
27. **Bapat, V. A. and Narayanaswamy, S.**, Mesocarp & endosperm culture of *Achras sapota* Linn. *in vitro, Indian J. Exp. Biol.,* 15, 294, 1977.
28. **Barker, W. G. and Collins, W.B.**, The blueberry rhizome: in vitro culture, *Can. J. Bot.,* 41, 1325, 1963.

28a. **Barlass, M. and Skene, K. G. M.**, In vitro propagation of grapevine (Vitis vinifera L.) from fragmented shoot apices, *Vitis,* 17, 335, 1978.

29. **Beauchesne, G. and Poulain, C. L.**, La culture de méristèmes, ses possibilités et ses limites actuelles pout les plantes ligneuses, in *Phytotronique et Prospective Horticole,* Chouard, P. and de Bilderling, Eds., Gauthier-Villars, Paris, 1972, 219.
30. **Béjaoui, M. and Pilet, P. E.**, Effects of fluoride on the oxygen uptake by *Rubus* tissues cultured in vitro, *C. R. Acad. Sci. Ser. D,* 280, 1457, 1975.
31. **Belkengren, R. O. and Miller, P. W.**, Culture apical meristems of *Fragaria vesca* strawberry plants as a method of excluding latent A virus, *Plant Dis. Rep.,* 46, 119, 1962.
32. **Benbadis, A.**, Culture de cellules de ronce sur des milieux non conditionnés renfermant de l'extrait de levure. Comparison avec les milieux simples conditionnes, *C. R. Acad. Sci. Ser. D.,* 264, 1612, 1967.
33. **Bennici, A., Buiatti, M., Togoni, F., Rosellini, D., and Giorgi, L.**, Habituation in *Nicotiana bigelovii* tissue cultures: different behaviour of two varieties, *Plant Cell Physiol.,* 13, 1, 1972.

33a. **Binding, H.**, Selektion in kalluskulturen mit haploiden zellen, *Z. Pflanzenzuecht.,* 67, 33, 1972.

33b. **Binding, H., Binding, K., and Straub, J.**, Selektion mit gewebekulturen mit haploiden zellen, *Naturwissenschaften,* 57, 138, 1970.

34. **Bini, G.**, Prove di colture "in vitro" di meristemi apicali di Vitis vinifera L., *Riv. Ortoflorofruttic. Ital.,* 60, 289, 1976.

35. **Bitters, W. P., Murashige, T., Rangan, T. S., Nauer, E.,** Investigations on established virus-free citrus plants through tissue culture, *Calif. Citrus Nurserymen's Soc.*, 9, 27, 1970.
36. **Bitters, W. P., Murashige, T. S., Rangan, T. S., and Nauer, E.,** Investigations on establishing virus-free citrus plants through tissue culture, in *Proc. 5th Conf. Int. Organization Citrus Virology, Price, W. C., Ed.*, University of Florida Press, Gainesville, 1972, 267.
37. **Blakely, L. M. and Steward, F. C.,** Growth and organized development of cultured cells. VII. Cellular variation, *Am. J. Bot.*, 51, 809, 1964.
38. **Borrod, G.,** Contribution à l'étude des cultures *in vitro* des tissus de châtaignier, *C. R. Acad. Sci. Ser. D*, 272, 56, 1971.
39. **Bouzid, S.,** Some aspects of the behaviour of *in vitro* cultures of *Citrus* cutting *C. R. Acad. Sci. Ser. D*, 280, 1689, 1975.
40. **Bouzid, S. and Lasram, M.,** Utilisation de cultures *in vitro* pour l'obtention de clones de *Citrus* homogenes et de bon etat sanitaire, *8th Congr. Int. Agrum. Mediter.*, vol. 2, 1, 1971.
41. **Bové, J.,** Le bouturage des *Citrus, Fruits*, 12, 253, 1957.
42. **Bové, J. and Morel, G.,** La culture de tissus de Citrus, *Rev. Gen. Bot.*, 64, 34, 1957.
43. **Boxus, P.,** La culture de meristemes de *Prunus.* Note preliminaire relative a l'espece *P. Pandora., Bull. Rech. Agron. Gembloux*, 6, 3, 1971.
44. **Boxus, P.,** La production de plants sains de fraisiers, *Acta Hortic.*, 30, 187, 1973.
45. **Boxus, P.,** The production of strawberry plants by *in vitro* micro-propagation, *J. Hortic. Sci.*, 49, 209, 1974.
46. **Boxus, P.,** Rapid production of virus-free strawberry by "in vitro" culture, *Acta Hortic.*, 66, 35, 1976.
47. **Boxus, P.,** The production of fruit and vegetable plants by in vitro culture — actual possibilities and perspectives, In Propagation of Higher Plants Through Tissue Culture: A Bridge Between Research and Application, Hughes, K. W., Henke, R., and Constantin, M., Eds., National Technical Information Service, U.S. Department of Energy, Springfield, Va., 1978, 44.
48. **Boxus, P. and Quoirin, M.,** La culture de méristèmes apicaux de quelques especès de Prunus, *Bull. Soc. R. Bot. Belg.*, 107, 91, 1974.
49. **Boxus, P. and Quoirin, M.,** Nursery behavior of fruit trees propagated by "in vitro", *Acta Hortic.*, 78, 373, 1977.
50. **Boxus, P., Quoirin, M., and Laine, J. M.,** Large scale propagation of strawberry plants from tissue culture, in *Plant Cell, Tissue, and Organ Culture,* Reinert, J. and Bajaj, Y. P. S., Eds., Springer-Verlag, Berlin, 1977, 130.
51. **Broertjes, C., Haccius, B., and Weidlich, S.,** Adventitious bud formation on isolated leaves and its significance for mutation breeding, *Euphytica*, 17, 321, 1968.
52. **Broome, O. C. and Zimmerman, R. H.,** In vitro propagation of blackberry, *HortScience*, 13, 151, 1978.
53. **Buckland, E. L.,** Cell Suspension Culture Studies of the *Coffea arabica* L. MS thesis, Department of Food Science, University of British Columbia, Vancouver, 1972.
54. **Burger, D. W. and Banks, M. S.,** Regeneration in *Citrus* tissue cultures, *HortScience*, 14 (Abstr.), 477, 1979a.
55. **Burger, D. W. and Banks, M. S.,** Citrus tissue culture, *HortScience*, 14 (Abstr.), 423, 1979b.
56. **Button, J. and Bornman, C. H.,** Development of nucellar plants from unpollinated and unfertilized ovules of the Washington Navel orange *in vitro, J. S. Afr. Bot* 37, 127, 1971a.
57. **Button, J. and Bornman, C. H.,** In vitro development of nucellar plants from unfertilized ovules of the Washington Navel orange, in *Morphogenesis in Plant Cell, Tissue, and Organ Cultures*, Johri, B. M. and Mohan Ram, H. Y., Eds., University of Delhi, India, 1971b, 43.
58. **Button, J. and Kochba, J.,** Tissue culture in the Citrus industry, in *Applied and Fundamental Aspects of Plant Cell, Tissue, and Organ Culture,* Reinert, J. and Bajaj, Y. P. S., Eds., Springer-Verlag, New York, 1977, 70.
59. **Button, J., Kochba, J., and Borman, C. H.,** Fine structure of and embryoid development from embryogenic ovular callus of "Shamouti" orange, *J. Exp. Bot.*, 25, 446, 1973.
60. **Bychenkova, E. A.,** An investigation of callus formation in certain trees and shrubs by the method of tissue culture in vitro, *Dokl. Akad. Nauk SSSR, Ser. Biol.*, 151, 1077, 1963.
60a. **Cameron, J. W. and Soost, R. K.,** Sexual and nucellar embryony in F_1 hybrids and advanced crosses of *Citrus* with *Poncirus, J. Am. Soc. Hortic. Sci.*, 104, 408, 1979.
61. **Campbell, R. A. and Durzan, D. J.,** Induction of multiple buds and needles in tissue cultures of *Picea glauca, Can. J. Bot.*, 53, 1652, 1975.
62. **Caponetti, J. D., Hall, G. C., and Farmer, R. E., Jr.,** In vitro growth of black cherry callus: effects of medium, environment, and clone, *Bot. Gaz. (Chicago)*, 132, 313, 1971.
63. **Caporali, L. and Weltzien, H. C.,** Structural anomalies and growth modifications caused by *Taphrina deformans* (Berk.) Tul. in tissues of *Prunus persica* L. cultivated *in vitro, Rev. Gen. Bot.*, 81, 85, 1974.

64. **Carlson, P. S.**, Production of auxotrophic mutants in ferns, *Genet. Res.*, 14, 337, 1969.

65. **Carlson, P. S.**, Induction and isolation of auxotrophic mutants in somatic cell cultures of *Nicotiana tabacum, Science,* 168, 487, 1970.

66. **Chaleff, R. S. and Carlson, P. S.**, Somatic cell genetics of higher plants, *Annu. Rev. Genet.*, 8, 267, 1974.

67. **Chaleff, R. S. and Carlson, P. S.**, Higher plant cells as experimental organisms, in *Modification of the Information Content of Plant Cells,* Markham, R., Davies, D. R., Hapwood, D. A., and Horne, R. W., Eds., North-Holland, Amsterdam, 1975, 197.

68. **Chaturvedi, H. C., Chowdhury, A. R., and Mitra, G. C.**, Morphogenesis in stem-callus tissue of *Citrus grandis* in long-term cultures — a biochemical analysis, *Curr. Sci.*, 43, 139, 1974.

69. **Chaturvedi, H. C. and Mitra, G. C.**, Clonal propagation of Citrus from somatic callus cultures, *HortScience,* 9, 118, 1974.

70. **Cheng, T. Y.**, Adventitious bud formation in culture of Douglas Fir (*Pseudotsuga menziesii* (Mirb.) Franco), *Plant Sci. Lett.*, 5, 97, 1975.

70a. **Cheng, T. Y.**, Micropropagation of clonal fruit tree rootstocks, *Compact Fruit Tree,* 12, 127, 1979.

71. **Cheng, T. Y.**, Clonal propagation of woody species through tissue culture techniques, *Proc. Int. Plant Prop. Soc.*, 28, 139, 1978.

72. **Cheng, T. Y. and Vogui-Dinh, T. H.**, Clonal propagation of selected deciduous trees through tissue culture, *HortScience,* 14 (Abstr.), 457, 1979.

73. **Chong, C. and Taper, C. D.**, *Malus* tissue cultures. I. Sorbitol (D-glucitol) as a carbon source for callus initiation and growth, *Can. J. Bot.*, 50, 1399, 1972.

74. **Chong, C. and Taper, C. D.**, *Malus* tissue cultures. II. Sorbitol metabolism and carbon nutrition, *Can. J. Bot.*, 52, 2361, 1974a.

75. **Chong, C. and Taper, C. D.**, Influence of light intensity on sorbitol metabolism, growth, and chlorophyll content of *Malus* tissue cultures, *Ann. Bot. (London),* 38, 359, 1974b.

76. **Coffin, R., Taper, C. D., and Chong, C.**, Sorbitol and sucrose as carbon source for callus culture of some species of the Rosaceae, *Can. J. Bot.*, 54, 547, 1976.

77. **Creasy, M. T. and Sommer, N. F.**, Growth of *Fragaria vesca* L. receptacles *in vitro* with reference to gibberellin inhibition by unfertilized carpels, *Physiol. Plant.*, 17, 710, 1964.

78. **Cutter, V. M. and Wilson, K. S.**, Effect of coconut endosperm and other growth stimulants upon the development in vitro of embryos of Cocos nucifera, *Bot. Gaz. (Chicago),* 115, 234, 1954.

79. **Dausend, B.**, Physiologische untersuchungen zur wurzelinduktion In Vitro an zwei verschiedenen Prunusarten (*Prunus avium* and *Prunus pseudocerasus*), Diss. Gieben, 1974.

80. **Dayton, D.**, Genetic heterogeneity in the histogenic layers of apple, *J. Am. Soc. Hortic. Sci.*, 94, 592, 1969.

81. **Dayton, D.**, New apple strains developed by forcing shoots on disbudded trees, *Ill. Res.*, 12, 10, 1970.

82. **De Bruijne, E., DeLanghe, D., and van Rijck, R.**, Actions of hormones and embryoid formation in callus cultures of *Carica papaya, Int. Symp. Fytofarm. Fytiat.*, 26, 637, 1974.

83. **De Capite, L.**, La colture dei frutti in vitro de fiori recisi de *Fragaria chiloensis* Ehrh × *F. virginica* Duch. var. Marshall e di *Pisum sativum* L. var. Zelka, *Ric. Sci.*, 25, 532, 1955.

84. **de Fossard, R. A.**, Tissue culture in horticulture — a perspective, *Acta Hortic.*, 78, 455, 1977.

85. **De Guzman, E. V.**, The growth and development of coconut "Makapuno" embryo *in vitro.* I. The induction of rooting, *Philipp. Agric.*, 53, 65, 1970.

86. **De Guzman, E. V., del Rosario, A. G., and Eusebio, E. C.**, The growth and development of "Makapuno" embryo *in vitro.* III. Resumption of root growth in high sugar media, *Philipp. Agric.*, 53, 566, 1971.

87. **Demetriades, S. D.**, Note preliminaire sur la culture *in vitro* des tissus de citronnier, *Ann. Inst. Phytopathol. Benaki,* 8, 103, 1954.

87a. **Diamantoglou, S. and Mitrakos, K.**, Sur la culture *in vitro* de l'embryon d'Olivier (*Olea europaea* L. *var.* Oleaster), *C. R. Acad. Sci. Ser. D,* 288, 1537, 1979.

88. **Drira, N. and Benbadis, A.**, Analysis of the androgenetic potentialities of two *Citrus* spp. (*Citrus medica* L. et *Citrus limon* L. Burm.) by anther culture, *C. R. Acad. Sci. Ser. D,* 281, 1321, 1975.

89. **Dutcher, R. D. and Powell, L. E.**, Culture of apple shoots from buds *in vitro, J. Am. Soc. Hortic. Sci.*, 97, 511, 1972.

90. **Eeuwens, C. J.**, Effects of organic nutrients and hormones on growth and development of tissue explants from coconut (*Cocos nucifera*) and date (*Phoenix dactylifera*) palms cultured *in vitro, Physiol. Plant.*, 42, 173, 1978.

91. **Eeuwens, C. J. and Blake, J.**, Culture of coconut and date palm tissue with a view to vegetative propagation, *Acta Hortic.*, 78, 277, 1977.

92. **Eichholtz, D. A., Robitaille, H. A., and Hasegawa, P. M.**, Morphogenesis in apple, *HortScience,* 14(Abstr.), 410, 1979.

92a. **Eichholtz, D. A., Robitaille, H. A., and Hasegawa, P. M.,** Adventive embryony in apple, *HortScience,* 14, 699, 1979.

93. **Einset, J. W.,** Citrus tissue culture — stimulation of fruit explant cultures with orange juice, *Plant Physiol.,* 62, 885, 1978.

94. **Elder, J. W., Benveniste, P., and Fonteneau, P.,** In vitro cyclization of squalene 2,3-Epoxide to α-amyrin by microsomes from bramble cell suspension cultures, *Phytochemistry,* 16, 490, 1977.

95. **Elliott, R. F.,** Axenic culture of meristem tips of *Rosa multiflora, Planta,* 95, 183, 1970.

96. **Elliott, R. F.** Axenic culture of shoot apices of apple, *N. Z. J. Bot.,* 10, 254, 1972.

97. **Epstein, E. and Lavee, S.,** Uptake and fate of IAA in apple callus tissue using IAA-1-^{14}C, *Plant Cell Physiol.,* 16, 553, 1975.

98. **Epstein, E., Klein, I., and Lavee, S.,** Uptake and fate of IAA in apple callus tissues: metabolism of IAA-2-^{14}C, *Plant Cell Physiol.,* 16, 305, 1975.

99. **Epstein, E., Kochba, J., and Neumann, H.,** Metabolism of indoleacetic acid by embryogenic and non-embryogenic callus lines of "Shamouti" orange (*Citrus sinensis* Osb), *Z. Pflanzenphysiol.,* 85, 263, 1977.

100. **Erner, Y., Reuveni, O., and Goldschmidt, E. E.,** Partial purification of a growth factor from orange juice which affects citrus tissue culture and its replacement by citric acid, *Plant Physiol.,* 56, 279, 1975.

101. **Fallot, J.,** Cultures de tissus de quelques espèces et variétiés du genre *Vitis* et leur comportement, *Bull. Soc. Hist. Nat. Toulouse,* 90, 163, 1955.

102. **Fallot, J.,** Cultures aseptiques de tiges de vigne prelévées juste avant et pendant le repos hivernal, *Bull. Soc. Hist. Nat. Toulouse,* 90, 173, 1955.

103. **Favre, J. M.,** Effects corrélatifs de facteurs internes et externes sur la rhizogenese d'un clone de vigne (*Vitis riparia × Vitis rupestris*) cultivé *in vitro, Rev. Gen. Bot.,* 80, 279, 1973.

104. **Favre, J. M.,** Influence de l'etat Physiologique de la Plante et de la Nature de l'Organe prelève sur l'obtention de Neoformations Caulinaires de Vigne, in Congr. Natl. Soc. Savantes, Lille, France, 1976.

105. **Favre, J. M.,** First results on *in vitro* production of bud-neoformations in grape vine, *Ann. Amelior. Plant.,* 27, 151, 1977.

106. **Feucht, W.,** Oxydativer phenolstoffwechsel bei kallus-kulturen von Prunus, *Mitt. Klosterneuburg,* 23, 131, 1973.

107. **Feucht, W.,** Effects of hormones on growth of *Prunus* tissue cultured *in vitro, Proc. 19th Int. Hortic. Congr.,* 1/A(Abstr.), 58, 1974.

108. **Feucht, W.,** Potential of cell division of stem explants from *Prunus* species with different growth vigor, *Gartenbauwissenschaft,* 40, 253, 1975.

109. **Feucht, W. and Dausend, B.,** Induktion von wurzelen bei oxytierenden Prunus-kalluskulturen unter dem gesichspunkt des auxinspiegels und der phenolsynthese, *Mitt. Klosterneuburg,* 22, 345, 1972.

110. **Feucht, W. and Dausend, B.,** Root induction in vitro of easy-to-root *Prunus pseudocerasus* and difficult-to-root *Prunus avium, Sci. Hortic.,* 4, 49, 1976.

111. **Feucht, W. and Johal, C. S.,** Effect of chlorogenic acids on the growth of excised young stem segments of Prunus avium, *Acta Hortic.,* 78, 109, 1977.

112. **Feucht, W. and Khan, Z.,** Effect of catechin on the growth of *Prunus* stem segments cultured *in vitro, Z. Pflanzenphysiol.,* 69, 242, 1973.

112a. **Feucht, W. and Nachit, M.,** Flavolans and growth-promoting catechins in young shoot tips of *Prunus* species and hybrids, *Physiol. Plant.,* 40, 230, 1977.

113. **Fisher, J. B. and Tsai, J. H.,** *In vitro* growth of embryos and callus of coconut palm, *In Vitro,* 14, 307, 1978.

114. **Fonteneau, P., Hartmann-Bouillon, M. A., and Benveniste, P.,** A 24-methylene lophenol C-28 methyltransferase from suspension cultures of bramble cells, *Plant Sci. Lett.,* 10, 147, 1977.

115. **Fowler, C. W.,** Feasibility studies for chromosome reduction in strawberry, Ph.D. thesis, Purdue University, West Lafayette, Ind., 1972.

116. **Fowler, C. W., Hughes, H. G., and Janick, J.,** Callus formation from strawberry anthers, *Hortic. Res.,* 11, 116, 1971.

117. **Fox, J. E.,** Growth factor requirements and chromosome number in tobacco tissue culture, *Physiol. Plant.,* 16, 793, 1963.

118. **Fujii, T. and Nito, N.,** Studies on the compatibility of grafting of fruit trees. I. Callus fusion between rootstock and scion, *J. Jpn. Soc. Hortic. Sci.,* 41, 1, 1972.

119. **Galzy, R.,** Confirmation de la nature virale du court-noué de la vigne par des essais de thermothérapie sur des cultures *in vitro, C. R. Acad. Sci.,* 253, 706, 1961.

120. **Galzy, R.,** Technique de thermothérapie des viroses de la vigne, *Ann. Epiphyt.,* 15, 245, 1964.

122. **Galzy, R.,** Action de la température 35°C sur *Vitis rupestris* atteint de court-noue, *Bull. Soc. Fr. Physiol. Veg.,* 12, 391, 1966.

123. **Galzy, R.**, Research on the growth of both fan leaf-infected and healthy *Vitis rupestris* Scheele cultivated *in vitro* at different temperatures, *Ann. Phytopathol.*, 1, 149, 1969a.
124. **Galzy, R.**, Remarques sur la croissance de Vitis rupestris cultivée in vitro sur différents milieux nutritifs, *Vitis*, 8, 191, 1969b.
125. **Galzy, R.**, La culture *in vitro* des apex de *Vitis rupestris*, *C. R. Acad. Sci. Ser. D*, 274, 210, 1972.
126. **Gamborg, O. L. and Eveleigh, D. E.**, Culture methods and detection of glucanases in suspension cultures of wheat and barley, *Can. J. Biochem.*, 46, 417, 1968.
127. **Garley, B., Stushnoff, C., Wildung, D., and Read, P. E.**, *In vitro* micropropagation of blueberry hybrids using single bud stem segments and stemtips, *HortScience*, 14 (Abstr.), 477, 1979.
128. **Gautheret, R. J.**, *Manuel Technique de Culture de Tissus Vegetaux*, Masson, Paris, 1942, 170.
129. **Gautheret, R. J.**, Sur la variabilité des proprietés physiologiques des cultures de tissus végétaux, *Rev. Gen. Bot.*, 62, 1, 1955a.
130. **Gautheret, R. J.**, The nutrition of plant tissue cultures, *Annu. Rev. Plant Physiol.*, 6, 433, 1955b.
131. **Gautheret, R. J.**, *La Culture des Tissus Vegetaux: Techniques et Realisations*, Masson, Paris, 1959, 23.
132. **Ghugale, D. D., Kulkarni, O. K., Narasimhan, R.**, Effect of auxins and gibberellic acid on growth and differentiation of *Morus alba* and *Populus nigra* tissues *in vitro*, *Indian J. Exp. Biol.*, 9, 381, 1971.
133. **Giladi, I., Altman, A., and Goren, R.**, Differential effects of sucrose, abscisic acid, and benzyladenine on shoot growth and callus formation in the abscission zone of excised Citrus buds, *Plant Physiol.*, 59, 1161, 1977.
134. **Gresshoff, P. M. and Doy, C. H.**, Development and differentiation of haploid *Lycopersicon esculentum* (tomato), *Planta*, 107, 161, 1972.
135. **Gresshoff, P. M. and Doy, C. H.**, Derivation of a haploid cell line from *Vitis vinifera* and the importance of the stage of meiotic development of anthers for haploid culture of this and other genera, *Z. Pflanzenphysiol.*, 73, 132, 1974.
136. **Grinblat, U.**, Differentiation of citrus stem *in vitro*, *J. Am. Soc. Hortic. Sci.*, 97, 599, 1972.
137. Gustafson, J. P., The evolutionary development of Triticale: the wheat-rye hybrid, in *Evolutionary Biology*, Vol. 9, Hecht, M. K., Steere, W. B., and Wallace, B., Eds., Plenum Press, New York, 107.
138. **Harada, H.**, *In vitro* organ culture of *Actinidia chinensis* Pl. as a technique for vegetative multiplication, *J. Hortic. Sci.*, 50, 81, 1975.
139. **Harn and Kim**, Induction of callus from anthers of Prunus armeniaca, *Korean J. Breeding*, 4, 49, 1971.
140. **Harper, P. C.**, Tissue culture propagation of blackberry and tayberry, *Hortic. Res.*, 18, 141, 1978.
141. **Harvey, A. E., Chakrovorty, A. K., Shaw, M., and Sctubb, L. A.**, Changes in ribonuclease activity in Ribes leaves and pine tissue culture with blister rust, Cronartium ribico La., *Peat Plant News*, 4, 359, 1974.
142. **Hawker, J. S., Downton, W. J. S., Wiskich, D., and Mullins, M. G.**, Callus and cell culture from grape berries, *HortScience*, 8, 398, 1973.
143. **Hedtrich, C. M.**, Differentiation of cultivated leaf discs of Prunus mahaleb, *Acta Hortic.*, 78, 177, 1977a.
144. **Hedtrich, C. M.**, Problems and possibilities of tissue culture, *Erwerbsobstbau*, 19, 147, 1977b.
145. **Heimer, Y. M. and Filner, P.**, Regulation of the nitrate assimilation pathway of cultured tobacco cells. II. Properties of variant cell line, *Biochim., Biophys. Acta*, 215, 152, 1970.
146. **Heintz, R. and Benveniste, P.**, Biosynthesis of methylene-24 cycloartenol starting from epoxide-2,3 of squalene by extracts of blackberry (*Rubus fruticosus*) microsome tissues cultivated in vitro, *C. R. Acad. Sci. Ser. D*, 274, 947, 1972.
147. **Heinz, D. J. and Mee, W. P.**, Morphologic cytogenetic, and enzymatic variation in *Saccharum* species hybrid clones derived from callus tissue, *Am. J. Bot.*, 58, 257, 1971.
148. **Heller, R.**, Sur l'emploi de papier filtre sans cendres comme support pour les cultures de tissus vegetaux, *C. R. Seances Soc. Biol. Paris*, 143, 335, 1949.
149. **Heller, R.**, Recherches sur la nutrition minerale des tissues vegetaux in vitro, *Ann. Sci. Nat. Bot., Biol. Veg.*, 14, 1, 1953.
150. **Hellman, E., Skirvin, R. M., and Otterbacher, A. G.**, Overcoming incompatibility associated with interspecific hybridization of red and black raspberries, *HortScience*, 14 (Abstr.), 423, 1979.
151. **Herman, E. B. and Haas, G. J.**, Clonal propagation of *Coffea arabica* L. from callus culture, *HortScience*, 10, 588, 1975.
152. **Heuser, C. W.**, Response of callus cultures of *Prunus persica, P. tomentosa*, and *P. besseyi* to cyanide, *Can. J. Bot.*, 50, 2149, 1972.
153. **Hildebrandt, A. C.**, Growth and differentiation of plant cell cultures, in *Control Mechanisms in the Expression of Cellular Phenotypes*, Padykul, H. A., Ed., Academic Press, New York, 1970.

154. **Hirabayashi, T., Kozaki, I., and Akihama, T.,** *In vitro* differentiation of shoots from anther callus in *Vitis, HortScience,* 11, 511, 1976.
155. **Hirsch, A. M.,** Identification et dosage des acides amines libres de fragments de tiges d'*Actinidia chinensis* Planchon. Mise en culture *in vitro* de ces fragments, *C. R. Acad. Sci. Ser. D.,* 270, 1462, 1970.
156. **Hirsch, A. M.,** Sur la culture de fruits d'*Actinidia chinensis* et la metabolisme des acides amines libres de fragments de tiges cultives *in vitro, C. R. Acad. Sci. Ser. D.,* 280, 1369, 1975.
157. **Hirsch, A. M., Bligny, D., and Tripathi, B. K.,** Biochemical properties of tissue cultures from different organs of Actinida chinensis, *Acta Hortic.,* 78, 75, 1977.
158. **Holsten, R. D., Burns, R. C., Hardy, R. W. F., and Hebert, R. R.,** Establishment of symbiosis between *Rhizobium* and plant cells *in vitro, Nature (London),* 232, 173, 1971.
159. **Huth, W.,** Culture of apple plants from apical meristems, *Gartenbauwissenschaft,* 43, 163, 1978.
160. **Huth, W.,** Culture of raspberry plants from apical meristems, *Gartenbauwissenschaft,* 44, 53, 1979.
161. **Jacquoit, C.,** Sur la culture in vitro de tissu cambial de Chataignier (*Castanea vesca* Gaertn.), *C. R. Acad. Sci.,* 231, 1080, 1950.
162. **Jacquoit, C.,** Observations sur l'histogénèse et la lignification dans les cultures *in vitro* de tissu cambial de certains arbres forestiers. Influence de quelques hydrazides sur ces phénomènes, *C. R. Acad. Sci.,* 236, 960, 1953.
163. **Jacquoit, C.,** Sur les besoins en auxine et les caractères morphologiques externes des cultures de tissu cambial de quelques espèces d'arbres, *C. R. Acad. Sci.,* 243, 510, 1956.
164. **Jacquoit, C.,** Contribution a l'Etude de l'Organogenese chez les des Vegetaux Lignaux Cultives In Vitro, in *Proc. 84th Congr. Soc.,* Savantes, Lille, France, 1959, 441.
164a. **Jalal, A. F. and Collin, H. A.,** Secondary metabolism in tissue cultures of *Theobroma cacao, New Phytol.,* 83, 343, 1979.
165. **James, D. J. and Newton, B.,** Auxin: cytokinin interactions in the in vitro micropropagation of strawberry plants, *Acta Hortic.,* 78, 321, 1977.
166. **Janick, J., Skirvin, M., and Janders, R. B.,** Comparison of in vitro and in vivo tissue culture systems in scented geranium, *J. Hered.,* 68, 62, 1977.
167. **Jona, R. and Webb, K. J.,** Callus and axilllary-bud culture of *Vitis vinifera* "Sylvaner Riesling", *Sci. Hortic.,* 9, 55, 1978.
168. **Jones, O. P.,** Effect of benzyl adenine on isolated apple shoots, *Nature (London),* 215, 1514, 1967.
169. **Jones, O. P.,** Effect of phloridzin and phloroglucinol on apple shoots, *Nature (London),* 262, 392, 1976; Erratum, id., 262, 724, 1976.
170. **Jones, O. P. and Hatfield, S. G. S.,** Root initiation in apple shoots cultured *in vitro* with auxins and phenolic compounds, *J. Hortic. Sci.,* 51, 495, 1976.
171. **Jones, O. P., Hopgood, M. E., and O'Farrell, D.,** Propagation *in vitro* of M.26 apple rootstocks, *J. Hortic. Sci.,* 52, 235, 1977.
171a. **Jones, O. P., Pontikius, C. A., and Hopgood, M. E.,** Propagation *in vitro* of five apple scion cultivars, *J. Hortic. Sci.,* 54, 155, 1979.
172. **Jones, O. P. and Taylor, E. C.,** Root initiation in apple shoots cultured *in vitro* with auxins and phenolic compounds, *J. Hortic. Sci.,* 51, 495, 1976.
173. **Jones, O. P. and Vine, S. J.,** The culture of gooseberry shoot tips for eliminating virus, *J. Hortic. Sci.,* 43, 289, 1968.
174. **Jordan, M.,** Multicellular pollen in *Prunus avium* after culture *in vitro, Z. Pflanzenzuecht.,* 71, 358, 1974.
175. **Jordan, M.,** *In vitro* culture of anthers from Prunus, Pyrus and Ribes, *Planta Med.,* Suppl. 1975, 59, 1975a.
176. **Jordan, M.,** 1975b. Histologische und physiologische untersuchungen zur Kapazität der androgenese bei *in vitro* kultivierten *Prunus-*, Pyrus- , und Ribes- antheren, Diss. Gieben 1976.
177. **Jordan, M. and Feucht, W.,** Processes of differentiation and metabolism of phenolics in anthers of *Prunus avium* and *Prunus*× "*Pandora*" cultivated *in vitro, Angew. Bot.,* 51, 69, 1977.
178. **Kano, Y. and Asahira, T.,** Effects of some growth regulators on the development of strawberry fruits *in vitro* culture, *J. Jpn. Soc. Hortic. Sci.,* 47, 195, 1978.
179. **Keller, H., Wanner, H., and Baumann, T. W.,** Synthesis in fruits and tissue cultures of *Coffea arabica, Planta,* 108, 339, 1972.
180. **Kester, D. E., Tabachnik, L., and Negueroles, J.,** Use of micropropagation and tissue culture to investigate genetic disorders in almond cultivars, *Acta Hortic.,* 78, 95, 1977.
181. **Kiss, F. and Zatykó, J.,** Vegetative propagation of *Rubus* species *in vitro, Bot. Kozl.,* 65, 65, 1978.
181a. **Kobayashi, S., Ikeda, I., and Nakatani, M.,** Studies on the nucellar embryogenesis in Citrus. II. Formation of nucellar embryo and development of ovule, *Bull. Fruit Tree Res. Stn. (Minist. Agric. Forest). Ser. E,* No. 2, 9, 1978.

181b. **Kobayashi, S., Ikeda, I., and Nakatani, M.,** Studies on nucellar embryogenesis in Citrus. II. Formation of the primordium cell of the nucellar embryo in the ovule of the flower bud, and its meristematic activity, *J. Jpn. Soc. Hortic. Sci.,* 48, 179, 1979.

182. **Knudson, L.,** Nutrient solutions for orchid seed germination, *Am. Orchid Soc. Bull.,* 12, 77, 1943.

183. **Kochba, J. and Button, J.,** The stimulation of embryogenesis and embryoid development in habituated ovular callus from the "Shamouti" orange (*Citrus sinensis*) as affected by tissue age and sucrose concentration, *Z. Pflanzenphysiol.,* 73, 415, 1974.

184. **Kochba, J., Button, J., Spiegel-Roy, P., Bornman, C. H., and Kochba, M.,** Stimulation of rooting of *Citrus* embryoids by gibberellic acid and adenine sulphate, *Ann. Bot. (London),* 38, 795, 1974.

185. **Kochba, J., Lavee, S., and Spiegel-Roy, P.,** Differences in peroxidase activity and isoenzymes in embryogenic and non-embryogenic "Shamouti" orange ovular callus lines, *Plant Cell Physiol.,* 18, 463, 1977.

186. **Kochba, J., and Spiegel-Roy, P.,** Effect of culture media on embryoid formation from ovular callus of "Shamouti" orange *(Citrus sinensis), Z. Pflanzenzuecht.,* 69, 156, 1973.

187. **Kochba, J. and Spiegel-Roy, P.,** Embryogenesis in gamma-irradiated habituated ovular callus of "Shamouti" orange as affected by auxin and tissue age, *Environ. Exp. Bot.,* 17, 151, 1977.

188. **Kochba, J. and Spiegel-Roy, P.,** The effects of auxins, cytokinins and inhibitors on embryogenesis in habituated ovular callus of the "Shamouti" orange *(Citrus sinensis), Z. Pfanzenphysiol.,* 81, 283, 1977.

189. **Kochba, J. and Spiegel-Roy, P.,** Cell and tissue culture for breeding and developmental studies of *Citrus, HortScience,* 12, 110, 1977.

190. **Kochba, J., Spiegel-Roy, P., Neumann, H., and Saad, S.,** Stimulation of embryogenesis in citrus ovular callus by ABA, ethephon, CCC, and alar and its supression by GA_3, *Z. Pflanzenphysiol.,* 89, 427, 1978.

190a. **Kochba, J., Spiegel-Roy, P., Saad, S., and Neumann, H.,** Tissue culture studies with Citrus: 1) the effect of several sugars on embryogenesis and 2) application of Citrus tissue cultures for selection of mutants, in *Production of Natural Compounds by Cell Culture Methods,* Alfermann, A. W. and Reinhard, E., Eds., Federal Ministry for Research and Technology, Munich, 1978, 223.

191. Kochba, J., **Spiegel-Roy, P.,** and **Safran, H.,** Adventive plants from ovules and nucelli in *Citrus, Planta,* 106, 237, 1972.

192. **Kordan, H. A.,** Proliferation of excised juice vesicles of lemon *in vitro, Science,* 129, 779, 1959.

193. **Kordan, H. A.,** Growth characteristics of Citrus fruit tissue *in vitro, Nature (London),* 198, 867, 1963.

194. **Kordan, H. A.,** Nucleolar activity and starch synthesis in lemon fruit tissue *in vitro, Bull. Torrey Bot. Club,* 92, 21, 1965.

195. **Kordan, H. A.,** Some morphological and physiological relationships of lemon fruit tissue *in vivo* and *in vitro, Bull. Torrey Bot. Club,* 92, 209, 1965b.

196. **Krul, W. R. and Worley, J. F.,** Formation of adventitious embryos in callus cultures of "Seyval", a French hybrid grape, *J. Am. Soc. Hortic. Sci.,* 102, 360, 1977.

197. **Kubicki, B., Telezynska, J., and Milewska-Pawliczuk, E.,** Induction of embryoid development from apple pollen grains, *Acta Soc. Bot. Pol.,* 44, 631, 1975.

198. **Lakshmi Sita, G., Singh, R., and Iyer, C. P. A.,** Plantlets through shoot-tip cultures in pineapple, *Curr. Sci.,* 43, 724, 1974.

199. **Lane, W. D.,** Regeneration of apple plants from shoot meristem-tips, *Plant Sci. Lett.,* 13, 281, 1978.

200. **Lane, W. D.,** Regeneration of pear plants from shoot meristem-tips, *Plant Sci. Lett.,* 16, 337, 1979.

201. **Latché, A., Pech, J. C., Codron, H., and Fallot, J.,** Study of some ripening characteristics of pear fruit disks *in vitro, Physiol. Veg.,* 17, 119, 1979.

202. **Lavee, S.,** Natural kinin in peach fruitlets, *Science,* 142, 583, 1963.

203. **Lavee, S.,** The growth potential of olive fruit mesocarp *in vitro* (Olea europea), *Acta Hortic.,* 78, 115, 1977.

204. **Lavee, S. and Hoffmann, M.,** The effect of potassium ions on peroxidase activity and its isozyme composition as related to apple callus growth *in vitro, Bot. Gaz. (Chicago),* 132, 232, 1971.

205. **Lee, E. C. M. and de Fossard, R. A.,** Regeneration of strawberry plants from tissue cultures, *Proc. Int. Plant Prop. Soc.,* 25, 277, 1975.

206. **Lee, E. C. M. and de Fossard, R. A.,** Some factors affecting multiple bud formation of strawberry (× Fragaria ananassa Duchesne) *in vitro, Acta Hortic.,* 78, 187, 1977.

206a. **Lescure, A. M.,** Selection of markers of resistance to base-analogues in somatic cell cultures of *Nicotiana tabacum, Plant Sci. Lett.,* 1, 375, 1973.

207. **Letham, D. S.,** Cultivation of apple-fruit tissue *in vitro, Nature (London),* 182, 473, 1958.

208. **Letham, D. S.,** Growth requirements of pome-fruit tissues, *Nature (London),* 188, 425, 1960.

209. **Lévi, E. and Maróti, M.,** Isolation of apple embryo and apple tissue cultures, *Bot. Kozl.,* 64, 75, 1977.

209a. **Lieberman, M., Wang, S. Y., and Owens, L. D.**, Ethylene production by callus and suspension cells from cortex tissue of postclimacteric apples, *Plant Physiol.*, 63, 811, 1979.

210. **Linsmaier, E. M. and Skoog, F.**, Organic growth factor requirements of tobacco tissue culture, *Physiol. Plant.*, 18, 100, 1965.

211. **Lis, P. K.**, Uptake and metabolism of sucrose-^{14}C and IAA-1-^{14}C in strawberry fruit explants cultivated *in vitro*, *Proc. 19th Int. Hortic. Congr.*, 1A (Abstr.), 61, 1974.

212. **Litz, R. E. and Conover, R. A.**, Tissue culture propagation of papaya, *Proc. Fla. State Hortic. Soc.*, 90, 245, 1977.

213. **Litz, R. E. and Conover, R. A.**, *In vitro* propagation of papaya, *HortScience*, 13, 241, 1978.

214. **Litz, R. E. and Conover, R. A.**, Recent advances in papaya tissue culture, *Proc. Fla. State Hortic. Soc.*, 91, 180, 1978.

215. **Lobov, V. P., Kalinin, F. L., and Korzh, L. P.**, Growth of a *Parthenocissus tricuspidata* gall tumor and *Rubus* tissue *in vitro* in relation to purine and pyrimidine bases and maleic hydrazide, *Physiol. Biochem. Cultiv. Plants (USSR)*, 1, 221, 1969.

216. **Lundergan, C.**, *In vitro* propagation of the apple, Ph.D. thesis, Purdue University, West Lafayette, Ind., 1978.

217. **Lundergan, C. and Janick, J.**, Low temperature storage of *in vitro* apple shoots, *HortScience*, 14, 514, 1979.

217a. **Lyrene, P. M.**, Blueberry callus and shoot-tip culture, *Proc. Fla. State Hortic. Soc.*, 91, 171, 1978.

217b. **Lyrene, P. M.**, In Vitro Propagation of Rabbiteye Blueberries, Proc. of the North American Blueberry Workers Conference, University of Arkansas, Fayetteville, 1979, 249.

217c. **Lyrene, P. M.**, Micropropagation of rabbiteye blueberries, *HortScience*, 15, 80, 1979.

218. **Ma, S. S. and Shii, C. T.**, Growing banana plantlets from adventitious buds, *J. Chin. Soc. Hortic. Sci.*, 20, 6, 1974.

219. **Maliga, P., Marton, L., and Sz-Breznovits, A.**, 5-bromodeoxy-uridine-resistant cell lines from haploid tobacco, *Plant Sci. Lett.*, 1, 119, 1973.

220. **Maliga, P., Sz-Breznovits, A., and Marton, L.**, Streptomycin-resistant plants from callus culture of haploid tobacco, *Nature (London) New Biol.*, 244, 29, 1973.

221. **Mapes, O.**, Tissue culture of Bromeliads, *J. Hawaii Agric. Exp. Stn.*, No. 1676, 1974.

222. **Maróti, M. and Lévi, E.**, Hormonal regulation of the organization from meristem cultures, in *Use of Tissue Cultures in Plant Breeding*, Novak, F. J., Ed., Czechoslovak Academy of Sciences, Prague, 1977, 337.

223. **Martinez, J., Hugard, H., and Jonard, R.**, The different grafting of shoot-tips realized *in vitro* between peach (*Prunus persica* Batsch) apricot (*Prunus armeniaca* L.) and myrobolan (*Prunus cesarifera* Ehrh.), *C. R. Acad. Sci. Ser. D*, 288, 759, 1979.

224. **Mathews, V. H., Rangan, T. S., and Narayanaswamy, S.**, Micro-propagation of *Ananas sativus in vitro*, *Z. Pflanzenphysiol.*, 79, 450, 1976.

225. **McGrew, J. R.**, Eradication of latent C virus in the Suwannee variety of strawberry by heat plus excised runner tip culture, *Phytopathology*, 55, 480, 1965.

225a. **McPheeters, K. and Skirvin, R. M.**, Chimeral manipulation of blackberry, *HortScience*, 15 (Abstr.), 415, 1980.

225b. **McPheeters, K., Skirvin, R. M., and Bly-Monnen, C. A.**, Culture of chestnut *(Castanea spp.) in vitro*, *HortScience*, 15 (Abstr.), 417, 1980.

226. **Medhi, A. A. and Hogan, L.**, Tissue culture of *Carica papaya*, *HortScience*, 11(Abstr.), 311, 1976.

226a. **Medora, R. S., Bikderback, D. E., and Mell, G. P.**, Effect of media on growth of papaya callus cultures, *Z. Pflanzenphysiol.*, 91, 79, 1979.

227. **Mehra, A. and Mehra, P. N.**, Organogenesis and plantlet formation *in vitro* in almonds, *Bot. Gaz. (Chicago)*, 135, 61, 1974.

228. **Messer, G. and Lavee, S.**, Studies on vigor and dwarfism of apple trees in an *in vitro* tissue culture system, *J. Hortic. Sci.*, 44, 219, 1969.

229. **Michellon, R., Hugard, J., and Jonard, R.**, Sur l'isolement de colonies tissulaires de Pêcher (*Prunus persica* Batsch, Cultivars Dixired et Nectared IV) et d'Amandier (*Prunus amygdalus* Stokes, cultivar Aï) à partir d'anthères cultivées *in vitro*, *C. R. Acad. Sci. Ser. D*, 278, 1719, 1974.

230. **Milewska-Pawliczuk, E. and Kubicki, B.**, Induction of androgenesis *in vitro* in *Malus domestica*, *Acta Hortic.*, 78, 271, 1977.

230a. **Miller, C. O.**, Kinetin and Kinetin-like compounds, in *Modern Methods in Plant Analysis*, Vol. 6, Paech, K. and Tracey, M. V., Eds., Springer-Verlag, Berlin, 1963, 194.

231. **Miller, C. O.**, Evidence for the natural occurrence of zeatin and derivatives: compounds from maize which promote cell division, *Proc. Natl. Acad. Sci. U.S.A.*, 54, 1052, 1965.

232. **Miller, C. O.**, Cytokinins in *Zea mays*, *Ann. N. Y. Acad. Sci.*, 144, 251, 1967.

233. **Miller, C. O. and Skoog, F.**, Chemical control of bud formation in tobacco stem segments, *Am. J. Bot.*, 40, 768, 1953.

234. **Miller, P. W. and Belkengren, R. O.**, Elimination of yellow edge, crinkle and veinbanding viruses and certain other virus complexes from strawberries by excision and culturing of apical meristems, *Plant Dis. Rep.*, 47, 298, 1963.
235. **Mitra, G. C. and Chaturvedi, H. C.**, Embryoids and complete plants from unpollinated ovaries and from ovules of in vitro-grown emasculated flower buds of *Citrus* spp., *Bull. Torrey Bot. Club*, 99, 184, 1972.
236. **Mohan Ram, H. Y. and Steward, F. C.**, The induction of growth in explanted tissue of the banana fruit, *Can. J. Bot.*, 42, 1559, 1964.
237. **Monaco, L. C., Söndahl, M. R., Carvalho, A., Crocomo, O. J., and Sharp, W. R.**, Application of tissue culture in the improvement of coffee, in *Applied and Fundamental Aspects of Plant Cell, Tissue, and Organ Culture*, Reinert, J. and Bajaj, Y. P. S., Eds., Springer-Verlag, New York, 1977, 107.
238. **Monnier, M.**, Culture *in vitro* de l'embryon immature de *Capsella bursa pastoris* Moench, *Rev. Cytol. Biol. Veg.*, 39, 1, 1976.
239. **Monsion, M. and Dunez, J.**, Young plants of *Prunus mariana* obtained from cuttings cultivated *in vitro*, *C. R. Acad. Sci. Ser. D*, 272, 1861, 1971.
240. **Morel, G.**, Sur le developpement de tissus de vigne cultives *in vitro*, *C. R. Seances Soc. Biol. Paris*, 138, 62, 1944.
241. **Morel, G.**, Action de l'acide indole-β-acétique sur la croissance des tissus de vigne, *C. R. Seances Soc. Biol. Paris*, 138, 93, 1944.
242. **Morel, G.**, Recherches sur la culture associée de parasites obligatoires de tissus vegetaux, *Ann. Epiphyt.*, 14, 123, 1948.
244. **Morel, G. and Martin, C.**, Guérison de dahlias attaints d'une maladie à virus, *C. R. Acad. Sci.*, 235, 1324, 1952.
245. **Morel, G. and Muller, J. F.**, La culture *in vitro* du meristeme apical de la pomme de terre, *C. R. Acad. Sci.*, 258, 5250, 1964.
246. **Mouras, A., Salesses, G., and Lutz, A.**, Protoplasts use in cytology: improvement of recent method in order to identify Nicotiana and Prunus mitotic chromosomes, *Caryologia*, 31, 117, 1978.
247. **Mullin, R. H. and Schlegel, D. E.**, Cold storage maintenance of strawberry meristem plantlets, *HortScience*, 11, 100, 1976.
248. **Mullin, R. H., Smith, S. H., Frazier, N. W., Schlegel, D. E., and McCall, S. R.**, Meristem culture frees strawberries of mild yellow edge, pallidosis, and mottle diseases, *Phytopathology*, 64, 1425, 1974.
249. **Mullins, M. G. and Srinivasan, C.**, Somatic embryos and plantlets from an ancient clone of the grapevine (cv. Cabernet-Sauvignon) by apomixis *in vitro*, *J. Exp. Bot.*, 27, 1022, 1976.
250. **Murashige, T.**, Plant propagation through tissue culture, *Annu. Rev. Plant Physiol.*, 25, 135, 1974.
251. **Murashiae, T.**, Clonal crops through tissue culture, in *Plant Tissue Culture and Its Bio-Technological Application*, Barz, W., Reinhard, E., and Zenk, M. H., Eds., Springer-Verlag, Berlin, 1977, 392.
252. **Murashige, T.**, The impact of plant tissue culture on agriculture, in *Frontiers of Plant Tissue Culture 1978*, Thorpe, T. A., Ed., University of Calgary, Canada, 1978, 15 and 518.
253. **Murashige, T., Bitters, W. P., Rangan, T. S., Nauer, E. M., Roistacher, C. N., and Holliday, P. B.**, A technique of shoot apex grafting and its utilization towards recovering virus-free *Citrus* clones, *HortScience*, 7, 118, 1972.
254. **Murashige, T., Nakano, R., and Tucker, D. P. H.**, Histogenesis and rate of nuclear change in *Citrus limon* tissue in vitro, *Phytomorphology*, 17, 469, 1967.
255. **Murashige, T. and Skoog, F.**, A revised medium for rapid growth and bioassays with tobacco tissue cultures, *Physiol. Plant.*, 15, 473, 1962.
256. **Murashige, T. and Tucker, D. P. H.**, Growth factor requirements of citrus tissue culture, *Proc. 1st Int. Citrus Symp.*, 3, 1155, 1969.
257. **Nabors, M. W., Daniels, A., Nadolny, L., and Brown, C.**, Sodium chloride tolerant lines of tobacco cells, *Plant Sci. Lett.*, 4, 155, 1975.
258. **Nachit, M. and Feucht, W.**, Die Eignung von phenolischen Sauren und aminosauren als selektionskriterien fur die wuchsigkeit von Malus-unterlagen, *Mitt. Klosterneuburg*, 26, 199, 1976.
259. **Navarro, L. and Juarez, J.**, Tissue culture techniques used in Spain to recover virus-free Citrus cultures, *Acta Hortic.*, 78, 425, 1977.
260. **Navarro, L., Roistacher, C. N., and Murashige, T.**, Improvement of shoot-tip grafting *in vitro* for virus-free Citrus, *J. Am. Hortic. Sci.*, 100, 471, 1975.
261. **Navarro, L., Roistacher, C. N., and Murashige, T.**, Effect of size and source of shoot tips on Psorosis-A and exocortis content of Navel orange plants obtained by shoot-tip grafting *in vitro*, in *7th Conf. Int. Organization Citrus Virology*, International Organization of Citrus Virology, Riverside, Calif., 1976, 194.
261a. **Negm, F. B. and Loesher, W. H.**, Detection and characterization of sorbitol dehydrogenase from apple callus tissue, *Plant Physiol.*, 64, 69, 1979.

262. **Nekrosova, T. V.**, The culture of isolated buds of fruit trees, *Sov. Plant Physiol.*, 11, 107, 1964.
263. **Nickell, L. G.**, Test-tube approaches to bypass sex, *Hawaii. Plant. Rec.*, 58, 293, 1973.
264. **Nickerson, N. L.**, Callus formation in lowbush blueberry fruit explants cultured *in vitro, Hortic. Res.*, 18, 85, 1978a.
265. **Nickerson, N. L.**, *In vitro* shoot formation in lowbush blueberry seedling explants, *HortScience*, 13, 698, 1978b.
266. **Nickerson, N. L. and Hall, I. V.**, Callus formation in stem internode sections of lowbush blueberry (*Vaccinium angustifolium* Ait.) cultured on a medium containing plant growth regulators, *Hortic. Res.*, 16, 29, 1976.
267. **Nitsch, C. and Nitsch, J. P.**, The induction of flowering in stem segments of *Plumbago indica* L. I. The production of vegetative buds, *Planta*, 72, 355, 1967.
268. **Nitsch, J. P.**, Growth and development *in vitro* of excised ovaries, *Am. J. Bot.*, 38, 566, 1951.
269. **Nitsch, J. P.**, Culture *in vitro* de tissus de fruits. I. Mesocarpe de pome, *Bull. Soc. Bot. Fr.* 106, 420, 1959.
270. **Nitsch, J. P.**, The *in vitro* culture of flowers and fruits, in *Plant Tissue and Organ Culture — A Symposium*, Maheshwari, P. and Ranga Swamy, N. S., Eds., International Society of Plant Morphologists, Delhi, 1963, 198.
271. **Nitsch, J. P.**, Culture *in vitro* de tissus de fruits. II. Orange, *Bull. Soc. Bot. Fr.*, 112, 20, 1965a.
272. **Nitsch, J. P.**, Culture *in vitro* de tissus de fruits. III. Mesocarpe et endocarpe de peche, *Bull. Soc. Bot. Fr.*, 112, 22, 1965.
273. **Nitsch, J. P.**, Haploid plants from pollen, *Z. Pflanzenzuecht.*, 67, 3, 1972.
274. **Nitsch, J. P., Asahira, T., Rossini, M. E., and Nitsch, C.**, Bases physiologiques de la production de chair de pomme et de poire *in vitro, Bull. Soc. Bot. Fr.*, 117, 479, 1970.
275. **Nitsch, J. P. and Nitsch, C.**, Haploid plants from pollen grains, *Science*, 163, 85, 1969.
276. **Nitsch, J. P., Nitsch, C., and Hamon, S.**, Réalisation experimentale de l'androgenese chez divers *Nicotiana, C. R. Seances Soc. Biol. Paris*, 162, 369, 1968.
277. **Nozeran, R.**, Multiple growth correlations in phanerogams, in *Tropical Trees as Living Systems*, Tomlinson, P. B. and Zimmeran, M. H., Eds., Cambridge University Press, New York, 1978a, 423.
278. **Nozeran, R.**, Reflexions sur les enchainements de fonctionnements au cours du cycle des vegetaux superieurs, *Bull. Soc. Bot. Fr.*, 125, 263, 1978.
279. **Nozeran, R. and Bancilhon, L.**, Les cultures *in vitro* en tant que techniques pour l'approche des problèmes poses par l'amélioration de plantes, *Ann. Amelior. Plant.*, 22, 167, 1972.
280. **Ohyama, K.**, Properties of 5-bromodeoxyuridine-resistant lines of higher plant cells in liquid culture, in *Genetic Manipulations with Plant Material*, Ledoux, L., Ed., Plenum Press, New York, 1974, 571.
281. **Ohyama, K.**, Properties of 5-bromodeoxyuridine-resistant lines of higher plant cells in liquid culture, *Exp. Cell Res.*, 89, 31, 1974.
282. **Ohyama, K. and Nitsch, J. P.**, Flowering haploid plants obtained from protoplasts of tobacco leaves, *Plant Cell Physiol.*, 13, 229, 1972.
283. **Oka, S. and Ohyama, K.**, Studies on in vitro culture of excised buds in mulberry tree. I. Effect of growth substances on the development of shoots and organ formation from winter buds, *J. Sericult. Sci. Jpn.*, 43, 230, 1974.
284. **Oka, S. and Ohyama, K.**, *In vitro* culture of excised mulberry bud and dormancy analysis with it, in Long Term Preservation of Favourable Germ Plasm in Arboreal Crops, Akihama, T. and Nakajima, K., Eds., Fruit Tree Research Station, Ministry of Agriculture and Forestry, Fujimoto, Japan, 1978, 136.
285. **Palmer, J. E. and Widholm, J. M.**, Characterization of carrot and tobacco cell cultures resistant to p-fluorophenylalanine, *Plant Physiol.*, 56, 233, 1975.
286. **Pannetier, C. and Lanaud, C.**, Divers aspects de l'utilisation possible des cultures "*in vitro*" pour la multiplication vegetative de l'*Ananas comosus* L. Merr, variete "Cayenne lisse," *Fruits*, 31, 739, 1976.
287. **Pech, J. C., Amboid, C., Latche, A., Diarra, A., and Fallot, J.**, Culture de tissus et de suspensions cellulaires de fruits. Moyen d'etude de certains aspects de la maturation, *Colloq. Int. C. N. R. S.*, 238, 211, 1975a.
288. **Pech, J. C., Latche, A., Austruy, M., and Fallot, J.**, *In vitro* growth of tissues and cells from apple fruits, *Bull. Soc. Bot. Fr.*, 122, 183, 1975b.
289. **Pelet, F., Hildebrandt, A. C., Riker, A. J., and Skoog, F.**, Growth in vitro of tissues isolated from normal stems and insect galls, *Am. J. Bot.*, 47, 186, 1960.
290. **Penas-Iglesias, A. and Ayuso, P.**, A new and accurate way of heat therapy of plants grown *in vitro* applied to the sanitary selection of spanish grape vine varieties, *Riv. Patol. Veg.*, 9, 172, 1973.
291. **Pence, V. C., Hasegawa, P. M., and Janick, J.**, Initiation of asexual embryogenesis in *Theobroma cacao* L., *HortScience*, 14(Abstr.), 477, 1979a.

292. **Pence, V. C., Hasegawa, P. M., and Janick, J.**, Asexual embryogenesis in *Theobroma cacao* L., *J. Am. Soc. Hortic. Sci.*, 104, 145, 1979b.

293. **Pieniazek, J.**, The growth *in vitro* of isolated apple-shoot tips from young seedlings on media containing growth regulators, *Bull. Acad. Pol. Sci. Ser. Sci. Biol.*, 16, 179, 1968.

294. **Pierik, R. L. M.**, Vegetative propagation of horticultural crops *in vitro* with special attention to shrubs and trees, *Acta Hortic.*, 54, 71, 1975.

295. **Pilet, P. E. and Roland, J. C.**, Effects of abscissic acid on the growth and ultra-structure of tissues cultivated *in vitro, Cytobiologie Z. Exp. Zellforsch.*, 4, 41, 1971.

296. **Popov, Yu. G. and Shchelkunova, S. E.**, Regeneration of *Rubus idaeus* L. stem apex *in vitro* in connection with presence of growth regulators in the nutrient medium, *Bot. Zh. (Leningrad)*, 58, 1515, 1973.

297. **Popov, Yu. G., Vysotskii, V. A., and Trushechkin, V. G.**, Cultivation of isolated stem apices of sour cherry, *Sov. Plant Physiol.*, 23, 435, 1976.

298. **Powell, L. E.**, The aseptic culture of excised apple buds, and their employment in dormancy studies, *HortScience*, 5(Abstr.), 326, 1970.

299. **Putz, C.**, Obtention de framboisiers (Var. Bois blancs) sans virus par la technique des cultures de meristems, *Ann. Phytopathol.*, 3, 493, 1971.

300. **Quak, F.**, Meristem culture and virus-free plants, in *Applied and Fundamental Aspects of Plant Cell, Tissue and Organ Culture*, Reinert, I. and Bajaj, Y. P. S., Eds., Springer-Verlag, Berlin, 1977, 598.

301. **Quoirin M.**, Premiers résultats obtenus dans la culture "in vitro" du méristème apical de subjets parte-greffe de pommier, *Bull. Rech. Agron. Gembloux*, 9, 189, 1974.

302. **Quoirin, M. and Lepoivre, P.**, Improved medium for *in vitro* culture of Prunus sp., *Acta Hortic.*, 78, 437, 1977.

303. **Quoirin, M., Boxus, P., and Gaspar, T.**, Root initiation and isoperoxidases of stem tip cuttings from mature *Prunus* plants, *Physiol. Veg.*, 12, 165, 1974.

304. **Quoirin, M., Gaspar, T., and Boxus, P.**, Changes in natural growth substances as related to survival and growth of the hybrid ornamental cherry *Prunus accolade* meristems grown *in vitro, C. R. Acad. Sci. Ser. D.*, 281, 1309, 1975.

305. **Radler, F.**, Versuche zur kulture isolierter beeren der rebe, *Vitis*, 4, 365, 1964.

306. **Radojević, L., Vujičić, R., and Nešković, M.**, Embryogenesis in tissue culture of *Corylus avellana* L., *Z. Pflanzenphysiol.*, 77, 33, 1975.

307. **Rajasekaren, K. and Mullins, M. G.**, Embryos and plantlets from cultured anthers of hybrid grapevines, *J. Exp. Bot.*, 30, 399, 1979.

308. **Rangan, T. S., Murashige, T., and Bitters, W. S.**, *In vitro* initiation of nucellar embryos in monoembryonic *Citrus, HortScience*, 3, 226, 1968.

309. **Rangan, T. S., Murashige, T., and Bitters, W. S.**, *In vitro* studies of zygotic and nucellar embryogenesis in citrus, *Proc. 1st Int. Citrus Symp.*, 1, 225, 1969.

310. **Ranga Swamy, N. S.**, Experimental studies on female reproductive structures of *Citrus microcarpa* Bunge, *Phytomorphology*, 11, 109, 1961.

310a. **Raveh, D., Huberman, E., and Galun, E.**, In vitro culture of tobacco protoplasts: Use of feeder techniques to support division of cells plated at low densities, *In Vitro*, 9, 216, 1973.

311. **Ravkin, A. S. and Popov, Y. G.**, Mitotic activity genome and chromosome variations in cells of apple cultures, *Tsitol. Genet.*, 7; 33, 1973; *Biol. Abstr.*, 57, 2361, 1973.

312. **Rekoslayskaya, N. I. and Gamburg, K. Z.**, Auxin induction of indolylacetyl aspartate synthesis in a culture of auxin-independent blackberry tissue, *Sov. Plant Physiol.*, 24, 970, 1977.

313. **Reuveni, O., Bar-Akiva, A., and Sagiv, J.**, Organic acids and potassium requirements of callus derived from orange peel, *Acta Hortic.*, 78(Abstr.), 123, 1977.

314. **Reuveni, O. and Licien-Kipnis, H.**, Studies of the *in vitro* culture of date palm (*Phoenix dactylifera* L.) tissues and organs, *Volcani Inst. Agric. Res. Div. Sci. Publ. Pam.*, 145, 3, 1974.

315. **Reynolds, J. F. and Murashige, T.**, Asexual embryogenesis in callus cultures of palms, *In Vitro*, 15, 383, 1979.

316. **Rilling, G. and Radler, F.**, Die kontrollierbare aufzucht der reblaus auf gewebekulturen von reben, *Naturwissenschaften*, 47, 547, 1960.

317. **Ringe, F. and Nitsch, J. P.**, Conditions leading to flower formation on excised Begonia fragments cultured *in vitro, Plant Cell Physiol.*, 9, 639, 1968.

317a. **Robles Moran, M. J.**, Multiplication vegetative, *in vitro*, des bourgeons axillaires de *Passiflora edulis* var. *flavicarpa* Degener et de *P. mollissima* Bailey, *Fruits*, 33, 693, 1978.

317b. **Robles, Moran, M. J.**, Potential morphogenetique des entrenoeuds de *Passiflora edulis* var. *flavicarpa* Degener et *P. mollissima* Bailey en culture "in vitro", *Turrialba*, 29, 224, 1979.

318. **Roistacher, C. N., Navarro, L., and Murashige, T.**, Recovery of Citrus selections free of several viruses, exocortis viroid, and *Spiroplasma citri* by shoot-tip grafting *in vitro*, in *7th Conf. Int. Organization Citrus Virology*, Calahan, E. C., Ed., International Organization of Citrus Virology, Riverside, Calif., 1976, 186.

319. **Roland, J. C. and Pilet, P. E.**, Production of free cells by tissues of bramble cultivated *in vitro* ultrastructural aspects, *C. R. Acad. Sci. Ser. D,* 274, 2969, 1972.
320. **Rosati, P., Devreux, M., and Laneri, U.**, Anther culture of strawberry, *HortScience,* 10, 119, 1975.
321. **Rosati, P., Marino, G., and Swierczewski, C.**, *In vitro* propagation of Japanese plum cv. "Calita", *HortScience,* 15, 383, 1979.
322. **Saad, A. T. and Boone, D. M.**, Nutritional and temperature studies on apple callus tissue *in vitro, Plant Physiol.,* Suppl. 39, 11(Abstr.), 1964.
323. **Saavedra, E.**, Experimentalle untersuchungen zur wachtums- und entwicklungsphysiologie von *Malus*-Unterlagen (*Malus sylvestris* Mill.), Diss. Gieben. 1971.
324. **Saavedra, E. and Feucht, W.**, Growth of callus cells and starch content of different apple rootstocks *in vitro, Gartenbauwiss,* 43, 167, 1978.
325. **Sabharwal, P. S.**, *In vitro* culture of nucelli and embryos of *Citrus aurantifolia* Swingle, in *Plant Embryology: A Symposium,* Maheshwari, P., Ed., Council of Scientific and Industrial Research, New Delhi, India, 1962, 239.
326. **Sabharwal, P. S.**, *In vitro* culture of ovules, nucelli and embryos of *Citrus reticulata Blanco Nagpuri,* in *Plant Tissue and Organ Culture — A Symposium,* Maheshwari, P. and Ranga Swamy, N. S., Eds., International Society of Plant Morphologists, Delhi, 1963, 265.

326a. **Salesses, G. and Mouras, A.**, The use of protoplasts for *Prunus* chromosome study, *Ann. Amelior. Plant.,* 27, 363, 1977.

327. **Schenk, R. U. and Hildebrandt, A. C.**, Medium and techniques for induction and growth of monocotyledonous and dicotyledonous plant cell cultures, *Can. J. Bot.,* 50, 199, 1972.
328. **Schneider, G. W., Lockard, T. G., and Cornelius, P. L.**, Growth controlling properties of apple stem callus *in vitro, J. Am. Soc. Hortic. Sci.,* 103, 634, 1978.
329. **Schroeder, C. A.**, Proliferation of mature fruit pericarp tissue slices *in vitro, Science,* 122, 601, 1955.
330. **Schroeder, C. A.**, Some morphological aspects of fruit tissues grown *in vitro, Bot. Gaz. (Chicago),* 122, 198, 1961.
331. **Schroeder, C. A.**, Longevity of fruit pericarp tissue *in vitro,* in *Morphogenesis in Plant Cell, Tissue and Organ Cultures,* Johri, B. M. and Mohan Ram, H. Y., Eds., University of Delhi, India, 1971, 77.
332. **Schroeder, C. A. and Kay, E.**, Temperature conditions and tolerance of avocado fruit tissue *in vitro, Calif. Avocado Soc. Yearb.,* 87, 1961.
333. **Schroeder, C. A. and Kay, E.**, Induced temperature tolerance of plant tissue *in vitro, Nature (London),* 200, 1301, 1963.
334. **Schroeder, C. A. and Spector, C.**, Effect of gibberellic acid and indoleacetic acid on growth of excised fruit tissue, *Science,* 126, 701, 1957.
335. **Schwabe, W. W.**, The long, slow road to better coconut plants, *Spectrum,* 103, 9, 1973.
336. **Scorza, R.**, *In vitro* flowering of *Passiflora suberosa* L., Ph.D. thesis, Purdue University, West Lafayette, Ind., 1979.
337. **Scorza, R. and Janick, J.**, Tissue culture in *Passiflora, 24th Ann. Congr. Am. Soc. Hortic. Soc. Trop. Reg.,* 179, 1976.
338. **Seirlis, G., Mouras, A., and Salesses**, *In vitro* culture of anthers and organ fragments of *Prunus, Ann. Amelior. Plant.,* 29, 145, 1979.
339. **Sharp, W. R., Caldas, L. S., Crocomo, O. J., Monaco, L. C., and Carvalho, A.**, Production of *Coffea arabica* callus of three ploidy levels and subsequent morphogenesis, *Phyton (Buenos Aires),* 31, 67, 1973.
340. **Shchelkunova, S. E. and Popov, Yu. G.**, Production of virus-free raspberry plants from cultures of isolated apexes, *Sov. Plant Physiol.,* 17, 513, 1970.

340a. **Sheridan, W. F.**, Plant regeneration and chromosome stability in tissue cultures, in *Genetic Manipulation with Plant Material,* Ledoux, L., Ed., Plenum, London, 1974, 263.

341. **Singh, V. P.** Raising nucellar seedlings of some *Rutaceae in vitro.* in *Plant Tissue and Organ culture,* Maheshwari, P. and Ranga Swamy, N., Eds., International Society of Plant Morphologists, Delhi, India, 1957, 275.
342. **Skene, K. G. M.**, Culture of protoplasts from grape vine pericarp callus, *Aust. J. Plant Physiol.,* 1, 371, 1974.
343. **Skene, K. G. M.**, Production of callus from protoplasts of cultured grape pericarp, *Vitis,* 14, 177, 1975.
344. **Skirvin, R. M.**, Fruit improvement through single-cell cultuure, *Fruit Var. J.,* 31, 82, 1977.
345. **Skirvin R. M.**, Natural and induced variation in tissue culture, *Euphytica,* 27, 241, 1978.
346. **Skirvin, R. M. and Chu, M. C.**, Tissue culture may revolutionize the production of peach shoots, *Ill. Res.,* 19, 18, 1977.
347. **Skirvin, R. M. and Chu, M. C.**, *Rubus* tissue culture, in Propagation of Higher Plants Through Tissue Culture, Hughes, K. W., Henke, R., and Constantin, M., Eds., National Technical Information Center, U. S. Department of Energy, Springfield 1978a, (Abstr.), 253.

348. **Skirvin, R. M. and Chu, M. C.**, Tissue culture of peach shoot tips, *HortScience,* 13(Abstr.), 359, 1978b.

348a. **Skirvin, R. M., Chu, M. C., and Gomez, E.**, *In vitro* propagation of trailing blackberries, *HortScience,* in press.

348b. **Skirvin, R. M., Chu, M. C., and Rukan, H.**, Tissue culture of peach, sweet and sour cherry, and apricot shoot tips, *Proc. Ill. State Hortic. Soc.,* 113, 30, 1980.

348c. **Skirvin, R. M., Chu, M. C., and Rukan, H.**, Rooting studies with *Prunus* spp. *in vitro, HortScience,* 15, 415, 1980.

348d. **Skirvin, R. M., Garrish, D. L., and Muldoon, D.**, The reduction of bacterial contamination from field-grown strawberry explants grown *in vitro, In Vitro,* in press.

349. **Skirvin, R. M. and Janick, J.**, Tissue culture-induced variation in scented *Pelargonium* spp., *J. Am. Soc. Hortic. Sci.,* 101, 281, 1976a.

350. **Skirvin, R. M. and Janick, J.**, "Velvet Rose" *Pelargonium,* a scented geranium, *HortScience,* 11, 61, 1976b.

351. **Skirvin, R. M. and Janick, J.**, Chromosome doubling and reduction in *in vitro* and *in vivo* systems, *HortScience,* 11(Abstr.), 329, 1976c.

352. **Skirvin, R. M. and Larson, D.**, Reduction of bacterial contamination in field-grown strawberry shoot-tip cultures, *HortScience,* 13(Abstr.), 349, 1978.

353. **Skoog, F. and Tsui, C.**, Chemical control of growth and bud formation in tobacco stem segments and callus cultures *in vitro, Am. J. Bot.,* 35, 782, 1948.

354. **Slabaeck-Szweykowska, A.**, On the conditions of anthocyanin formation in the Vitis vinifera tissue cultivated in vitro, *Acta. Soc. Bot. Pol.,* 21, 537, 1952.

355. **Smith, R. H. and Murashige, T.**, In vitro development of the isolated shoot apical meristem of angiosperms, *Am. J. Bot.,* 57, 562, 1970.

356. **Sommer, N. F., Bradley, M. V., and Creasy, M. T.**, Peach mesocarp explant enlargement and callus production in vitro, *Science,* 136, 264, 1962.

356a. **Sommer, N. F. and Creasy, M. T.**, Peach fruit tissue culture, *Am. J. Bot.,* 49(Abstr.), 663, 1962.

356b. **Söndahl, M. R., Salisbury, J. L., and Sharp, W. R.**, SEM characterization of embryogenic tissue and globular embryos during high frequency somatic embryogenesis in coffee callus cells, *Z. Pflanzenphysiol.,* 94, 185, 1979.

357. **Söndahl, M. R. and Sharp, W. R.**, High frequency induction of somatic embryos in cultured leaf explants of *Coffea arabica* L., *Z. Pflanzenphysiol.,* 81, 395, 1977.

357a. **Söndahl, M. R., Spahlinger, D. A., and Sharp, W. R.**, A histological study of high frequency and low frequency induction of somatic embryos in cultured leaf discs of *Coffea arabica* L., *Z. Pfanzenphysiol.,* 94, 101, 1979.

358. **Spiegel-Roy, P. and Kochba J.**, Stimulation of differentiation in orange (*Citrus simensis*) ovular callus in relation to irradiation of the media, *Radiat. Bot.,* 13, 97, 1973.

359. **Spiegel-Roy, P. and Kochba, J.**, Mutation breeding in *Citrus,* in Induced Mutations in Vegetatively propagated plants, Vol. 339, International Atomic Energy Association, United Nations Organization, Vienna, 1973b, 91.

360. **Spiegel-Roy, P. and Kochba, J.**, Production of solid mutants in Citrus, utilizing new approaches and techniques, in Improvement of Vegetatively Propagated Plants Through Induced Mutations, International Atomic Energy Association, United Nations Organization, Vienna, 1975, 117.

361. **Srinavasan, C. and Mullins, M. G.**, Control of flowering in the grapevine (*Vitis vinifera* L). Formation of inflorescences *in vitro* by isolated tendrils, *Plant Physiol.,* 61, 127, 1978.

362. **Startisky, G.**, Embryoid formation in callus tissues of coffee, *Acta Bot. Neerl.,* 19, 509, 1970.

363. **Staut, G., Borner, H. G., Becker, H.**, Studies on callus growth of di- and tetra- ploid grapes *in vitro, Vitis,* 11, 1, 1972.

364. **Stone, O. M.**, Factors affecting the growth of carnation shoot apices, *Ann. Appl. Biol.,* 52, 199, 1963.

365. **Syono, K. and Furuya, T.**, Induction of auxin-nonrequiring tobacco calluses and its reversal by treatments with auxins, *Plant Cell Physiol.,* 15, 7, 1974.

366. **Tabachnik, L. and Kester, D. K.**, Shoot culture for almond — peach hybrid clones *in vitro, HortScience,* 12, 545, 1977.

367. **Takatori, F. H., Murashige, T., and Stillman, J. I.**, Vegetative propagation of asparagus through tissue culture, *HortScience,* 3, 20, 1968.

368. **Teo, C. K. H.**, Plant tissue culture and its potential contributions to Malaysia agriculture, *Planter (Malaya),* 49, 491, 1973.

369. **Teo, C. K. H.**, Clonal propagation of pineapple by tissue culture, *Planter (Malaya),* 50, 58, 1974.

370. **Tisserat, B. and Murashige, T.**, Probable identity of substances in Citrus that repress asexual embryogenesis, *In Vitro,* 13, 785, 1977a.

371. **Tisserat, B. and Murashige, T.**, Repression of asexual embryogenesis in vitro by some plant growth regulators, *In Vitro,* 13, 799, 1977b.

372. **Townsley, P. M.**, Chocolate aroma from plant cells, *Can. Inst. Food Sci. Technol. J.*, 7, 76, 1974a.
373. **Townsley, P. M.**, Production of coffee from plant cell suspension cultures, *Can. Inst. Food Sci. Technol. J.*, 7, 79, 1974b.
374. **Trippi, V. S.**, Studies on ontogeny and senility in plants. II. Seasonal variation in proliferative capacity *in vitro* of tissues from juvenile and adult zones of *Aesculus hippocastanum* and *Castanea vulgaris, Phyton (Buenos Aires)*, 20, 146, 1963a.
375. **Trippi, V. S.**, Studies on ontogeny and senility in plants. III. Changes in the proliferative capacity in vitro during ontogeny in *Robinia pseudoacacia* and *Castanea vulgaris* and in adult and juvenile clones of *R. pseudoacacia, Phyton (Buenos Aires)*, 20, 153, 1963b.
376. **Tucker, D. P. H. and Murashige, T.**, High temperature growth effects on *Citrus limon* fruit tissue as studied *in vitro, J. Hortic. Sci.*, 43, 453, 1968.
377. **Tukey, H. B.**, Artificial culture of sweet cherry embryos, *J. Hered.*, 24, 7, 1933.
378. **Van de Voort, F. and Townsley, P. M.**, A gas chromotographic comparison of the fatty acids of the green coffee bean, Coffea arabica, and the submerged coffee cell culture, *Can. Inst. Food Sci. Technol. J.*, 7, 82, 1974.
379. **Vardi, A.**, Studies on isolation and regeneration of orange protoplasts, in *Production of Natural Compounds by Cell Culture Methods*, Proc. Int. Symp. Plant Cell Culture, Alfermann, A. W. and Reinhard, E., Eds., Federal Ministry for Research and Technology, Munich, 1978, 234.
380. **Vardi, A. and Ravek, D.**, Cross-feeder experiments between tobacco and orange protoplasts, *Z. Pflanzenphysiol.*, 78, 350, 1976.
381. **Vieitez, A. M., Gonzalez, M. L., and Vieitez, E.**, In vitro culture of cotyledon tissue of *Castanea sativa* Mill., *Sci. Hortic.*, 8, 243, 1978.
382. **Vine, S. J.**, Improved culture of apical tissues for production of virus-free strawberries, *J. Hortic. Sci.*, 43, 293, 1968.
382a. **Volk, R., Harel, E., and Mayer, A. M.**, Repression by ethionine of catechol oxidase activity in apple fruit suspension cultures, *Ann. Bot. (London)*, 43, 787, 1979.
383. **Walkey, D. G.**, Production of apple plantlets from axillary-bud meristems, *Can. J. Plant Sci.*, 52, 1085, 1972.
384. **Wallner, S. J.**, Apple fruit explant responses *in vitro* and textural characteristics of the derived tissue cultures, *J. Am. Soc. Hortic. Sci.*, 102, 743, 1977.
385. **Westphalen, H. H. and Billen, W.**, Erzeugung von erdbeerpflenzen in groben mengen durch sprobspitzenkultur, *Erwerbs-Obstau*, 18, 49, 1976.
386. **Wetmore, R. H. and Rier, J. P.**, Experimental induction of vascular tissues in callus of angiosperms, *Am. J. Bot.*, 50, 418, 1963.
387. **Wetmore, R. H. and Sorokin, S.**, On the differentiation of xylem, *J. Arnold Arbor. Harv. Univ.*, 36, 305, 1955.
388. **White, P. R.**, *A Handbook of Plant Tissue Culture*, The Jacques Catlell Press, Lancaster, Pa., 1943.
389. **White, P. R.**, *The Cultivation of Animal and Plant Cells*, 2nd ed., The Ronald Press, New York, 1963.
390. **Widholm, J. M.**, Anthranilate synthetase from 5-methyltrytophan susceptible and resistant cultured *Daucus carota* cells, *Biochim. Biophys. Acta*, 279, 48, 1972a.
391. **Widholm, J. M.**, Cultured *Nicotiana tabacum* cells with an altered anthranilate synthetase which is less sensitive to feedback inhibition, *Biochim. Biophys. Acta*, 261, 52, 1972b.
392. **Widholm, J. M.**, Selection and characteristics of biochemical mutants of cultured plant cells, in *Tissue Culture and Plant Science 1974*, Street, H. E., Ed., Academic Press, London, 1974a, 287.
393. **Widholm, J. M.**, Cultured carrot cell mutants: 5-methyl-tryptophan resistance trait carried from cell to plant and back, *Plant Sci. Lett.*, 3, 323, 1974b.
394. **Widholm, J. M.**, Selection and characterization of cultured carrot and tobacco cells resistant to lysine, methionine, and proline analogs, *Can. J. Bot.*, 54, 1523, 1976.
395. **Willemot, C. and Boll, W. G.**, Changed response of excised tomato roots to pyridoxine deficiency in prolonged sterile culture, *Can. J. Bot.*, 40, 1107, 1959.
396. **Williams, E.**, A hybrid between *Trifolium repens* and *T. ambiguum* obtained with the aid of embryo culture, *N. Z. J. Bot.*, 16, 499, 1978.
396a. **Wittenbach, V. A. and Bukovac, M. J.**, *In vitro* culture of Sour cherry fruits, *J. Am. Soc. Hortic. Sci.*, 105, 277, 1980.
397. **Yie, S. T. and Liaw, S. I.**, Plant regeneration from shoot tips and callus of papaya, *In Vitro*, 13, 564, 1977.
398. **Yokoyama, T. and Takeuchi, M.**, Organ and plantlet formation from callus in Japanese persimmon *(Diospyros kaki), Phytomorphology*, 26, 273, 1976.
399. **Youtsey, C. O.**, A method for virus-free propagation of Citrus — Shoot-tip grafting, *Citrus Ind.*, 59, 39, 1978.
400. **Zatykó, J. M. and Simon, I.**, In vitro culture of ovaries of Rubus species, *Acta Agron. Acad. Sci. Hung.*, 24, 277, 1975.

401. **Zatykó, J. M., Simon, I., and Szabó, Cs.**, Induction of polyembryony in cultivated ovules of red currant, *Plant Sci. Lett.*, 4, 281, 1975.

402. **Zdruikovskaya-Rikhter, A. I.**, Culture of early ripening embryos of peach *in vitro, Tr. Gos. Nikitsk. Bot. Sad,* 36, 213, 1962.

403. **Zenk, M. H.**, Haploids in physiological and biochemical research, in *Haploids in Higher Plants, Advances and Potential,* Kasha, K. J., Ed., University of Guelph, Ontario, 1974, 339.

404. **Zimmerman, R. H.**, Tissue culture of fruit trees and other fruit plants, *Proc. Int. Prop. Soc.*, 28, 539, 1978.

405. **Zimmerman, R. H., and Broome, O. C.**, *In vitro* propagation of apple cultivars, *HortScience,* 14 (Abstr.), 478, 1979a.

406. **Zimmerman, R. H., and Broome, O. C.**, Propagation of blueberries through tissue culture, *Hort-Science,* 14(Abstr.), 477, 1979b.

407. **Zwagerman, D. and Zilis, M.**, Propagation of *Prunus cerasifera* "Thundercloud" by tissue culture, *HortScience,* 14(Abstr.), 478, 1979.

408. **Broome, O.**, Personal communication, Science and Education Administration, Beltsville Agricultural Research Center-West, U.S. Department of Agriculture, Beltsville, Md. 1979.

409. **Raghavan, V.**, Diets and culture media for plant embryos, in *CRC Handbook Series in Nutrition and Food,* Vol. 4, Rechcigl, M., Jr., Ed., CRC Press, Boca Raton, Fla., 1977, 364.

Chapter 4

VEGETABLE CROPS*

Paul J. Bottino

TABLE OF CONTENTS

* Scientific article no. A2658, Contribution No. 5699 of the University of Maryland, Agriculture Experiment Station. This work was supported by project No. MDJ107 from the Maryland Agriculture Experiment Station.

I. INTRODUCTION

There are several possible applications of tissue culture technology. The most notable are the use of tissue culture as a tool to genetically modify plants for plant breeding[1-4] and the application of tissue culture techniques to the in vitro propagation of plants.[5-8] The major area where tissue culture techniques are presently being successfully applied is in the clonal multiplication of plants. The excellent reviews by Murashige[5-8] clearly establish the background methods and success in this area for ornamental plants. The approach is economically sound, and there is now a rapidly increasing number of commercial operations devoted exclusively to in vitro propagation of ornamental plants.

The subject of the present chapter, however, is not ornamentals, but vegetable crops. The techniques of in vitro propagation have application with vegetable crops for the same reasons they apply to ornamentals. These include: (1) production of virus-free stock plants, (2) production of large numbers of genetically identical plants, and (3) propagation of large numbers of plants where seed production is difficult. The purpose of this review is to explore the techniques and approaches used in the in vitro propagation of vegetable crops and to survey and establish the current state of the art for each major crop species.

II. BACKGROUND

The excellent reviews by Murashige[5-8] will provide the reader with a more general background into in vitro propagation. This section will investigate specific aspects of the process as they apply directly to vegetable crops.

A. Pathways of In Vitro Clonal Multiplication

According to Murashige[6,7] there are three pathways by which clonal multiplication can occur in vitro. These are somatic cell embryogenesis, adventitious shoots, and enhanced axillary branching.

Somatic embryos are identical to embryos arising from the zygote except that their origin is a somatic cell. The result is an intact plant. The details of in vitro embryogenesis have been very adequately reviewed.[9] Embryo formation can occur directly on the explant as has been reported for cauliflower[10] and potato[11] or more commonly from callus as in celery,[12,15] potato,[14,15] pumpkin,[16,17] carrot,[18,19] and asparagus.[20] According to Wetherell[9] embryogenesis is initiated by effectively reducing the auxin levels in medium on which the tissues are being cultured.

Adventitious shoots or roots arise in tissues which originate at sites other than normal leaf axils. This process is much more common than somatic embryogenesis. Shoots or roots may arise directly on the explant as in lettuce[21,22] and cabbage,[23] or more commonly the shoots may arise from callus derived from the primary explant. This is the case in lettuce,[22,24] tomato,[25,28] onion,[29] garlic,[30] chive,[31] cabbage,[23,32] carrot,[33] broccoli,[34,35] pepper,[36] cauliflower,[37,38] brussels sprout,[39] potato,[40,41] kale,[42,46] sweet potato,[47] and asparagus.[48,49] One precaution should be made whenever callus is the source of the somatic cells. The problem of genetic instability may be enhanced in plants arising by adventitious organogenesis. This may be particularly true when repeated subculturing of callus is carried out. Also, over time the callus may lose its potential for organogenesis or embryogenesis as a result of accumulated chromosomal abnormalities. The evidence with cabbage,[32,50] broccoli,[35] asparagus,[48,49,51] and carrot[52] would seem to bear this out.

Finally, enhanced axillary branching might occur from shoots originating at their normal position such as stem tips and lateral buds. Very little callus is produced, and once started, successive subculturing results in increased proliferation of more shoots. This process is quite applicable in several vegetable crop plants (asparagus,[53,58] sweet potato,[59] potato,[60] brussels sprout,[61] cauliflower,[38,62] cucumber,[63] broccoli,[64] cabbage,[65,66] garlic,[30] tomato,[27] and lettuce[22]. One major advantage of cloning plants through axillary shoot formation is the reduced or total lack of genetic abnormalities which occur.[7]

B. Stepwise Sequence in Clonal Multiplication

The sequential steps involved in clonal propagation are now very well outlined and have been discussed in detail by Murashige.[5-8] On a commercial level, a stepwise procedure must be identified and conditions for optimum growth and multiplication for each stage must be clearly identified. Much attention has been given to this in vegetable crops. Stage I has for its main objective the establishment of an aseptic tissue culture of the plant. The primary aim should be prolonged survival of the explant. During this period visible enlargement of buds or stem tips occurs, or adventitious organs begin to appear from the explant. In Stage II there is a rapid increase in structures which will ultimately give rise to plants. This is the multiplication step in the process. It may involve repeated subculture to realize maximum production of plants, and as long as no genetic abnormalities appear, this stage may be carried on for very long periods of time. Stage III is a step which prepares the plants for reestablishment in the soil. A high frequency of success is absolutely necessary and the details of this stage are frequently overlooked. It has been clearly established that tissue-culture derived plants differ considerably from conventionally produced seedlings. Grout and Aston[67] reported that tissue-culture produced cauliflower plants have a modified leaf anatomy, where the palisade mesophyll tissue is absent or minimal. In addition, poor vascular connections occur between the root and shoot, and little epicuticular wax is present. These plants also differ physiologically in that they have lower levels of chlorophyll, Hill activity, and carbon dioxide fixation than have plants of a comparable age grown from seed.[67] Full photosynthetic competence did not develop until the plants had been established in the soil. The main goal of Stage III, then, is to root shoots which have developed, to harden the plants to moisture stress, and to convert the plant to a more autotrophic form of growth.

Murashige[8] has recently defined Stage IV as the reestablishment of the plants in the soil. Water loss upon transplanting to the soil is particularly acute due to the reduced epicuticular wax. Therefore, the initial stages of establishment in the soil must occur under conditions of high humidity. This has produced good survival with kale[45] and asparagus.[57,68] Also, water loss in newly transplanted cauliflower plants has been slowed by spraying the plants with polyvinyl resin S600.[69] This had the effect of increasing cuticular resistance to water loss, but not impairing stomatal function.

C. Special Considerations

There are three areas of special consideration which are important in the development of an optimal in vitro system for clonal propagaion of plants. These are explant source, nutrient requirements of the culture, and environmental conditions under which the culture is grown. These have been very adequately described in the reviews of Murashige,[5-8] and comprehensive treatment of these subjects will not be made here. However, these considerations as they apply to cloning vegetable crops will be specifically discussed.

Even though totipotency is a characteristic of all plant cells, its expression may be limited to specific cells. This is why Murashige believes that the state of the initial explant is the most important factor determining success in clonal propagation.[7] The most important areas of consideration are organ source, physiological and developmental state of the organ, season, size and overall physiological state of the plant from which the explant comes. As can be seen from Table 1,* the sources of explant material with vegetable crops are quite diverse. They include cotyledons, hypocotyls, leaf sections, axillary buds, shoot tips and stem apices, or other meristematic regions. The stem tip apices and meristems usually include several leaf primordia. Rarely, if ever, does the apical dome alone serve as the explant source. It is interesting that in many cases the explant originates from young seedling tissue. Young tissue clearly has greater regenerative potential than older tissue. The smaller the explant the better the chance is for the tissue being free of pathogens. However, the frequency of survival decreases as the explant gets smaller, and unless the objective is specifically to obtain virus-free clones, no advantage is gained by using extremely small explants.

The next most important consideration is the nutrient medium used for initiating and maintaining the cultures. Since this subject has been specifically dealt with,[70,71] only observations relating to vegetable crops will be made. Medium components are normally divided into three major groups: (1) inorganic salts, (2) organic components, and (3) natural complexes. It can be seen in Table 1, that with few exceptions, the inorganic salt formulation of choice for vegetable crops has been that of Murashige and Skoog.[72] There has also been use of Gamborg's B5,[73] Miller's[74] or White's[75] media, either as originally described or with slight modifications. Among the organic constituents, the carbohydrates, vitamins, amino acids, and inositol are used at concentrations similar to those described by Murashige and Skoog.[72] The formation of shoots, roots, and embryos clearly is under the influence of an auxin and cytokinin. The results for vegetable crops agree with those originally described by Skoog and Miller[76] who determined that the balance between the auxin and the cytokinin regulates the growth pattern of the culture. Table 1 lists the specific hormone combination for each crop. The most commonly used auxins are indoleacetic acid (IAA) and napthalene acetic (NAA) in the range 0.5 to 5.0 mg/ℓ. Kinetin, benzyladenine (BA), and N_6-iospentenyladenine(2ip) are the most commonly used cytokinins, also in the range 0.5 to 5.0 mg/ℓ. Specific hormones and their concentrations in the medium can be found in Table 1 and in the discussion of each specific crop. Either yeast extract, malt extract, or casein hydrolysate and 10% coconut water in varying amounts are sometimes used. These substances are only added to the medium as a last resort to initiate cell growth or to stimulate organogenesis. The predominant case is for the medium to be solidified with 0.8 to 1.0% agar. There are a few examples of the successful use of liquid suspensions and filter-paper bridge cultures. Normally the medium is the same for both Stage I and Stage II. The plants are usually transferred to a medium with no hormones or a lowered hormone level than previously used for rooting in Stage III.

Detailed studies on the environmental conditions favorable to mass propagation with vegetable crops are lacking. In general, they follow very closely those summarized for ornamentals by Murashige.[6] These include a 16-hr photoperiod with a light intensity of at least 1000 1x. Temperature ranges used extend from 20 to 27°C with a constant 25 to 27°C being the most common.

III. SUMMARY OF IN VITRO PROPAGATION OF SPECIFIC VEGETABLE CROPS

This section will summarize the literature and current state of in vitro propagation

* Table 1 will be found at the end of the text.

for each vegetable crop. The original list of crops for consideration came from the U.S. Department of Agriculture, Washington, D.C., 1978 Annual Summary of Acreage, Yield, Production, and Value, in which the principal vegetable crops in the U.S. are listed. In addition, other minor crops on which in vitro propagation has been successful are also included. This discussion is intended to supplement the basic information on each crop species in Table 1.

A. Asparagus (*Asparagus officinalis* L.)

Becasue it is clearly an advantage to propagate superior genetically identical asparagus plants in large numbers, a great deal of work has occurred with this crop. Yang[70] has very recently summarized the published literature in this area. As a result only the essential features of in vitro propagation of asparagus will be summarized here.

A number of studies have demonstrated plant regeneration from callus cultures. Wilmar and Hellendoorn[48] induced shoot production from callus on medium without exogenously supplied hormones, and Takatori et al.[49] induced plants from callus on MS medium[72] containing 0.5 mg/ℓ NAA, 50 mg/ℓ adenine sulfate, and 15% coconut water. Even though callus cells showed some aneuploidy, all regenerated plants were diploid. On the other hand, Malnassy and Ellison[51] observed a shift to tetraploidy in the plants regenerated from callus. Steward and Mapes[77] clearly demonstrate the feasibility of propagation of asparagus from callus. However, Yang[70] believes the use of callus cultures results in too much genetic instability to be used as the standard in vitro method of propagation and would be useful only for producing genetic variability for breeding purposes.

Shoot apices have consistently proved a good explant source for diploid plants. Murashige et al.[53] and Hasegawa et al.[54] report successful propagation of plants from shoot apices on MS medium[72] containing 0.3 mg/ℓ NAA, 0.1 mg/ℓ kinetin, 40 mg/ℓ adenine sulfate, 500 mg/ℓ Difco Bacto® malt extract, and 170 mg/ℓ $NaH_2PO_4 \cdot H_2O$. In 80 to 90% of the explants, both multiple spear initiation and rooting occurred on the same medium, however, medium without NAA allowed more extensive root formation. All plants produced in this manner were diploid, and Haseqawa et al.[54] estimate that 300,000 transplantable asparagus plants could be produced from a single shoot apex culture in 1 year.

Yang and Clore[55] have approached propagation of asparagus in a slightly different way with remarkable success. They first established aseptic stock plants from lateral bud spears. The medium used to maintain the sterile stock plants is MS medium[72] containing 0.05 to 0.1 mg/ℓ NAA and 0.05 mg/ℓ kinetin. Clone increase occurs when shoots containing buds from the stock plants are used as explants on Ms medium[72] containing 0.1 mg/ℓ NAA and 0.1 mg/ℓ kinetin. Buds from various portions of the stock plants are used. However, more plants are produced from basal buds than middle or apical buds. When stem segments with branched shoots are used, additional plantlets are formed. Yang[78] has also reported that low levels of benomyl (10 to 50 mg/ℓ) will promote vigorous shoot development in these cultures. The apical dome less than 0.1 mm in height and free of leaf primordia was used by Yang[79] to successfully rid plants of their viruses. Finally, protoplasts have been produced in asparagus,[20,80] and embryoids form when callus is cultured first on MS medium[72] containing 8×10^{-6} *M* zeatin and 3×10^{-6} *M* NAA or 5×10^{-6} *M* BA, 5×10^{-6} *M* NAA and 40 mg/ℓ adenine for 6 to 8 weeks and then transferred to hormone-free medium. No cytological investigations were carried out to determine the chromosomal composition of the callus.

Yang[70] has evaluated the rate of plant production through the method of aseptic stock plants. He estimates about 70,000 plants can be produced by one person working 200 days in a year.

B. Broccoli (*Brassica oleracea* L.)

There are two major studies on in vitro propagation of broccoli. Anderson and Carstens[64] used flower buds as the explant material. On MS medium[72] containing 80 mg/ℓ adenine sulfate, 170 mg/ℓ $NaH_2PO_4 \cdot H_2O$, 1 mg/ℓ IAA, and 4 mg/ℓ 2iP, three shoots were produced from each flower bud in 5 weeks. A subculture on the same medium resulted in a 7.5-fold increase in shoots. Root development occurred on medium with 0.2 mg/ℓ IAA replacing the adenine sulfate and the 2iP. At 7 weeks the plants were transplanted to the soil with 96% survival. When stock plants were grown at 16°C prior to taking the explants, flower and lateral buds produced multiple shoots in 4 weeks. Regeneration from leaf blade and midrib tissue did not occur under any condition. Johnson and Mitchell[34] obtained roots and shoots from leaf and stem explant material on MS medium[72] with 3 to 6 mg/ℓ kinetin and 8 to 9 mg/ℓ IAA, with leaf explants providing the best results.

Anther culture in broccoli has also been undertaken. Quazi[35] plated anthers from young flower buds on a complex medium containing MS salts,[72] B5 amino acids,[73] 2.5 g/ℓ yeast extract, 2×10^{-6} *M* BA, 2×10^{-6} *M* 2,4-dichlorophenoxyacetic acid (2,4-D), and 10% coconut water. Callus developed first, followed by root and shoot development. Chromosome counts of root cells showed primarily the tetraploid chromosome number. The in vitro propagated plants consistently produced more shoots than those grown from seed.

In the only study of its kind on any vegetable crop species, Anderson and Meagher[81] conducted a cost evaluation of propagating broccoli plants through tissue culture. This study provides a guide for estimating propagation costs and found tissue culture propagation to cost 11.6¢ per plant compared with greenhouse seed production which costs 15.4¢ per plant.

C. Brussels Sprout (*Brassica oleracea* L.)

It has been pointed out by Clare and Collin[39] that micropropagation is ideal for plants that are difficult to propagate by seed. Such is the case with brussels sprouts. Using inner buds or leaf petioles, good callus resulted on MS medium[72] with 0.2 mg/ℓ 2,4-D and 0.5 mg/ℓ kinetin. Shoot organogenesis occurred on MS medium[72] with 2 mg/ℓ IAA, 0.5 mg/ℓ kinetin, 1 mg/ℓ indolebuteric acid (IBA), and 10% coconut water. No chromosome number changes occurred in any of the regenerated plants. Clare and Collin[61] also have induced shoot and root development from shoot apices on hormone-free MS medium[72] containing 100 mg/ℓ of only thiamine, inositol, and casamino acids.

D. Bulbs (*Allium* sp.)

Even though onions (*Allium cepa*) rank third in acreage in the U.S., very little in the way of in vitro propagation has been done. Friedborg[29] using inner scales of aerial bulbs obtained first callus on B5 medium[73] with 5×10^{-6} *M* 2,4-D in either solid or liquid medium. Shoot formation occurred when 2iP at either 5×10^{-6} *M* or 5×10^{-7} *M* replaced the 2,4-D. Regeneration again occurred in both solid or liquid medium, and the callus remained diploid for over a year.

Other members of the genus *Allium* have also been propagated in tissue culture. Kehr and Schaeffer[30] using garlic, *Allium sativum* L., regenerated plants starting on MS medium[72] containing 1 mg/ℓ IAA, 1 mg/ℓ 2,4-D and 25 mℓ of deproteinized coconut water and then transferring to the medium of Nitsch and Nitsch.[82] Debergh and Standaert-De Metsenaere[83] regenerated bulbils of leek (*Allium porrum* L.) from callus on half-strength MS medium[72] containing 10 mg/ℓ IBA and 0.1 mg/ℓ BA. Finally, Zee et al.[31] using leaf tissue from seedlings obtained plant regeneration from callus on MS medium[72] containing 0.5 mg/ℓ IBA.

E. Cabbage (*Brassica oleracea* L.)

Tissue culture propagation has been widespread and successful in all the members of the genus *Brassica*. In one of the most comprehensive studies on cabbage, Bajaj and Nietsch[23] carried out detailed experiments on plant regeneration from several parts of the plant. The original explant material came from 3- to 5-week-old seedlings. MS medium[72] was used in all studies. With root explants, 0.5 mg/ℓ IAA, 0.5 mg/ℓ kinetin, and 500 mg/ℓ casein hydrolysate produced callus which developed buds in 4 to 6 weeks or could be induced to develop shoots on 1 mg/ℓ IAA, 0.5 mg/ℓ kinetin, and 10% coconut water. Hypocotyl segments produced buds directly on 1 mg/ℓ IAA, 2 mg/ℓ kinetin, and 500 mg/ℓ casein. Cotyledons and leaves produced abundant buds on 2 mg/ℓ IAA and 2 mg/ℓ kinetin. Plants could be transplanted into soil and developed into normal plants. Miszke and Skucinska[65] found that stump segments containing buds would develop shoots on MS medium[72] with the organic addenda of Horak et al.,[43] 0.2 mg/ℓ IAA, and 2.0 mg/ℓ kinetin. Rooting of the plantlets took place on MS medium[72] with 0.1 mg/ℓ NAA and 2.0 mg/ℓ kinetin. Kuo and Tsay[66] could induce axillary buds to develop on one-third strength MS medium[72] containing 1 mg/ℓ BA. Rooting occurred with 1 mg/ℓ NAA or 1 mg/ℓ IBA. They concluded that good shoot growth occurred only when the ratio of cytokinin to auxin was greater than one. Mascarenhas et al.[32] recently observed organogenesis from many parts of the plant including axillary buds from full grown heads on MS medium[72] containing 2 mg/ℓ kinetin and 0.4 mg/ℓ NAA, or MS medium[72] and 20% coconut water and 2 mg/ℓ NAA. However, the morphogenetic capacity of callus was lost after three subcultures. Genetic instability is always a concern with tissue culture propagation. Usually the problem is one of polyploidization of the culture. However, Maryakhina and Butenko[50] observed the reverse. Octoploid callus cultures after two subcultures reverted back to the diploid state. Only 30% of the cells were still octoploid after the second subculture. Finally, Kameya and Hinata[84] reported the induction of haploid plants from anther-derived callus using Nitsch salts and vitamins,[82] and 1 mg/ℓ each NAA and kinetin.

F. Carrot (*Daucus carota* L.)

Since the original discoveries by Steward et al.[33] and Reinert[85] that carrot plants could be regenerated from callus, totipotency of plant cells has been firmly established. Since that time the carrot has served as a model system for numerous studies into plant regeneration from unorganized tissues. Clearly the methods are available for mass in vitro propagation of the carrot. In the early experiments of Steward et al.,[33] coconut milk was an integral part of the medium. However, it was shown very soon that shoot regeneration or embryogenesis would occur under defined conditions. Kato and Takeuchi[18] using tap root sections as the explant observed both organogenesis and embryogenesis from callus on White's medium,[75] which contained 0.1% Difco yeast extract and 1.0 or 10.0 mg/ℓ IAA. Halperin and Wetherell[86] also obtained embryos from callus if 0.1 mg/ℓ 2,4-D replaced the coconut milk. Furthermore, Halperin[87] and Halperin and Wetherell[88] regenerated plants from embryos derived from callus on a medium consisting of the basal salts of Lin and Staba[89] and 2,4-D. When the 2,4-D concentration was higher than 0.1 mg/ℓ, only callus growth occurred. However, when the 2,4-D concentration was lowered to 0.01 mg/ℓ, embryos rapidly developed. The state of embryogenesis in carrots has been reviewed by Steward et al.,[90] and it is not the intent of this discussion to cover completely all aspects of this enormous field. The reader is directed to the recent review by Steward et al.[91] for a current treatment.

The possibility for rapid clonal propagation of the carrot is clearly at hand based on the extensive work done on this plant. One precaution is necessary, however. Smith and Street[52] showed that on a medium containing 2,4-D, the embryogenic potential of carrot suspension cultures decreased over time and subculturing. This drop in embryogenic potential correlated directly with a rise in the occurrence of tetraploid cells. This observation, which probably would be applicable to most tissue cultures, could be an explanation of the inability for plant regeneration from long-term callus.

G. Cassava (*Manihot utilissium* Rohl.)

Plants have been regenerated from meristems,[92,93] young shoots,[93] and whole buds.[93] In one study with meristems, Kartha et al.[92] cut stakes into sections with two nodes each, sealed the upper cut ends with paraffin, and planted them in vermiculite in 9-in. pots. The pots were transferred to a growth cabinet and incubated at 26°C, 60% RH, and a light/dark cycle of 16/18 hr with a light intensity of 4000 lx. The cuttings sprouted within 5 to 7 days, and meristems were taken from the shoot apices. After sterilization, the domes were plated on an agar (0.6%) medium containing the macro- and micro-elements of MS medium[72] with the vitamins of B5 medium[73] and 2% sucrose.

Plants were regenerated in either a two-phase or a one-phase system. For the one-phase system, BA, gibberellic acid (GA), and NAA were combined in the culture medium at molar concentrations of 5×10^{-7}, 10^{-7}, and 10^{-6}, respectively. Within 25 days, complete plants with well-developed shoot and root systems could be produced. When plants were at the 3- to 4-leaf stage, they were transferred to pots, watered with half-strength Hoagland's solution, and grown in growth chambers at 80% RH.

Bajaj[93] recently described in vitro clonal propagation of cassava from stem segments, buds, and meristem tips. Stakes were cut into 10 cm long pieces and planted in pots. After the sprouting of lateral buds and shoot development, these were excised and surface sterilized with 0.63% sodium hypochlorite (12% commercial bleach). One cm long shoot segments were plated on basal White's[75] medium. Roots were produced at the basal end. With the addition of 1 mg/ℓ NAA, 0.2 mg/ℓ kinetin, and 0.5 mg/ℓ GA callus arose at the cut end of the segments and at the base of the bud and meristem. Shoots formed on the explants within 2 to 3 weeks. The resulting plantlets were transferred to pots where they continued to grow. These results show that in vitro propagation is very effective from preformed structures in the absence of extensive callus.

H. Cauliflower (*Brassica oleracea* L.)

Cauliflower probably has been studied more than any member of the genus *Brassica*. Pow[37] cultured curds on filter-paper bridge cultures in LS medium[94] containing 0.0256 mg/ℓ kinetin and 8 mg/ℓ IAA and observed the development of roots after 1 week and shoots in 3 to 4 weeks. Plants were transplanted into the soil at 6 to 7 weeks. Walkey and Woolfitt[38] using floret tissue from the mature curd in a liquid shake culture obtained shoots in LS medium[94] with 8 mg/ℓ IAA and 2.56 mg/ℓ kinetin. They also observed that lowering the kinetin favored root development. Crisp and Walkey[62] using the meristem portion from the curd obtained large numbers of leafy shoots on liquid stationary cultures in the medium of Walkey and Woolfitt.[38] They estimate 10,000 apices per curd which are potential plants through in vitro culture. No genetic instabilities have been observed in all the plants regenerated from these apices. This method has also worked for freeing cauliflower of virus infection. Using the method of either Pow[37] or Walkey and Woolfitt,[38] Walkey et al.[95] found that 32% of the plants regenerated from apices were free from turnip mosaic and cauliflower mosaic viruses that infected the stock plants. Embryogenesis was induced in leaf callus by Pareek and Chandra.[10] Embryoids developed directly from the callus on MS medium[72] containing 1 mg/ℓ IAA and 0.5 mg/ℓ kinetin. Once embryoids began to form, the callus was transferred to lower auxin levels (0.1 to 0.01 mg/ℓ IAA) where embryogenesis was accelerated.

Since Grout and Aston[67,96] have shown that cauliflower plants from culture differ from conventionally produced seedlings in a number of morphological and physiological features, some concern has been displayed for protecting these plants upon transplanting in soil. According to Grout[97] poor survival of plants regenerated from meristem culture when transplanted to field and greenhouse conditions is due primarily to uncontrolled water loss. This is due primarily to differences in the quantity of wax on leaf surfaces. Grout and Aston[98] have reported a decrease in epicuticular wax and incomplete vascular development between roots and shoots of the regenerated plants, and in addition, they[96] have also shown that meristem-cultured plants differed physiologically from plants of a comparable age grown from seed. A solution to the water loss problem in meristem-cultured plants was found by Wardle et al.[69] who showed that cuticular resistance to water loss could be increased 2.5-, to 3-fold by a single application of polyvinyl resin S600.

I. Celery (*Apium graveolens* L.)

Celery cultures seem to regenerate plants exclusively through embryogenesis. Williams and Collin[12] using petiole sections on MS medium[72] with 0.5 mg/ℓ 2,4-D and 0.6 mg/ℓ kinetin produced callus. Subculturing the callus produced embryoids. Lowering the 2,4-D concentration and increasing the kinetin further stimulated embryoid formation. In suspension culture, embryoids develop best in the absence of all hormones. New complete plants developed 10 weeks after starting. Al-Abta and Collin[13] made a study of the process of embryogenesis in celery. They suggest that embryogenesis is similar to that in the carrot. The cultures show no reduction in embryogenic potential over 4 years as long as 2,4-D is omitted from the medium. The loss of embryogenic potential occurs immediately upon the use of 2,4-D in the medium.

J. Cucumber (*Cucumis sativus* L.)

A very recent report has appeared on plant regeneration in the cucumber. Handley and Chambliss,[63] using axillary buds 1- to 3-mm long, could get large numbers of plants to regenerate. The best medium was MS[72] with 0.1 mg/ℓ NAA and 0.1 mg/ℓ kinetin. On this medium there was production of both roots and shoots. They could not get plant regeneration from callus.

K. Eggplant (*Solanum melongena* L.)

In a recent report Kamat and Rao[99] carried out an extensive study on morphogenesis from hypocotyl sections of 4-week-old seedlings. On a modified MS medium,[72] extensive plant regeneration occurred at several combinations of auxins and cytokinins. The best overall success occurred with 5.7 μM IAA and 4.4 μM BA. Regenerated plants could be easily established in soil in pots. They eventually flowered and set fruit.

L. Kale (*Brassica oleracea* L.)

Kale is one of the brassicas in which plant regeneration has been possible for several years. Lustinec and Horak[42] observed plant regeneration from callus from virtually every part of the plant. In all cases MS medium[72] was used. The best hormone combination was 0.2 to 1.0 mg/ℓ 2,4-D and 0.5 to 3.0 mg/ℓ kinetin. The morphogenetic potential of the callus decreased with time. Horak et al.[45] noted the production of polyploid plants using the same medium. Horak[44] then tested 17 plants regenerated from tissue of a single diploid plant and found that only three plants corresponded in chromosome number to the original diploid plant. The rest were tetraploid. Also, Horak et al.[45] using pith tissues produced callus on a medium free of 2,4-D (MS[72] + 0.5 mg/ℓ kinetin, 2 mg/ℓ NAA, and 10% coconut water). When the callus was transferred to medium free of NAA, shoots developed and root formation increased. The plants were transplanted to the soil. Only 8% of the regenerated plants were diploid, and 76% were tetraploid. The remaining 16% had thick stems and petioles, deformed leaves, and poor survival in the field. The dominant chromosome number in these plants was octoploid. More recently Zee and Hui[46] obtained explants from hypocotyls and cotyledons of 10- to 12-day-old seedlings. On MS medium[72] with 1 mg/ℓ NAA and 1 mg/ℓ kinetin, roots developed from the hypocotyl explant in 6 days and shoots in 12 to 16 days. Roots and shoots developed from cotyledons on MS medium[72] with 4 mg/ℓ NAA and 0.5 mg/ℓ kinetin. Plants were easily established in the soil when they were watered wtih Hoagland's solution containing 10 μg/mℓ gentamicin and 2 mg/ℓ NAA. Unfortunately, they do not report any information on the chromosome composition of the regenerated plants. Since callus growth was kept to a minimum, and in the absence of 2,4-D, the plants may not have been polyploid as in the earlier studies.

M. Legumes (*Phaseolus vulgaris* L., *Pisum sativum* L.)

Success in in vitro propagation in the legumes has been very limited. There has generally been lack of success in plant regeneration from in vitro cultures in the entire family. In the legumes, root morphogenesis seems to be easily controllable. But shoot morphogenesis has been much more difficult. The most notable success with *Phaseolus vulgaris* so far is that of Crocomo et al.[100] Starting with leaves, a callus could be developed quite easily on a medium containing only IAA and kinetin. However, with the addition of a bean seed extract to the basal medium containing 2 mg/ℓ IAA, 1 mg/ℓ NAA, and 0.2 mg/ℓ kinetin, shoots were induced to develop from callus. The results were only two plantlets in over 400 cultures, but some progress was made in the direction of shoot morphogenesis.

More progress has been made with the pea, *Pisum sativum* L. Gamborg et al.[101] excised apices 2 to 3 mm in length from 3- to 4-day-old shoots and macerated them into cell masses that were placed on B5 medium[73] and MS medium[72] with varying concentrations of growth regulators. Better results were obtained wtih B5 medium. Callus initially formed followed by shoot formation at several concentrations of BA. NAA was not necessary and appeared to repress shoot formation. Additional nitrogen above that found in regular B5 medium[73] was beneficial to shoot development. Root formation occurred randomly. Kartha et al.[102] used only apical domes from peas measuring 200 to 300 μm as the explant. On B5 medium,[73] BA alone at 5×10^{-7} *M* or with 10^{-6} *M* NAA induced shoot development from the apex. The shoots could be induced to root in 7 to 10 days on half-strength B5 medium containing 10^{-6} *M* NAA.

Recently Kartha et al.[103] have made another development in in vitro propagation of peas. They report plant regeneration from meristems which had been frozen at liquid nitrogen temperatures for 1 hr. The maximum regeneration of 73% occurred when the meristems had been precultured for 48 hr prior to freezing on B5 medium[73] supplemented with 0.5 μ*M* BA and 5% dimethylsulfoxide (DMSO). Apparently the preculturing of the meristems in DMSO protected the cells from freezing injury without effecting their morphogenetic potential. These results represent a significant contribution in the storage of valuable germplasm by freezing, with subsequent plant regeneration from the thawed material.

N. Lettuce (*Lactuca sativa* L.)

Adventitious shoot formation in lettuce, *Lactuca sativa* L., was reported in 1967 by Doeschug and Miller.[21] Cultures were established from cotyledons, hypocotyls, and roots of seedlings. Prolific shoot formation occurred from cotyledons on a medium of their own design containing 0.5 mg/ℓ kinetin and 5.0 mg/ℓ IAA. They also reported that adenine could replace kinetin in inducing bud formation. Similar results were reported by Kadkade and O'Connor[104] with almost the same medium. Kadkade and Seibert[24] then reported that organogenesis was enhanced by red light. A 2- to 2.5-fold increase in the number of shoots and a 1.5-fold increase in callus weight compared to the controls were observed. They suggest that the light regulation of shoot production occurs at the initiation of buds. Finally, Koevary et al.[22] using explants from stem and leaf tissue observed large amounts of shoot development. Stem material from apical segments produced shoots in 14 days; nonexpanded buds remained dormant or produced only callus as did pith. Leaf material from near the margin produced callus, followed by shoots and roots in 14 days. Overall root formation was highly unpredictable but could be induced with IBA. The procedure that works best, according to Koevary et al.,[22] is to culture the original explant at 22.5°C for 2 weeks on Miller's basal medium[74] with 5 mg/ℓ IAA and 0.5 mg/ℓ kinetin to induce shoot formation and transfer to the same medium containing 1 mg/ℓ IBA for root induction. Plants could be transplanted into pots 4 weeks after initiating the cultures and to the field within an additional 3 to 5 weeks. Heads appeared, and viable seed was produced.

O. Pepper (*Capsicum annuum* L.)

Gunay and Rao[36] have reported shoot formation in red pepper using cotyledons and hypocotyls from 4-week-old germinated seedlings. The most effective medium consisted of MS salts[72] with 2 mg/ℓ BA and 1 mg/ℓ IAA. When NAA was substituted for IAA, shoot formation was completely suppressed. They also observed slight differences in varietal responses to this medium. George and Narayanaswamy[105] have also obtained haploid plantlets in pepper through anther culture. Anther cultures are started on medium consisting of LS salts[94] with kinetin and IAA, which initiates the development of proembryo-like multicellular structures. The culture is then transferred to LS medium[94] containing 0.5 mg/ℓ IAA and 400 mg/ℓ casein hydrolysate where further development of the haploid embryo continues. Ultimately, only a few complete haploid plants were obtained.

P. Potato (*Solanum tuberosum* L.)

A number of different approaches have been successfully used in the regeneration of potato plants from in vitro cultures. Mellor and Stace-Smith[60] have rooted excised buds in filter-paper wick culture. The medium was MS[72] with no exogenous hormones. Lam[11] induced embryoids on tuber discs on MS medium,[72] Nitsch and Nitsch[82] organic addenda, 0.4 mg/ℓ IAA, 0.4 mg/ℓ GA, 0.8 mg/ℓ kinetin, and 1.0 g/ℓ casein. Shoots develop when the embryoids are transferred to medium containing 0.4 mg/ℓ BA. This procedure yielded many "abnormal" shoots that were reddish in color and had unexpanded leafy structures resembling bracts. Skirvin et al.[106] removed the abnormal shoots and cultured them on a modified White's medium[75] with no growth regulators. In 3 to 4 days the shoots turned green, roots formed, and in 4 weeks the plants were transplanted to the soil. Finally, Roest and Bokelmann[40] explanted stem sections on an MS medium[72] containing 10 mg/ℓ GA, 1.0 mg/ℓ BA, and 1.0 mg/ℓ IAA and observed plant regeneration.

Anther culture has been used to produce both haploid and dihaploid plants. Dunwell and Sunderland[14] cultured anthers on MS medium[72] with 0.01 to 5.0 mg/ℓ kinetin and NAA above 1.0 mg/ℓ. Some callus formed, followed by embryoids developing from the callus. They estimate only 0.05% of the pollen responded. Foroughi-Wehr et al.[107] induced monohaploid plants to develop from anthers on LS medium[94] containing 1.0 mg/ℓ IAA and 1.0 mg/ℓ BA. Hormones were clearly required for the process, and a differential response among several different genotypes occurred. Sopory et al.[108] obtained 37% haploid embryoids from anthers cultured on MS medium[72] with 6×10^{-6} *M* IAA and 4×10^{-6} *M* BA. Also, 5.0 g/ℓ of activated charcoal stimulated haploid development. Finally, Sopory[41] cultured anthers of dihaploid clones with successful production of diploid and tetraploid plants. Nitsch[82] medium with 10^{-6} *M* BA, 10^{-5} *M* BA, and 10^{-5} *M* NAA was the best for callus growth. There was some genotypic effects. Of the 40 clones tested only 22 produced callus, and when transferred to MS[72] with 10^{-5} *M* zeatin, shoot production occurred.

There are also good examples of plant regeneration from protoplasts of potato. Shepard and Totten[109] starting with mesophyll protoplasts were able to regenerate a callus and induce both shoot and root organogenesis. The medium was basically that used by Lam[11] with 0.1 mg/ℓ IAA, 0.5 mg/ℓ zeatin, and 40 mg/ℓ adenine sulfate. This same medium would promote shoot development from callus derived from tuber sections. Binding et al.,[110] also starting with mesophyll protoplasts, obtained shoot development. Two media were effective in promoting shoot formation: MS[72] with 15 μM kinetin and 5 μM IAA or B5[73] with either 2.5 μM BA alone or 5% coconut water, 20 μM zeatin, and 5 μM IAA. Rooting occurred best on hormone-free medium. Grun and Chu[15] also carried out an extensive study on culture and plant regeneration from mesophyll protoplasts. After culturing the protoplasts and callus formation, embryoids developed on the medium of Lam[11] or on the medium of Upadhya[111] using the hormone composition of Lam.[11] On these media shoots and roots developed within 1 or 2 months. According to Grun and Chu[15] and Upadhya,[111] the potato is particularly sensitive to media containing ammonium ions and only grows well in media that is low or lacking in ammonium ions.

Q. Pumpkin (*Cucurbita pepo* L.)

Embryogenesis from callus has been reported by Jelaska.[16,17] Originally,[16] MS medium[72] containing 0.3 mg/ℓ 2,4-D, 1 mg/ℓ IBA, and 10% watermelon sap was used to induce embryos to form from callus. Later,[17] embryogenesis occurred from callus with just 1 mg/ℓ IBA or 1 mg/ℓ NAA and 13.5 mg/ℓ adenine. The cultures have retained their embryogenic properties for 3 years.

R. Sweet Potato (*Ipomoea batatas* L.)

Shoot tips and axillary buds were used by Litz and Conover[59] for propagating the sweet potato. They used MS medium[72] with 10 g/ℓ activated charcoal. In 2 weeks callus developed, followed by roots and shoots. Optimum shoot regeneration from "White Star" explants occurred with 1 mg/ℓ BA and from P1315343 with 1 mg/ℓ kinetin and 1 mg/ℓ IAA. Sehgal[47] used anthers to produce a callus which could be induced to develop plants. The callus grew best on MS medium[72] with 0.1 mg/ℓ kinetin and 1 mg/ℓ 2,4-D. When the callus was transferred to medium without hormones, plant regeneration occurred. Since the callus originated from the anther wall, all plants regenerated were diploid.

S. Tomato (*Lycopersicon esculentum* Mill.)

A number of papers report plant regeneration in tomato. Behki and Lesley[25] report callus, shoot, and root formation from discs of young expanded leaves of 15 mutant clones. Shoot formation occurred in 12 clones on MS medium[72] with 5×10^{-7} *M* NAA and 1×10^{-5} *M* BA. Root formation occurred upon transfer to MS medium[72] with 2.5×10^{-6} *M* NAA and 1×10^{-6} *M* BA. Kartha et al.[26] reported both root and shoot formation from leaf explants on medium consisting of MS salts,[72] B5 vitamins,[73] and 0.1 μM zeatin and 10 μM IAA or 1 μM zeatin and 0.1 μM IAA. Also, BA or zeatin alone at 5 ^-M induced complete plant regeneration from callus. Rooting in general was best on hormone-free medium. Kartha et al.,[27] again starting with shoot tips, observed plant differentiation from callus at all concentrations of BA or zeatin between 0.1 to 10 μM. The process occurred in 35 days. They observed excellent uniformity in the plants regenerated from the callus. De Langhe and De Bruijne[28] reported shoot formation from stem-derived callus on LS medium[72] with 20% coconut water and 10^{-5} M zeatin. The shoot forming ability of the explants increased greatly when the plants in the greenhouse were pretreated wtih Chlormequat prior to taking the original explants.

Haploid plants have been produced from anther culture in tomato. Gresshoff and Doy[112] tested 43 races for haploid plant production. Using either the mineral salts of Gamborg and Eveleigh[113] with 2 mg/ℓ NAA and 5 mg/ℓ kinetin or the salts of Blaydes[114] with 2 mg/ℓ NAA and 1.0 mg/ℓ kinetin, haploid callus differentiated from the anthers followed by regeneration of haploid plants. The callus remained haploid after 1 year of culture. Ramulu et al.[115] produced haploid plants in the same way as Gresshoff and Doy,[112] but also produced plants from anthers and stem internodes. They used an MS medium[72] with 10^{-6} M NAA, 10^{-5} M GA_3 10^{-5} M abscisic acid (ABA), 15% coconut water, and 5 mg/ℓ glycine. Some differential genotypic responses were observed.

T. Yam (*Dioscorea alata* L.)

Two studies have reported successful plant regeneration in the yam. Mascarenhas et al.[32] obtained callus from the swollen tuberous region of seedlings. Multiple shoots differentiated from the callus on MS medium[72] containing 10% coconut water. Mantell et al.[116] investigated the two major food yams of the caribbean, *D. alata* L. and *D. rotundata* Poir. Nodal segments of shoots were used as the explant. Shoot development occurred on MS medium[72] with either 10 mg/ℓ NAA and 0.5 mg/ℓ or 1.0 mg/ℓ BA. They also reported that a 16-hr photoperiod was four times more effective than a 12-hr photoperiod in promoting shoot formation.

IV. CONCLUSIONS AND PROSPECTS

It should be evident from this review that a great deal of progress has been made on the in vitro propagation of vegetable crops, with good success for most of the major crops grown in the U.S. While in vitro techniques might not be as uniformly applicable to vegetable crops as they are to ornamental crops, the technique has application for mass propagation of many vegetable species. The method has already been demonstrated to be economically feasible in some cases, and it is being used for large scale nursery production of potato. According to Boxus[117] in vitro methods are clearly justified for any vegetatively propagated crop. In addition, in vitro techniques certainly are useful for producing uniform desirable genotypes for use in breeding programs. Single selections could be increased through this method to produce enough plants for use in a breeding program. For example, it has been used in breeding asparagus for supermale production to be used in the creation of entirely male F_1 male hybrids.[117] Also, plants which would be difficult to maintain because of high self-incompatibility could be easily maintained and increased through in vitro methods. Murashige[8] suggests some additional unexpected benefits from in vitro propagation such as greater vigor, earlier flowering, fuller branching, and higher yield. The technique is already established as a valuable method for freeing valuable stock plants from viruses, and aseptic cultures can be used as a method to transport germplasm across international boundaries.[118]

In summary, methods of in vitro propagation of vegetable crops are well developed and can be carried out with relative ease with most major vegetable species. Each crop represents specific problems for the breeder, but in vitro techniques can clearly be applied to maintenance and propagation of most vegetable crop species.

ACKNOWLEDGMENTS

The author wishes to thank Dr. Amul Purohit for his help in conducting a computer search of the literature. This paper is dedicated to the memory of the late Arnold H. Sparrow who provided the author with both the opportunity and the encouragement to enter the field of plant tissue culture.

Table 1
VEGETABLE CROPS WHICH CAN BE PROPAGATED THROUGH IN VITRO TECHNIQUES

Crop plant	Explant	Type of regeneration	Medium	Ref.
Asparagus	Hypocotyl-derived callus	Plants	LS (no hormones)	48
	Spear-derived callus	Plants	MS + 0.5 mg/ℓ NAA, 50 mg/ℓ adenine sulfate, and 15% CW	49
	Shoot apices	Plants	MS + 3.0 mg/ℓ NAA, 0.1 mg/ℓ kinetin, 40 mg/ℓ adenine sulfate, 500 mg/ℓ malt extract, and 170 mg/ℓ $NaH_2PO_4 \cdot H_2O$	53, 54
	Shoot apices	Rooting	Same as above without NAA	53, 54
	Shoots from aseptic stock plants	Plants	MS + 0.1 mg/ℓ NAA and 0.1 mg/ℓ kinetin	55, 58
	Protoplast-derived callus	Embryoids	MS + 8×10^{-6} *M* zeatin 3×10^{-6} *M* NAA or 5×10^{-6} *M* BA, 5×10^{-6} *M* NAA, 40 mg/ℓ adenine, and transfer to hormone-free medium	20, 80
Broccoli	Flower buds	Shoots	MS + 1.0 mg/ℓ IAA, 4.0 mg/ℓ 2iP, 80 mg/ℓ adenine sulfate, and 170 mg/ℓ $NaH_2PO_4 \cdot H_2O$	64
	Flower buds	Roots	MS + 0.2 mg/ℓ IAA	64
	Leaf and stem segments	Roots/shoots	MS + 3—6 mg/ℓ kinetin, and 8—9 mg/ℓ IAA	34
	Anthers	Tetraploid plants	MS salts + B5 amino acids, 2×10^{-6} *M* BA, 2×10^{-6} *M* 2,4-D, 2.5 g/ℓ yeast extract, and 10% CW	35
Brussels sprout	Inner buds or mature leaf petioles	Shoots	MS + 2.0 mg/ℓ IAA, 0.5 mg/ℓ kinetin, 1.0 mg/ℓ IBA, and 10% CW	39

Table 1 (continued)
VEGETABLE CROPS WHICH CAN BE PROPAGATED THROUGH IN VITRO TECHNIQUES

Crop plant	Explant	Type of regeneration	Medium	Ref.
	Shoot apex	Shoots/roots	MS + 100 mg/ℓ casamino acids	61
Bulbs				
Onion	Inner scale of aerial bulb	Shoots	B5 + 5.0×10^{-6} or 10^{-7} *M* 2,4-D	29
Garlic	Meristems	Shoots	MS + 1.0 mg/ℓ IAA, 25% CW, and then transfer to medium of Nitsch	30, *82*
Leek	Bulb-derived callus	Bulbils	MS (½ strength) + 10 mg/ℓ IBA and 0.1 mg/ℓ BA	83
Chive	Seedling leaf	Shoots/roots	MS + 0.5 mg/ℓ IBA	31
Cabbage	Seedling roots	Shoots	MS + 0.5 mg/ℓ IAA, 0.5 mg/ℓ kinetin, and 500 mg/ℓ casein hydrolysate or MS + 1.0 mg/ℓ IAA, 0.5 mg/ℓ kinetin, and 10% CW	23
	Seedling hypocotyl	Buds	MS + 1.0 mg/ℓ IAA, 2.0 mg/ℓ kinetin, and 500 mg/ℓ casein	23
	Cotyledon and leaves	Buds	MS + 2.0 mg/ℓ IAA and 2.0 mg/ℓ kinetin	23
	Stump segments containing buds	Shoots	MS + Horak et al. organics, 0.2 mg/ℓ IAA, and 2.0 mg/ℓ kinetin	*43*, 65
	Stump segments containing buds	Roots	MS + 0.1 mg/ℓ NAA and 2.0 mg/ℓ kinetin	43, 65
	Axillary buds	Shoots	MS (⅓ strength) + 1.0 mg/ℓ BA	66
	Axillary buds	Roots	MS (⅓ strength) + 1.0 mg/ℓ NAA or 1.0 mg/ℓ IBA	66
	Roots, stems, leaves,	Plants	MS + 0.4 mg/ℓ NAA and 2.0	32

	axillary buds		mg/ℓ kinetin or MS + 2.0 mg/ℓ NAA and 20% CW	
	Anthers	Haploid plants	Nitsch salts + 1.0 mg/ℓ NAA and 1.0 mg/ℓ kinetin	*82*, 84
Carrot	Tap root	Shoots/embryos	White's + 1.0 or 10.0 mg/ℓ IAA and 0.1% Difco yeast extract	18, *75*
	Petiole-derived callus	Embryos	Lin-Staba salts + 0.01 mg/ℓ 2,4-D	19, 86, 87, *89*
Cassava	Meristem tip	Shoots/roots	MS salts, B5 vitamins, + 5 × 10^{-7} *M*BA, 10^{-7} *M*GA, and 10^{-6} *M*NAA	92
	Stem segments/buds	Roots, callus/shoots 2—3 weeks later	White's + 1 mg/ℓ NAA, 0.2 mg/ℓ kinetin, and 0.5 mg/ℓ GA	75, 93
Cauliflower	Curds	Roots and shoots	LS + 8.0 mg/ℓ IAA and 0.0256 mg/ℓ kinetin	37
	Floret tissue from mature curd	Shoots	LS + 8.0 mg/ℓ IAA and 2.56 mg/ℓ kinetin	38
	Floret tissue from mature curd	Roots	Lowered kinetin	38
	Meristem portion of curd	Shoots	Walkey and Woolfitt	*38*, 62
	Leaf-derived callus	Embryoids	MS + 1.0 mg/ℓ IAA and 0.5 mg/ℓ kinetin	10
Celery	Petiole sections	Embryoids	MS + 0.6 mg/ℓ kinetin and 0.5 mg/ℓ 2,4-D	12, 13, 119
Cucumber	Axillary buds	Plants	MS + 0.1 mg/ℓ NAA and 0.1 mg/ℓ kinetin	63
Eggplant	Seedling hypocotyls	Plants	MS + 5.7 μ*M*IAA and 4.4 μ*M* BA	99
Kale	Roots, hypocotyls, cotyledons, leaves, pith	Roots/shoots	MS + 0.2—1.0 mg/ℓ 2,4-D and 0.5—3.0 mg/ℓ kinetin	42
	Pith-derived callus	Roots/shoots	Same as Lustinec and Horak without 2,4,D	*42*, 43
	Stem pith	Plants	Same as Horak et al.	*43*, 44
	Pith-derived callus	Polyploid roots and shoots	MS + 0.5 mg/ℓ kinetin and 10% CW	45

Table 1 (continued)
VEGETABLE CROPS WHICH CAN BE PROPAGATED THROUGH IN VITRO TECHNIQUES

Crop plant	Explant	Type of regeneration	Medium	Ref.
	Hypocotyls	Roots/shoots	MS + 1.0 mg/ℓ NAA and 1.0 mg/ℓ kinetin	46
	Cotyledons	Roots/shoots	MS + 4.0 mg/ℓ NAA and 0.5 mg/ℓ kinetin	46
Legumes				
Bean	Seedling leaves	Plants	Medium of Veliky et al. + bean seed extract	100, *120*
Pea	Apical dome	Shoot development	B5 + 5×10^{-7} *M* BA alone or with 10^{-6} *M* NAA	102
	Shoot apex	Shoots	B5 + 0.2—5.0 μ*M* BA	101
Lettuce	Cotyledons	Buds/shoots	Doeschug & Miller salts + 0.5 mg/ℓ kinetin and 5.0 mg/ℓ IAA	21
	Cotyledons	Shoots	Miller + 1.0 mg/ℓ kinetin and 40 mg/ℓ adenine sulfate	*74*, 104
	Cotyledons	Roots	Miller + 0.1 mg/ℓ kinetin	*74*
	Cotyledons	Roots/shoots	MS + 1.0 mg/ℓ IAA, 0.5 mg/ℓ kinetin, and 40 mg/ℓ adenine sulfate	24
	Apex segments	Shoots	Miller + 5.0 mg/ℓ IAA and 0.5 mg/ℓ kinetin	22, *74*
	Apex segments	Roots	Miller + 1.0 mg/ℓ IBA	22, *74*
Pepper	Cotyledons, hypocotyls	Shoots	MS salts + 1.0 mg/ℓ IAA and 2.0 mg/ℓ BA	36
	Anthers	Haploid plants	Start culture on LS + 3.0 mg/ℓ IAA and 0.3 mg/ℓ kinetin or LS + 1.0 mg/ℓ IAA 3.0 mg/ℓ kinetin Transfer to LS + 0.5 mg/ℓ IAA and 400 mg/ℓ casein hydrolysate	105

Potato	Excised buds	Rooted plants	MS (no hormones)	60
	Tuber disc	Embryoids, "abornormal shoots"	Start culture on MS + 0.4 mg/ℓ IAA, 0.4 mg/ℓ GA, 0.8 mg/ℓ kinetin, and 1.0 g/ℓ casein Transfer to same medium + 0.4 mg/ℓ BA	11
	"Abnormal shoots" of Lam	Normal plants	Modified White's medium (no hormones)	*11*, *75*, 106
	Stem sections	Shoots	MS + 1.0 mg/ℓ IAA, 1.0 mg/ℓ BA, and 10.0 mg/ℓ GA	40
	Anthers	Haploid embryoids	MS + 1.0 mg/ℓ NAA and 0.01—5.0 mg/ℓ kinetin	14
	Anthers	Plants	LS + 1.0 mg/ℓ IAA and 1.0 mg/ℓ BA	107
	Anthers	Embryoids	MS + 6×10^{-6} *M* IAA and 4×10^{-6} *M* BA, also 5.0 g/ℓ activated charcoal	108
	Anthers	Plants	Nitsch + 10^{-6} *M* BA 10^{-5} *M* NAA, and 10^{-5} *M* BA	41, *82*
	Callus from mesophyll protoplasts	Plants	Lam + 0.1 mg/ℓ IAA, 0.5 mg/ℓ zeatin, and 40 g/ℓ adenine sulfate	*11*, 109
	Callus from mesophyll protoplasts	Shoots	MS + 5 μ*M* IAA, and 15 μ*M* kinetin or B5 + 2.5 μ*M* BA or B5 + 5 μ*M* IAA, 20 μ*M* zeatin, and 5% CW	110
	Callus from mesophyll protoplasts	Roots	MS (hormone-free)	110
	Callus from mesophyll protoplasts	Embryoids	Lam or Upadhya with hormones from Lam	*11*, 15, *111*
Pumpkin	Hypocotyls, cotyledons	Embryos	MS + 0.3 mg/ℓ 2,4-D, 1.0 mg/ℓ IBA, and 10% watermelon sap	16
	Hypocotyl-derived callus	Embryos	MS + 1.0 mg/ℓ IBA or 1.0 mg/ℓ NAA and 13.5 mg/ℓ adenine	17
Sweet potato	Anther-derived diploid callus	Plants	MS (no hormones)	47

Table 1 (continued)
VEGETABLE CROPS WHICH CAN BE PROPAGATED THROUGH IN VITRO TECHNIQUES

Crop plant	Explant	Type of regeneration	Medium	Ref.
	Shoot tips and axillary buds	Roots/shoots	MS + 1.0 mg/ℓ BA or 1.0 mg/ℓ kinetin, 1.0 mg/ℓ IAA, and 10 g/ℓ activated charcoal	59
Tomato	Leaf discs	Shoots	MS + 5×10^{-7} *M*NAA and 1×10^{-5} *M*BA	25
	Leaf discs	Roots	MS + 2.5×10^{-6} *M*NAA and 1.0×10^{-6} *M*BA	25
	Leaf	Shoots	MS, B5 vitamins + 0.1 μ*M* zeatin and 10 μ*M*IAA or 5.0 μ*M*BA and 5.0 μ*M*zeatin or 1.0 μ*M*zeatin and 0.1 μ*M*IAA	26
	Shoot tips	Shoots	MS, B5 vitamins + 0.1—10 μ*M*BA	27
	Stem sections	Shoots	LS + 10^{-5} *M*zeatin and 20% CW	28
	Anthers	Shoots, haploid callus, haploid plants	Gamborg & Eveleigh salts + 2.0 mg/ℓ NAA and 5.0 mg/ℓ kinetin or Blaydes salts + 2.0 mg/ℓ NAA and 1.0 mg/ℓ kinetin	112, *113*, *114*
Yams	Seedling tubers	Shoots	MS + 10% CW	32
	Nodes	Shoots	MS + 10 mg/ℓ NAA and 0.5 mg/ℓ BA	116

Note: Abbreviations: LS = Linsmaier and Skoog;[94] MS = Murashige and Skoog;[72] B5 = Gamborg, Miller, and Ojima;[73] CW = Coconut water; NAA = Naphthaleneacetic acid; BA = N_6-benzyladenine; 2,4-D = 2,4-dichlorophenoxyacetic acid; IBA = Indole-3-butyric acid; IAA = Indole-3-acetic acid; and GA = Gibberellic acid. Reference numbers that are italicized are medium types. These medium types are rarely used.

REFERENCES

1. **Bottino, P. J.**, The potential of genetic manipulation in plant cell cultures for plant breeding, *Radiat. Bot.*, 15, 1, 1975.
2. **Bajaj, Y. P. S.**, Potentials of protoplast culture work in agriculture, *Euphytica*, 23, 623, 1974.
3. **Gamborg, Ol L., Ohyama, K., Pelcher, L. E., Fowke, L. C., Kartha, K., Constabel, F., and Kao, K.**, Genetic modification of plants, in *Plant Cell and Tissue Culture*, Sharp, W. R., Larsen, P. O., Paddock, E. F., and Raghaven, V., Eds., Ohio State University Press, Columbus, 1979, 371.
4. **Kleinhofs, A. and Behke, R.**, Prospects for plant genome modification by nonconventional methods, *Annu. Rev. Genet.*, 11, 79, 1977.
5. **Murashige, T.**, Plant propagation through tissue cultures, *Annu. Rev. Plant Physiol.*, 25, 135, 1974.
6. **Murashige, T.**, Manipulation of organ initiation in plant tissue culture, *Bot. Bull. Acad. Sin.*, 18, 1, 1977.
7. **Murashige, T.**, Clonal crops through tissue culture, in *Plant Tissue Culture and Its Bio-Technological Application: Proceedings*, Barz, W., Reinhard, E., and Zenk, M. H., Eds., Springer-Verlag, Berlin, 1977, 392.
8. **Murashige, T.**, Principles of rapid propagation, in Propagation of Higher Plant Through Tissue Culture: A Bridge Between Research and Application, Hughes, K. W., Henke, R., and Constantin, M. J., Eds., National Technical Information Service, U.S. Department of Energy, Springfield, Va., 1978, 14.
9. **Wetherell, D. F.**, In vitro embryoid formation in cells derived from somatic plant tissue, in Propagation of Higher Plants Through Tissue Culture: A Bridge Between Research and Application, Hughes, K. W., Henke, R., and Constantin, M. J., Eds., National Technical Information Service, U.S. Department of Energy, Springfield, Va., 1978, 102.
10. **Pareek, L. K. and Chandra, N.**, Somatic embryogenesis in leaf callus from cauliflower (*Brassica oleraces* var. Botrytis), *Plant Sci. Lett.*, 11, 311, 1978.
11. **Lam, S.-L.**, Shoot formation in potato tuber discs in tissue culture, *Am. Potato J.*, 52, 103, 1975.
12. **Williams, L. and Collin, H. A.**, Embryogenesis and plantlet formation in tissue cultures of celery, *Ann. Bot. (London)*, 40, 32, 1976.
13. **Al-Abta, S. and Collin, H. A.**, Cell differentiation in embryoids and plantlets of celery tissue cultures, *New Phytol.*, 80, 517, 1978.
14. **Dunwell, J. M. and Sunderland, N.**, Anther culture of *Solanum tuberosum* L., *Euphytica*, 22, 317, 1973.
15. **Grun, P. and Chu, L.-J.**, Development of plants from protoplasts of Solanum (solanaceae), *Am. J. Bot.*, 65, 538, 1978.
16. **Jelaska, S.**, Embryoid formation by fragments of cotyledons and hypocotyls in *Cucurbita pepo*, *Plants*, 103, 278, 1972.
17. **Jelaska, S.**, Embryogenesis and organogenesis in pumpkin explants, *Physiol. Plant.*, 31, 257, 1974.
18. **Kato, H. and Takeuchi, M.**, Morphogenesis in vitro starting from single cells of carrot root, *Plant Cell Physiol.*, 4, 243, 1963.
19. **Halperin, W.**, Alternative morphogenetic events in cell suspensions, *Am. J. Bot.*, 43, 443, 1966.
20. **Ha, D. B.-D., Norrell, B., and Masset, A.**, Regeneration of *Asparagus officinalis* L. through callus cultures derived from protoplasts, *J. Exp. Bot.*, 26, 263, 1975.
21. **Doeschug, M. R. and Miller, C. O.**, Chemical control of adventitious organ formation in *Lactua sativa* explants, *Am. J. Bot.*, 54, 410, 1967.
22. **Koevary, K., Rappaport, L., and Morris, L. L.**, Tissue culture propagation of head lettuce, *HortScience*, 13, 39, 1978.
23. **Bajaj, Y. P. S. and Nietsch, P.**, In vitro propagation of red cabbage (*Brassica oleracea* L. var. Capitata), *J. Exp. Bot.*, 26, 883, 1975.
24. **Kadkade, P. and Seibert, M.**, Phytochrome-regulated organogenesis in lettuce tissue cultures, *Nature (London)*, 270, 49, 1977.
25. **Behki, R. M. and Lesley, S. M.**, In vitro plant regeneration from leaf explants of *Lycopersicon esculentum* (tomato), *Can. J. Bot.*, 54, 2409, 1976.
26. **Kartha, K. K., Gamborg, O. L., Shyluk, J. P., and Constabel, F.**, Morphogenetic investigations on in vitro leaf cultures of tomato (*Lycopersicon esculentum* Mill. cv. Starfire) and high frequency plant regeneration, *Z. Pflanzenphysiol.*, 77, 292, 1976.
27. **Kartha, K. K., Champous, S., Gamborg, O. L., and Phal, K.**, In vitro propagation of tomato by shoot apical meristem culture, *J. Am. Soc. Hortic. Sci.*, 102, 236, 1977.
28. **De Langhe, E. and De Bruijne, E.**, Continuous propagation of tomato plants by means of callus cultures, *Sci. Hortic.*, 4, 221, 1976.
29. **Friedborg, G.**, Growth and organogenesis in tissue cultures of *Allium cepa* var. Proliferum, *Physiol. Plant.*, 25, 436, 1971.

30. **Kehr, A. E. and Schaeffer, G. W.**, Tissue culture and differentiation of garlic, *HortScience*, 11, 422, 1976.
31. **Zee, S. Y., Fung, A., and Yue, S. B.**, Tissue culture and differentiation of Chinese chive, *HortScience*, 12, 264, 1977.
32. **Mascarenhas, A. F., Hendre, R. R., Nadgir, A. L., Durga, D., Barve, M., and Jagannathan, V.**, Differentiation in tissue culture of cabbage, *Indian J. Exp. Biol.*, 16, 122, 1978.
33. **Steward, F. C., Mapes, M. O., and Mears, K.**, Growth and organized development of cultured cells. II. Organization in cultures from freely suspended cells, *Am. J. Bot.*, 45, 705, 1958.
34. **Johnson, B. B. and Mitchell, E. D., Jr.**, In vitro propagation of broccoli from stem, leaf, and leaf rib explants, *HortScience*, 13, 246, 1978.
35. **Quazi, H. M.**, Regeneration of plants from anthers of broccoli (*Brassica oleracea* L.), *Ann. Bot. (London)*, 42, 473, 1978.
36. **Gunay, A. L. and Rao, P. S.**, In vitro plant regeneration from hypocotyl and cotyledon explants of red pepper *(Capsicum)*, *Plant Sci. Lett.*, 11, 365, 1978.
37. **Pow, J. J.**, Clonal propagation in vitro from cauliflower curd, *Hortic. Res.*, 9, 151, 1969.
38. **Walkey, D. G. A. and Woolfitt, J. M.**, Rapid clonal multiplication of cauliflower by shake culture, *J. Hortic. Sci.*, 45, 205, 1970.
39. **Clare, M. and Collin, H. A.**, The production of plantlets from tissue cultures of brussels sprout (*Brassica oleracea* L. var. Gemmifera D. C.), *Ann. Bot. (London)*, 38, 1067, 1974.
40. **Roest, S. and Bokelmann, G. S.**, Vegetative Propagation of *Solanum tuberosum* L., *Potato Res.*, 19, 173, 1976.
41. **Sopory, S.**, Differentiation in callus from cultured anthers of dihaploid clones of *Solanum tuberosum*, *Z. Pflanzenphysiol.*, 82, 88, 1977.
42. **Lustinec, J. and Horak, J.**, Induced regeneration of plants in tissue cultures of *Brassica oleracea*, *Experientia*, 26, 919, 1970.
43. **Horak, J., Landa, Z., and Lustinec, J.**, Production of polyploid plants from tissue cultures of *Brassica oleracea* L., *Phyton (Buenos Aires)*, 28, 7, 1971.
44. **Horak, J.**, Ploidy chimeras in plants regenerated from the tissue culture of *Brassica oleracea* L., *Biol. Plant.*, 14, 423, 1972.
45. **Horak, J., Lustinec, L., Mesicek, J., Kaminek, M., and Polackova, D.**, Regeneration of diploid and polyploid plants from the stem pith explants of diploid marrow stem kale (*Brassica oleracea* L.), *Ann. Bot. (London)*, 39, 571, 1975.
46. **Zee, S. Y. and Hui, L. H.**, In vitro plant regeneration from hypocotyl and cotyledons of Chinese kale (*Brassica alboglabra* Bailey), *Z. Pflanzenphysiol.*, 82, 440, 1977.
47. **Sehgal, C. B.**, Regeneration of plants from anther cultures of sweet potato (*Ipomoes batatas* Poir), *Z. Pflanzenphysiol.*, 88, 349, 1978.
48. **Wilmar, C. and Hellendoorn, M.**, Growth and morphogenesis of asparagus cells cultured in vitro, *Nature (London)*, 217, 369, 1968.
49. **Takatori, F. H., Murashige, T., and Stillman, J. I.**, Vegetative propagation of asparagus through tissue culture, *HortScience*, 3, 20, 1968.
50. **Maryakhina, I. Ya. and Butenko, R. G.**, Somatic reduction in a cabbage tissue culture, *Cytol. Genet.*, 8, 74, 1974.
51. **Malnassy, P. and Ellison, J. H.**, Asparagus tetraploids from callus tissue, *HortScience*, 5, 444, 1970.
52. **Smith, S. M. and Street, H. E.**, The decline of embryogenic potential as callus and suspension cultures of carrot (*Daucus carota* L.) are serially subcultured, *Ann. Bot. (London)*, 38, 223, 1974.
53. **Murashige, T., Shabde, M. N., Hasegawa, P. M., Takatori, F. H., and Jones, J. B.**, Propagation of asparagus through shoot apex culture. I. Nutrient medium for formation of plantlets, *J. Am. Soc. Hortic. Sci.*, 97, 158, 1972.
54. **Hasegawa, P. M., Murashige, T., and Takatori, F. H.**, Propagation of asparagus through shoot apex culture. II. Light and temperature requirements, transplantability of plants and cyto-histological characteristics, *J. Am. Soc. Hortic. Sci.*, 98, 143, 1973.
55. **Yang, H.-J. and Clore, W. J.**, Rapid vegetative propagation of asparagus through lateral bud culture, *HortScience*, 8, 141, 1973.
56. **Yang, H.-J. and Clore, W. J.**, Development of complete plantlets from moderately vigorous shoots of stock plants of asparagus in vitro, *HortScience*, 9, 138, 1974.
57. **Yang, H.-J. and Clore, W. J.**, Improving the survival of aseptically-cultured asparagus plants in transplanting, *HortScience*, 9, 235, 1974.
58. **Yang, H.-J. and Clore, W. J.**, In vitro reproductiveness of asparagus stem segments with branch-shoots at a node, *HortScience*, 10, 411, 1975.
59. **Litz, R. E. and Conover, R. A.**, In vitro propagation of sweet potato, *HortScience*, 13, 659, 1978.
60. **Mellor, F. C. and Stace-Smith, R.**, Development of excised potato buds in nutrient cultures, *Can. J. Bot.*, 47, 1617, 1969.
61. **Clare, M. V. and Collin, H. A.**, Meristem culture of brussels sprouts, *Hortic. Res.*, 13, 111, 1974.

62. **Crisp, P. and Walkey, D. G. A.**, The use of aseptic meristem culture in cauliflower breeding, *Euphytica*, 23, 305, 1974.
63. **Handley, L. W. and Chambliss, O. L.**, In vitro propagation of *Cucumis sativus* L., *HortScience*, 14, 22, 1979.
64. **Anderson, W. C. and Carstens, J. B.**, Tissue culture propagation of broccoli *Brassica oleracea* (Italica group) for use in F_1 hybrid seed production, *J. Am. Soc. Hortic. Sci.*, 102, 69, 1977.
65. **Miszke, W. and Skucinska, B.**, In vitro vegetative propagation of *Brassica oleracea* V. Capitata L. f. alba, *Z. Pflanzenphysiol.*, 76, 81, 1976.
66. **Kuo, C. G. and Tsay, J. S.**, Propagating Chinese cabbage by axillary bud culture, *HortScience*, 12, 456, 1977.
67. **Grout, B. W. W. and Aston, M. J.**, Transplanting of cauliflower plants regenerated from meristem culture. II. Carbon dioxide fixation and the development of photosynthetic ability, *Hortic. Res.*, 17, 65, 1978.
68. **Yang, H.-J.**, Tissue culture technique developed for asparagus propagation, *HortScience*, 12, 140, 1977.
69. **Wardle, K., Quinlan, A., and Simpkins, I.**, Abscisic acid and the regulation of water loss in plantlets of *Brassica oleracea* L. var. Botrytis regenerated through apical meristem culture, *Ann. Bot. (London)*, 43, 745, 1979.
70. **Gamborg, O. L., Murashige, T., Thorpe, T. A., and Vasil, I. K.**, Plant tissue culture media, *In Vitro*, 12, 473, 1976.
71. **Murashige, T.**, Nutrition of plant cells and organs in vitro, *In Vitro*, 9, 81, 1973.
72. **Murashige, T. and Skoog, F.**, A revised medium for rapid growth and bioassays with tobacco tissue cultures, *Physiol. Plant.*, 15, 473, 1962.
73. **Gamborg, O. L., Miller, R. A., and Ojima, K.**, Nutrient requirements of suspension cultures of soybean root cells, *Exp. Cell Res.*, 50, 148, 1968.
74. **Miller, C. O.**, Kinetin and kinetin-like compounds, in *Modern Methods of Plant Analysis*, Vol. 6, Linskens, H. F. and Tracey, M. V., Eds., Springer-Verlag, Berlin, 1963, 194.
75. **White, P. R.**, Nutrient deficiency studies and an improved inorganic nutrient for cultivation of excised tomato roots, *Growth*, 7, 53, 1943.
76. **Skoog, F. and Miller, C. O.**, Chemical regulation of growth and organ formation in plant tissues cultured in vitro, *Symp. Soc. Exp. Biol.*, 11, 118, 1957.
77. **Steward F. C. and Mapes, M. O.**, Morphogenesis and plant propagation in aseptic cultures of asparagus, *Bot. Gaz. (Chicago)*, 132, 70, 1971.
78. **Yang, H.-J.**, Effect of benomyl on *Asparagus officinalis* L. shoot and root development in culture media, *HortScience*, 11, 473, 1976.
79. **Yang, H.-J.**, Obtaining virus-free plants of *Asparagus officinalis* L. by culturing shoot tips and apical meristems, *HortScience*, 11, 474, 1976.
80. **Ha, D. B.-D. and Mackenzie, I. A.**, The division of protoplasts from *Asparagus officinalis* L. and their growth and differentiation, *Protoplasma*, 78, 215, 1973.
81. **Anderson, N. C. and Meagher, G. W.**, Cost of propagating broccoli plants through tissue culture, *HortScience*, 12, 543, 1977.
82. **Nitsch, J. P. and Nitsch, C.**, Haploid plants from pollen grains, *Science*, 163, 85, 1969.
83. **Debergh, P. and Standaert-De-Metsenaere, R.**, Modification of bulbils in *Allium porrum* L. cultured in vitro, *Sci. Hortic.*, 5, 11, 1976.
84. **Kameya, T. and Hinata, K.**, Induction of haploid plants from pollen grain of brassica, *Jpn. J. Breed.*, 20, 82, 1970.
85. **Reinert, J.**, Über die Kontrolle der Morphogenese und die Induktion von Adventive-Embryonea an Gewebekulturen aus Karotten, *Planta*, 4, 318, 1959.
86. **Halperin W. and Wetherell, D. F.**, Adventive embryony in tissue culture of the wild carrot, *Daucus carota*, *Am. J. Bot.*, 51, 274, 1964.
87. **Halperin, W.**, Morphogenetic studies with partially synchronized cultures of carrot embryos, *Science*, 146, 408, 1964.
88. **Haplerin, W. and Wetherell, D. F.**, Ontogeny of adventive embryos of wild carrot, *Science*, 147, 756, 1965.
89. **Lin, M.-L. and Staba, E. J.**, Peppermint and spearmint tissue culture. I. Callus formation and submerged culture, *Lloydia*, 24, 139, 1961.
90. **Steward, F. C., Mapes, M. O., Kent, A. E., and Holsten, R. D.**, Growth and development of cultured plant cells, *Science*, 143, 20, 1964.
91. **Steward, F. C., Israel, H. W., Mott, R. L., Wilson, H. J., and Krikorian, A. D.**, Observations on growth and morphogenesis in cultured cells of carrot (*Daucus carota* L.), *Philos. Trans. R. Soc. London Ser. B*, 273, 33, 1975.
92. **Kartha, K. K., Gamborg, O. L., Constabel, F., and Shyluk, J. P.**, Regeneration of cassava plants from apical meristems, *Plant Sci. Lett.*, 2, 107, 1974.

93. **Bajaj, Y. P. S.**, *In vitro* clonal propagation of cassava, *Curr. Sci.*, 47, 971, 1978.
94. **Linsmaier, E. M. and Skoog, F.**, Organic growth factor requirements of tobacco tissue cultures, *Physiol. Plant.*, 18, 100, 1965.
95. **Walkey, D. G. A., Cooper, V. C., and Crisp, P.**, The production of virus-free cauliflowers by tissue culture, *J. Hortic. Sci.*, 49, 273, 1974.
96. **Grout, B. W. W. and Aston, M. J.**, Modified leaf anatomy of cauliflower plantlets regenerated from meristem culture, *Ann. Bot. (London)*, 42, 993, 1978.
97. **Grout, B. W. W.**, Wax development of leaf surfaces of *Brassica oleracea* var. Currawong cauliflower regenerated from meristem culture, *Plant Sci. Lett.*, 5, 401, 1975.
98. **Grout, B. W. W. and Aston, M. J.**, Transplanting of cauliflower plants regenerated from meristem culture. I. Water loss and water transfer related to changes in leaf wax and to xylem regeneration, *Hortic. Res.*, 17, 1, 1977.
99. **Kamat, M. G. and Rao, P. S.**, Vegetative multiplication of eggplants *(Solanum melongena)* using tissue culture techniques, *Plant Sci. Lett.*, 13, 57, 1978.
100. **Crocomo, O. J., Sharp, W. R., and Peters, J. E.**, Plantlet morphogenesis and the control of callus growth and root induction of *Phaseolus vulgaris* with the addition of a bean seed extract, *Z. Pflanzenphysiol.*, 78, 456, 1976.
101. **Gamborg, O. L., Constabel, F., and Shyluk, J. P.**, Organogenesis in callus from shoot apices of *Pisum sativum, Physiol. Plant.*, 30, 125, 1974.
102. **Kartha, K. K., Gamborg, O. L., and Constabel, F.**, Regeneration of pea (*Pisum sativum* L.) plants from shoot apical meristems, *Z. Pflanzenphysiol.*, 72, 172, 1974.
103. **Kartha, K. K., Leung, N. L., and Gamborg, O. L.**, Freeze-preservation of pea meristems in liquid nitrogen and subsequent plant regeneration, *Plant Sci. Lett.*, 15, 7, 1979.
104. **Kadkade, P. G. and O'Connor, H. J.**, Interactive effects of growth regulators on organogenesis in lettuce tissue culture, *Plant Physiol.*, 57, 75, 1976.
105. **George, L. and Narayanaswamy, S.**, Haploid *Capsicum* through experimental androgenesis, *Protoplasma*, 78, 467, 1973.
106. **Skirvin, R. M., Lam, S. L., and Janick, J.**, Plantlet formation from potato callus in vitro, *HortScience*, 10, 413, 1975.
107. **Foroughi-Wehr, B., Wilson, H. M., Mix, G., and Gaul, H.**, Monohaploid plants from anthers of a dihaploid genotype of *Solanum tuberosum* L., *Euphytica*, 26, 361, 1977.
108. **Sopory, S. K., Jacobsen, E., and Wenzel, G.**, Production of monohaploid embryoids and plantlets in cultured anthers of *Solanum tuberosum, Plant Sci. Lett.*, 12, 47, 1978.
109. **Shepard, J. F. and Totten, R. E.**, Mesophyll cell protoplasts of potato; isolation, proliferation, and plant regeneration, *Plant Physiol.*, 60, 313, 1977.
110. **Binding, H., Nehls, R., Schieder, O., Sopory, S. K., and Wenzel, G.**, Regeneration of mesophyll protoplasts isolated from dihaploid clones of *Solanum tuberosum, Physiol. Plant.*, 43, 52, 1978.
111. **Upadhya, M. D.**, Isolation and culture of mesophyll protoplasts of potato (*Solanum tuberosum* L.), *Potato Res.*, 18, 438, 1975.
112. **Gresshoff, P. M. and Doy, C. H.**, Development and differentiation of haploid *Lycopersicon esculentum* (tomato), *Planta*, 107, 161, 1972.
113. **Gamborg, O. L. and Eveleigh, D.**, Culture methods and detection of glucanases in suspension cultures of wheat and barley, *Can. J. Biochem.*, 46, 417, 1968.
114. **Blaydes, D. F.**, Interaction of kinetin and various inhibitors in the growth of soybean tissue, *Physiol. Plant.*, 19, 748, 1966.
115. **Ramulu, K. S., Devreau, M., Ancora, G., and Laneri, U.**, Chimerism in *Lycopersicum peruvianum* plants regenerated from in vitro cultures of anthers and stem internodes, *Z. Pflanzenphysiol.*, 76, 299, 1976.
116. **Mantell, S. H., Haque, S. Q., and Whitehall, A. P.**, Clonal multiplication of *Dioscorca alata* L. and *Dioscorea rotundata* Poir. yams by tissue culture, *J. Hortic. Sci.*, 53, 95, 1978.
117. **Boxus, P.**, The production of fruit and vegetable plants by in vitro culture, in Propagation of Higher Plants Through Tissue Cultures: A Bridge Between Research and Application, Hughes, K. W., Henke, R., and Constantin, M. J., Eds., National Technical Information Service, U.S. Department of Energy, Springfield, Va., 1978, 44.
118. **Kahn, R. P.**, International exchange of genetic stocks, in Propagation of Higher Plants Through Tissue Culture: A Bridge Between Research and Application, Hughes, K. W., Henke, K. W., Henke, R., and Constantin, M. J., Eds., National Technical Information Service, U.S. Department of Energy, Springfield, Va., 1978, 233.
119. **Al-Abta, S. and Collin, H. A.**, Control of embryoid development in tissue cultures of celery, *Ann. Bot. (London)*, 42, 180, 1978.
120. **Veliky, I. A. and Martin, S. M.**, A fermenter for plant cell suspension cultures, *Can. J. Microbiol.*, 16, 223, 1970.

Chapter 5

AGRONOMIC CROPS

B. V. Conger

TABLE OF CONTENTS

I. INTRODUCTION

Agronomic crops supply the greater proportion of man's food either for direct or indirect consumption through livestock products. The most important agronomic crops include the cereal grains, seed legumes, and forage crops. Cereal grains, which are grown for their edible starchy seeds, are by far the most important source of concentrated carbohydrates for man and beast.[1] They are the main items in the diet of much of the world's population and about 70% of the harvested acreage in the world is devoted to growing them.[2] Recent statistics show that only three species, wheat, rice, and corn, each account for essentially one quarter of the total cereal supply.[3] Seed legumes are higher in protein than the cereals and have evolved symbiotic associations with bacteria for the fixation of gaseous nitrogen. Forage crops, which include both annual and perennial grasses and legumes, contribute 25 to 30% of the typical American's food supply through livestock products and more acreage is devoted to growing them than all other crops combined.[4]

Recent progress in plant cell and tissue culture has created much excitement and speculation concerning the use of this technology in the improvement of agriculturally important plants including agronomic crops. Potential in vitro manipulations that are of importance include:

1. Induction of haploid plantlets from anther and pollen culture
2. Protoplast culture — DNA uptake, parasexual hybridization, and genetic engineering
3. Induction and selection of mutants
4. Cloning — rapid and mass propagation of specific genotypes

Although the first three may be of most interest and potentially have the most significant application in agronomic crops, the latter, i.e., in vitro propagataion is the primary subject of this book and will be emphasized in this chapter. However, numerous references are also given for haploid and protoplast culture of certain agronomic species.

This chapter will present a classification of agronomic crops according to use, the current status of in vitro propagation, technology for culture of somatic tissue in a general sense and for specific crops, problems of using in vitro techniques for cloning, and potential applications and future prospects of this technology for improving agronomic crops.

II. CLASSIFICATION OF AGRONOMIC CROPS

Agronomic crops may be classified in several ways, e.g., on a taxonomic basis or on the basis of a morphological similarity of plant parts. Classification may also be made on a commodity basis, i.e., on the basis of use. This is a convenient and practical method of classification and will be used in this chapter even though it is recognized that some crops have more than one use. For example, flax may be classified as both a fiber and oil seed crop; corn and soybeans are used for oil as well as for whole seed consumption and many crops can be and are used for forage. The classification scheme used here comes primarily from Martin et al.[3]

1. Cereal or grain crops — The true cereals are all members of the Gramineae (grass) family, but seeds from plants of other families, e.g., buckwheat, are also sometimes included since their seeds are used similarly. The product of the true cereals is the grain or caryopsis, a fruit in which the ovary wall turns hard and fuses with the single seed. True cereals include rice (*Oryza*), wheat (*Triticum*), corn (*Zea*), barley (*Hordeum*), oats (*Avena*), rye (*Secale*), sorghum (*Sorghum*), and various millets (*Panicum, Pennisetum, Setaria, Paspalum, Eleusine,* etc.).
2. Seed legumes — The chief legumes grown for their seed are soybean (*Glycine*), peanut (*Arachis*), field bean (*Phaseolus*), field pea (*Pisum*), chickpea (*Cicer*), broad bean (*Vicia*), lentil (*Lens*), and pigeon pea (*Cajanus*).
3. Forage crops — These crops are grown for their vegetable matter, fresh or preserved, utilized as feed for animals. They primarily include both annual and perennial grasses and legumes. Both grass and legume forage species commonly have small seeds as compared to the cereals and seed legumes. There are several genera of both grasses and legumes, and in some cases many species within each genera, which could be listed as economically important. Some of those commonly grown are listed below. Forage grasses and legumes will be treated separately for the remainder of this chapter.
   a. Forage grasses — Orchardgrass (*Dactylis*), fescue (*Festuca*) bluegrass (*Poa*), bromegrass (*Bromus*), timothy (*Phleum*), ryegrass (*Lolium*), canarygrass (*Phalaris*), wheatgrass(*Agropyron*), bluestem (*Andropogon*), indiangrass (*Sorghastrum*), switchgrass (*Panicum*), bahiagrass (*Paspalum*), bermudagrass (*Cynodon*), and forage sorghums including sudangrass (*Sorghum*).
   b. Forage legumes — Alfalfa (*Medicago*), sweetclover (*Melilotus*), true clovers (*Trifolium*), birdsfoot trefoil (*Lotus*), vetch (*Vicia*), lespedeza (*Lespedeza*), and lupines, (*Lupinus*).
4. Sugar crops — Crops grown for their juice from which sucrose is extracted and cystallized include sugar beet (*Beta*) and sugarcane (*Saccharum*).

5. Oil seed crops — These crops are grown for their oily seeds and include sunflower (*Helianthus*), rape (*Brassica*), flax (*Linum*), crambe (*Crambe*), safflower (*Carthamus*), sesame (*Sesame*), and castor bean (*Ricinus*). Other crops grown for oil which are already listed above include corn and soybean.
6. Fiber crops — Of the plants grown for their natural fiber, cotton (*Gossypium*) is probably the most important. Other fiber crops include hemp (*Cannabis*), jute (*Corchorus*), and flax (*Linum*). The latter genus is also listed above under oil seed crops.
7. Drug and miscellaneous crops — These include tobacco (*Nicotiana*), mint (*Mentha*), buckwheat (*Fagopyrum*), and numerous other species which are only of minor importance for food, feed, or fiber.

Root and tuber crops such as potato, sweet potato, and cassava are covered in the chapter on Vegetable Crops.

III. CURRENT STATUS OF IN VITRO PROPAGATION

Of the various in vitro manipulations available to plant scientists and growers of technology and actual application is most advanced in the area of clonal propagation. This is especially true for the classes of agricultural plants covered in other chapters of this book. Murashige[5,6] lists several agronomic crop species whose propagation through cell and tissue culture might be possible. However, it may be stated at this point that in vitro techniques for cloning or mass propagation are currently being used very little, if at all, in agronomic crops.

Reasons for this include:

1. Most important food and feed crops are normally propagated by seed and not vegetatively.
2. Usually very large acreages are grown. This is especially true of the cereals, seed legumes, and forage crops. Presently, it would seem difficult and probably unfeasible to vegetatively propagate an entire wheat or soybean field or a tall fescue pasture consisting of hundreds of hectares.
3. Many agronomic species such as most of the small grains and grain legumes are naturally self-fertilizing; thus, inbred lines are easily developed and maintained by seed. In a cross pollinating species such as maize both self and cross pollinations can be easily made and controlled, again making inbred lines easy to develop and maintain by seed. Thus, in species such as these, cloning by in vitro methods, or otherwise, has not been considered necessary to maintain selected genotypes.
4. Technology for in vitro culture of the major food and feed crop species is not as advanced as that for species in other classes of agricultural plants and many agronomic species do not appear to be as amenable to in vitro culture. A notable exception is tobacco which is not an important food and feed crop. However, even in tobacco, which has probably been the most extensively cultured plant species, in vitro techniques are not commonly used for mass propagation.
5. There are several problems involved with culture of agronomic species, including low regeneration frequency and genetic instability, which hinder, or even preclude, the use of in vitro techniques for cloning. Some of these problems have been previously reviewed[7] and will be dealt with in more detail later in this chapter.

Nevertheless, the potential does exist for in vitro propagation of a certain number of agronomic crops and there are certain crop species in which in vitro propagation might be advantageous over current methods for cloning desirable genotypes.

According to Murashige,[8] multiplication in tissue culture can occur through (1) enhanced formation of axillary shoots followed by rooting of individual shoots; (2) production of adventitious shoots also followed by rooting of individual shoots; and (3) somatic cell embryogenesis. Production of plants by the axillary shooting method has an advantage in that aberrant plants are produced at a frequency no higher than the spontaneous rate. Even though it is slower than other tissue culture methods, it enables multiplication rates that are a million times faster than traditional methods.[8]

Adventitious shoots or asexual embryos can arise directly from the explant or can originate in callus.[8] Both methods offer potentially much higher multiplication rates than axillary shooting. However, there are problems maintaining genetic stability in callus cultures (to be discussed in more detail later) and problems of handling somatic embryos. Also, plantlet formation from callus has not yet been reported for many agronomic crops including some of those classified above. Cotton and soybean are important species in this category. However, in soybean, adventitious bud development from hypocotyl sections has been reported.[9]

Agronomic crop species in which plantlet formation from callus, meristem tip culture, and adventitious budding has been reported and appropriate references are listed in Table 1. Meristem tip culture may have more application in obtaining disease-free plants than for cloning, however, since it represents plantlet formation from somatic tissue some references are included here. Reports of successful haploid culture and plantlet regeneration from protoplasts and cell suspension cultures are also given for information. Although the list of references is extensive it is not intended to be exhaustive. For many of the species only one or a few reports of successful culture have been published and it is hoped that most of these are included. On the other hand, it would be difficult and perhaps superfluous to list all the references for plantlet regeneration in tobacco.

IV. TECHNOLOGY AND PROCEDURE

Current technology and procedures will be discussed below for specific classes of crops and crop species. This discussion will be limited to information available for induction of callus from somatic tissue, the formation of plantlets from callus, or directly from other somatic tissue, e.g., meristem culture or adventitious budding and the establishment of actively growing plants from these cultures. Since anther and protoplast culture have less relevance for in vitro propagation or cloning they will not be discussed further. For additional information on these subjects the reader is invited to consult specific references listed in Table 1 or the following reviews.[191-200]

A. Sterilization of Plant Material

General procedures concerning sterile technique have been discussed in other chapters of this book and elsewhere,[201-205] and therefore, will not be dealt with in depth here. Published procedures for surface disinfection of explants in specific crops will be given later. Since plants and plant parts are contaminated with a wide range of fungi and bacteria it is necessary to disinfect tissues with a minimum amount of cellular damage to the host tissue. Fresh plant tissue is usually less contaminated, especially with spores, than dormant tissue such as seeds. The latter, however, will withstand more rigorous disinfection treatments without damage. Surface sterilization of explants is usually done with ethanol, dilute sodium or calcium hypochlorite, detergent,

Table 1
REPORTS OF PLANTLET FORMATION FROM IN VITRO CULTURE OF AGRONOMIC PLANTS

Class, Genus-species	Common name	Somatic tissue derived callus, meristem tips, or adventitious budding[a]	Anther, microspore, or haploid culture	Protoplast or cell suspension culture
Cereals				
Avena sativa L.	Oat	10—12		
Hordeum vulgare L.	Barley	13—17	18—23	
Oryza sativa L.	Rice	24—35	36—49	
Secale cereale L.	Rye		45—49	
Sorghum bicolor L. Moench	Sorghum	34, 55—58		
Triticum aestivum L.	Bread wheat	11, 34, 59—65, 258	66—72	
T. durum Desf.	Durum wheat	73		
Zea mays L.	Corn or maize	11, 34, 74—78	79—80	
Panicum miliaceum L.	Common millet	81		
Paspalum scrobiculatum L.	Koda millet	82		
Eleusine coracane Gaertn.	Finger millet	82		
Pennisetum typhoideum Pers.	Bulrush millet	82		
Setaria italica (L.) Beauv.	Foxtail millet		83	
Pennisetum americanum (L.) K. Schum.	Pearl millet			
× *Triticosecale* Wittmack	Triticale		85—88	
Aegilops	Aegilops		84	
Triticum — Agropyron			89	
Hordeum — Secale			90	
Phragmites communis L.		91		
Legume seeds				
Phaseolus vulgaris L.	Bean	92		
P. aureus Roxb.	Green gram	93		
P. mungo L.	Black gram	93		
Pisum sativum L.	Pea	93—97		
Cajanus cajan (L.) Mills. P.	Pigeon pea	98		
Cicer arietinum L.	Chickpea	93		
Glycine max (L.) Merrill	Soybean	9		

Vicia faba L.	Broad bean	99		
Lens esculentum Moench.	Lentil	93		
Forage grasses				
Andropogon gerardii Vitman	Big bluestem	100		
Sorghastrum nutans L.	Indiangrass	101		
Dactylis glomerata L.	Orchardgrass or Cocksfoot	102, 103		
Festuca arundinacea Schreb.	Tall fescue	103, 104	105	
Lolium multiflorum Lam. × *Festuca arundinacea* Schreb.	Ryegrass-tall, fescue hybrids	106	107	
Lolium multiflorum Lam.	Italian ryegrass	103, 108, 109	110	
Lolium multiflorum Lam. × *L. perenne* L.	Ryegrass hybrids	111, 112		
Bromus inermis Leyss.	Bromegrass			113, 114
Phleum pratense L.	Timothy	103		
Forage legumes				
Medicago sativa L.	Alfalfa	115—119		120
Trifolium repens L.	White clover	121—123		
T. pratense L.	Red clover	124—126		
T. incarnatum L.	Crimson clover	126		
T. alexandrinum L.	Berseem clover	127	127	
Lotus corniculatus L.	Birdsfoot trefoil	128, 129		
Stylosanthes hamata L.	Caribbean stylo	130		
Sugar crops				
Beta vulgaris L.	Sugar beet	131—134	135	
Saccharum officinarium L.	Sugarcane	136—144		145
Oil seeds				
Brassica napus L.	Rape	146	50, 54, 147—151	152
B. campestris L.	Rape		153, 154	
Crambe maritima L.	Crambe	155		
Helianthus annuus L.	Sunflower	156		
Linum usitatissimum L.	Flax	157—160	159	
Fiber crops				
Linum usitatissium L.	Flax	(See oil seeds)		
Drug and miscellaneous crops				
Nicotiana spp.	Tobacco	161—167	168—178	176, 179—190
Fagopyrum esculentum Moench.	Buckwheat	259		

[a] Also includes other types of plantlet formation directly on explants, e.g., axillary shooting.

or a combination of these. An important factor is the ease with which these agents can be removed from the tissue since the retention of such noxious chemicals may seriously affect callus growth.[202] After sterilization, the explant must be handled and all further operations conducted under completely aseptic conditions. The development of sterile laminar air-flow cabinets has contributed greatly to the ease of conducting plant cell and tissue culture experiments and these cabinets are now in routine use for this work.

B. Media Composition and Preparation

The basis for all plant tissue culture media is a mixture of mineral salts combining the essential macro- and micronutrient elements together with a carbon source, usually sucrose.[202] Various organic constituents, mainly vitamins and hormones, are also usually added to the medium. Numerous plant tissue culture media and modifications have been described in the literature. Fossard[205] and Gamborg et al.[206] have listed the components of some of the most commonly used media. Many of the media described were developed for tobacco culture. However, they are also commonly used, sometimes with various modifications, for other plant species including all major agronomic crops. The inorganic nutrients (salt composition) of various commonly used media and the original references are presented in Table 2. The organic constituents (vitamins, auxins, cytokinins, etc.) of the same media are presented in Table 3. Modifications of these for specific crops will be mentioned below where applicable and pertinent.

One of the most widely used defined media developed early is that by White[207] and is included here because of its historical significance. This medium has been superseded, however, by several more recently developed media. Perhaps the most widely used and cited medium is that published by Murashige and Skoog.[208] This medium, developed for tobacco culture and referred to as MS medium, has been used for a wide range of plant genera and species including many agronomic crops. The inorganic constituents of the LS medium developed by Linsmaier and Skoog[209] are identical to those of the MS medium, but the LS medium contains four times as much thiamine and omits pyroxidine, nicotinic acid, and glycine. The LS medium is purported by the authors to give high yields and excellent vigor of callus and organ development and is superior to the earlier MS medium for plant regeneration in vitro.[209]

Eriksson's[210] medium is similar to the MS and LS media, but contains twice the amount of phosphate and only one tenth the concentration of most of the micronutrients. The B5 medium of Gamborg et al.[211] and Blaydes[212] medium were developed for soybean culture. Blaydes medium contains lower amounts of all ions than MS with the exception of phosphorus, which is lower in MS, and iodine which is about the same in the two media. The Schenk and Hildebrandt (SH)[213] medium which was developed for culture of a wide range of plant genera, including both monocotyledonous and dicotyledonous species, is similar in composition to the B5 medium.

The PC-L2 medium developed by Phillips and Collins[124] and the P2-mod. medium by Potrykus et al.[214] are relatively new and have not yet received widespread use. In fact, they both may require further development and modification before optimum results are obtained. They are included here since they represent attempts to develop specific media for the two most important classes of agronomic crops, i.e., the legumes (PC-L2) and cereals (P2-mod.). The PC-L2 medium was optimized for culture of red clover, but is also broadly supportive of other legume species, including alfalfa, soybeans, and jack bean (*Canavalia ensiformus* L.).[124] This medium contains no nicotinic acid, that was found inhibitory to legume cell and callus cultures, and utilizes 4-amino-3,5,6-trichloropicolinic acid (picloram) as the auxin and 6-benzylaminopurine (BAP) as the cytokinin. The P2-mod. cereal culture medium[214] was developed for protoplast

Table 2
INORGANIC NUTRIENTS (SALT COMPOSITION) OF VARIOUS MEDIA USED FOR CELL AND TISSUE CULTURE OF AGRONOMIC CROPS

	White[207]		Murashige and Skoog (MS)[208]		Linsmaier and Skoog (LS)[209]	
Macronutrient element	mg/ℓ	m*M*	mg/ℓ	m*M*	mg/ℓ	m*M*
NH_4NO_3	—	—	1650	20.6	1650	20.6
KNO_3	80	0.79	1900	18.8	1900	18.8
$Ca(NO_3)_2 \cdot 4H_2O$	300	1.27	—	—	—	—
$CaCl_2 \cdot 2H_2O$	—	—	440	3.00	440	3.00
KCl	65	0.87	—	—	—	—
$MgSO_4 \cdot 7H_2O$	720	2.92	370	1.50	370	1.50
Na_2SO_4	200	1.41	—	—	—	—
KH_2PO_4	—	—	170	1.25	170	1.25
$NaH_2PO_4 \cdot H_2O$	16.5	0.12	—	—	—	—
Micronutrient elements	mg/ℓ	μ*M*	mg/ℓ	μ*M*	mg/ℓ	μ*M*
KI	0.75	4.52	0.83	5.00	0.83	5.00
H_3BO_3	1.5	24.3	6.2	100	6.2	100
$MnSO_4 \cdot 4H_2O$	7	31.4	22.3	100	22.3	100
$ZnSO_4 \cdot 7H_2O$	3	10.4	8.6	29.9	8.6	29.9
$Na_2MoO_4 \cdot 2H_2O$	—	—	0.25	1.0	0.25	1.0
$CuSO_4 \cdot 5H_2O$	0.001	0.004	0.025	0.1	0.025	0.1
$CoCl_2 \cdot 6H_2O$	—	—	0.025	0.1	0.025	0.1
$Na_2 \cdot EDTA$	—	—	37.23	100	37.23	100
$FeSO_4 \cdot 7H_2O$	—	—	27.95	100	27.95	100
$Fe_2(SO_4)_3$	2.5	6.25	—	—	—	—
MoO_3	0.0001	0.0007	—	—	—	—
pH	5.5		5.7		5.6	

	Schenk and Hildebrandt (SH)[213]		Eriksson (ER)[210]		Gamborg et al. (B5)[211]	
Macronutrient elements	mg/ℓ	m*M*	mg/ℓ	m*M*	mg/ℓ	m*M*
NH_4NO_3	—	—	1200	15.0	—	—
KNO_3	2500	24.7	1900	18.8	2500	24.7
$CaCl_2 \cdot 2H_2O$	200	1.36	440	3.00	150	1.02
$MgSO_4 \cdot 7H_2 \cdot 7H_2O$	400	1.62	370	1.50	250	1.01
$(NH_4)_2SO_4$	—	—	—	—	134	1.01
KH_2PO_4	—	—	340	2.50	—	—
$NaH_2PO_4 \cdot H_2O$	—	—	—	—	150	1.09
$NH_4H_2PO_4$	300	2.61	—	—	—	—
Micronutrient elements	mg/ℓ	μ*M*	mg/ℓ	μ*M*	mg/ℓ	μ*M*
KI	1.0	6.02	—	—	0.75	4.52
H_3BO_3	5.0	80.7	0.63	10.2	3.0	48.5
$MnSO_4 \cdot 4H_2O$	—	—	2.23	10.0	—	—
$MnSO_4 \cdot H_2O$	10.0	59.2	—	—	10.0	59.2
$ZnSO_4 \cdot 7H_2O$	1.0	3.48	—	—	2.0	6.96
$Zn\text{-}Na_2EDTA$	—	—	15	37.0	—	—
$Na_2MoO_4 \cdot 2H_2O$	0.1	0.41	0.025	0.10	0.25	1.03
$CuSO_4 \cdot 5H_2O$	0.2	0.80	0.0025	0.01	0.025	0.10

Table 2 (continued)
INORGANIC NUTRIENTS (SALT COMPOSITION) OF VARIOUS MEDIA USED FOR CELL AND TISSUE CULTURE OF AGRONOMIC CROPS

	Schenk and Hildebrandt (SH)[213]		Eriksson (ER)[210]		Gamborg et al. (B5)[211]	
Macronutrient elements	mg/ℓ	m*M*	mg/ℓ	m*M*	mg/ℓ	m*M*
$CoCl_2 \cdot 6H_2O$	0.1	0.42	0.0025	0.01	0.025	0.11
$Na_2 \cdot EDTA$	20.0	53.7	37.3	100	37.3	100
$FeSO_4 \cdot 7H_2O$	15.0	53.9	27.85	100	27.8	100
pH	5.8		5.8		5.5	

	Blaydes[212]		Phillips and Collins (PC-L2)[124]		Potrykus et al. (P2—mod.)[214]	
Macronutrient elements	mg/ℓ	m*M*	mg/ℓ	m*M*	mg/ℓ	m*M*
NH_4NO_3	1000	12.5	1000	12.5	—	—
KNO_3	1000	9.89	2100	20.8	505	5.00
$Ca(NO_3)_2 \cdot 4H_2O$	—	—	—	—	2360	10.0
$Ca(NO_3)_2$	347	2.11	—	—	—	—
$CaCl_2 \cdot 2H_2O$	—	—	600	4.08	6235	42.4
KCl	65	0.87	—	—	—	—
$MgSO_4$	35	0.29	—	—	—	—
$MgSO_4 \cdot 7H_2O$	—	—	435	1.75	246	1.00
KH_2PO_4	300	2.20	325	2.39	—	—
$NaH_2PO_4 \cdot H_2O$	—	—	85	0.62	—	—
Na_2HPO_4	—	—	—	—	21	0.15
Micronutrient elements	mg/ℓ	μ*M*	mg/ℓ	μ*M*	mg/ℓ	μ*M*
KI	0.8	4.82	1.0	6.02	—	—
H_3BO_3	1.6	25.9	5.0	80.9	10.0	161.7
$MnSO_4 \cdot 4H_2O$	—	—	—	—	25.0	112.1
$MnSO_4 \cdot H_2O$	—	—	15.0	88.8	—	—
$MnSO_4$	4.4	29.1	—	—	—	—
$ZnSO_4 \cdot 7H_2O$	—	—	5.0	17.4	—	—
$ZnSO_4 \cdot 4H_2O$	—	—	—	—	10.0	42.8
$ZnSO_4$	1.5	9.29	—	—	—	—
$Na_2MoO_4 \cdot 2H_2O$	—	—	0.4	1.65	0.25	1.03
$CuSO_4 \cdot 5H_2O$	—	—	0.1	0.40	0.025	0.10
$CoCl_2 \cdot 6H_2O$	—	—	0.1	0.42	—	—
$Na_2 \cdot EDTA$	—	—	—	—	37.3	100
Na-Fe EDTA	32.0	87.2	—	—	—	—
$FeSO_4 \cdot 7H_2O$	—	—	25.0	89.9	27.8	100
pH	6.0		5.8		5.8	

culture of corn, and therefore, does not contain agar. Agar at the rate of 6 to 8 g/ℓ could, however, be added for a solid medium. This medium contains a high amount of calcium chloride and the primary nitrogen source is calcium nitrate. The medium also contains a number of organic compounds not found in the others. It is not known if the composition of this medium would provide optimal gowth of a wide range of

Table 3

ORGANIC CONSTITUENTS OF VARIOUS NUTRIENT MEDIA USED FOR CELL AND TISSUE CULTURE OF AGRONOMIC CROPS

	White[207]		Murashige and Skoog (MS)[208]		Linsmaier and Skoog (LS)[209]	
Compound	mg/ℓ	μM	mg/ℓ	μM	mg/ℓ	μM
Inositol	—	—	100	555.1	100	555.1
Nicotinic acid	0.5	4.06	0.5	4.06	—	—
Pyroxidine · HCl	0.1	0.49	0.5	2.43	—	—
Thiamine″HCl	0.1	0.30	0.1	0.30	0.4	1.18
Glycine	3.0	40.0	2.0	26.6	—	—
IAA[a]	—	—	1—30	5.71—171	1—30	5.71—171
Kinetin	—	—	0.04—10	0.18—46.5	0.001—10	0.005—46.5
Sucrose[b]	20 g	—	30 g	—	30 g	—
Agar[b]	5 g	—	10 g	—	10 g	—

	Schenk and Hildebrandt (SH)[213]		Eriksson (ER)[210]		Gamborg et al. (B5)[211]	
Compound	mg/ℓ	μM	mg/ℓ	μM	mg/ℓ	μM
Inositol	1000	5506	—	—	100	555.1
Nicotinic acid	5.0	40.6	0.5	4.06	1.0	8.12
Pyroxidine · HCl	0.5	2.4	0.5	2.43	1.0	4.86
Thiamine″HCl	5.0	14.8	0.5	1.48	10.0	29.6
Glycine	—	—	2.0	26.6	—	—
NAA[c]	—	—	1.0	5.37	—	—
2,4-D[d]	0.5	2.3	—	—	0.1—1.0	0.45—4.52
p-CPA[e]	2.0	10.7	—	—	—	—
Kinetin	0.1	0.5	0.02	0.09	0.1	0.46
Sucrose[b]	30 g	—	40 g	—	20 g	—
Agar[b]	6 g	—	7 g	—	6—8 g	—

	Blaydes[212]		Phillips and Collins (PC-L2)[124]		Potrykus et al. (P2—mod.)[214]	
Compound	mg/ℓ	μM	mg/ℓ	μM	mg/ℓ	μM
Inositol	—	—	250	1388	200	1110
Nicotinic acid	0.5	2.43	—	—	5.0	40.6
Pyroxidine · HCl	0.1	0.49	0.5	2.43	0.5	2.43
Thiamine · HCl	0.1	0.30	2.0	5.92	0.5	1.48
Glycine	2.0	26.6	—	—	2.0	26.6
Glutamine	—	—	—	—	300	2053
Serine	—	—	—	—	100	951.6
2,4-D[d]	—	—	—	—	5.0	26.6
IBA[f]	—	—	—	—	2.5	12.3
Picloram[g]	—	—	0.06	0.25	—	—
IAA[a]	5.0	28.5	—	—	—	—
Kinetin	0.5	2.32	—	—	—	—
BAP[h]	—	—	0.10	0.44	—	—
Folic acid	—	—	—	—	0.5	1.13
Biotin	—	—	—	—	0.05	0.20
Yeast extract[i]	—	—	—	—	25	—
Soluble starch[j]	—	—	—	—	50	—
Sucrose[b]	30 g	—	25 g	—	20 g	—

Table 3 (continued)
ORGANIC CONSTITUENTS OF VARIOUS NUTRIENT MEDIA USED FOR CELL AND TISSUE CULTURE OF AGRONOMIC CROPS

	Blaydes[212]		Phillips and Collins (PC-L2)[124]		Potrykus et al. (P2—mod.)[214]	
Compound	mg/ℓ	μM	mg/ℓ	μM	mg/ℓ	μM
Mannitol[b]	—	—	—	—	36.43 g	—
Sorbitol[b]	—	—	—	—	36.43 g	—
Agar[b]	10 g	—	8 g	—	—	—

[a] Indole-3-acetic acid.
[b] g/ℓ.
[c] α-Naphthaleneacetic acid.
[d] 2,4-Dichlorophenoxyacetic acid.
[e] p-Chlorophenoxyacetic acid.
[f] Indole-3-butyric acid.
[g] 4-Amino-3,5,6-trichloropicolinic acid.
[h] 6-Benzylaminopurine.
[i] Merck 3753.
[j] Merck 1252.

cereal and other grass species. In fact, it was recently reported by the same authors that this medium was not effective for inducing protoplast divisions in corn and a new medium (NN67) was developed for such purpose.[215]

The preparation of media has been described in detail by Gamborg[204] and Fossard.[205] Basically, the chemicals are dissolved in glass-distilled water. The micronutrient elements and organic constituents are usually made up as stock solutions and then added to the medium. There are several ways to prepare stock solutions. Gamborg[204] lists the ingredients for stock solutions for the B5, MS, and Eriksson's media and the amounts that are to be added to make up the final medium. As an example, the following procedure for preparation of the B5 medium is from Gamborg.[204]

Microelements	mg/100 mℓ	Vitamins	mg/100 mℓ
$MnSO_4 \cdot H_2O$	1,000	Nicotinic acid	100
H_3BO_3	300	Thiamine·HCl	1,000
$ZnSO_4 \cdot 7H_2O$	200	Pyroxidine·HCl	100
$Na_2MoO_4 \cdot 2H_2O$	25	*Myo*-inositol	10,000
$CuSO_4 \cdot 5H_2O$	2.5		
$CoCl_2 \cdot 6H_2O$	2.5		

After preparation, inorganic stock solutions should be stored at 2 to 4°C (refrigerator) and the organic stock solutions at −20°C (freezer).

Calcium chloride ($CaCl_2 \cdot 2H_2O$) solution is prepared by adding 15 g/100 mℓ and potassium iodide (KI) solution is prepared with 75 mg/100 mℓ. The KI solution should be stored in an amber bottle in the refrigerator.

Auxins such as 2,4-dichlorophenoxyacetic acid (2,4-D) and α-naphthalene-acetic acid (NAA) are dissolved in ethanol with slight heating and gradually diluted to 100 mℓ with water. The initial amount to be added depends on the auxin concentration desired. Kinetin is dissolved in a small amount of 0.5 *N* HCl with slight heating and then diluted to 100 mℓ with distilled water. Auxin and cytokinin stock solutions should be stored in the refrigerator after preparation.

The amounts per liter contained in the final medium are as follows: $NaH_2PO_4 \cdot H_2O$, 150 mg; KNO_3, 2500 mg; $(NH_4)_2SO_4$, 134 mg; $MgSO_4 \cdot 7H_2O$, 250 mg; ferric EDTA, 40 mg; sucrose, 20 g; $CaCl_2 \cdot 2H_2O$ stock solution, 1.0 mℓ; micronutrient elements stock solution, 1.0 mℓ; KI stock solution, 1.0 mℓ. Vitamins stock solutions to be added to the final solution would depend on the concentration of the stock solution and the final amount of these ingredients desired in the medium. The final pH is adjusted to 5.5 with 0.2 *N* KOH or 0.2 *N* HCl. If a solid medium is desired, 6 to 8 g/ℓ of agar is added. The agar can be dissolved in an autoclave or by heating while stirring. Another outline of procedure for preparing stock solutions is given by Fossard.[205]

In our laboratory, stock solutions are prepared as groups of compounds regardless of whether they are macro- or micronutrient elements, i.e., sulfate, phosphate, and nitrate stock solutions are prepared separately. Glass-distilled deionized water is used for preparing all stock solutions. The preparation of LS medium would be as follows:

Sulfate stock solution — Weigh out the following: $MgSO_4 \cdot 7H_2O$, 3.7 g; $MnSO_4 \cdot H_2O$, 0.169 g; and $ZnSO_4 \cdot 7H_2O$, 0.863 g. $CuSO_4 \cdot 5H_2O$ is added by weighing 0.0025 g and diluting the 100 mℓ. Ten mℓ of this solution are added to the other ingredients and water is added to bring to a total volume of 200 mℓ. Twenty mℓ of this stock solution are used in preparing 1ℓ of medium.

Phosphate stock solution — Add 1.70 g of KH_2PO_4 to water to bring to 200 mℓ vol. Twenty mℓ are used to prepare 1ℓ of medium.

Nitrate stock solution — Add 19.0 g of KNO_3 and 16.50 g of NH_4NO_3 to glass-distilled deionized water and bring to 200 mℓ total volume. Use 20 mℓ for 1 ℓ of final medium.

Calcium chloride stock solution — Add 4.414 g of $CaCl_2 \cdot 2H_2O$ and bring to a volume of 200 mℓ as above. Use 20 mℓ for 1ℓ of final medium.

Potassium iodide stock solution — Weigh 0.0083 g of KI and bring to a volume of 200 mℓ as above. Use 20 mℓ for 1ℓ of final medium.

Other microelements stock solution — Weigh out 0.0250 g of $Na_2MoO_4 \cdot 2H_2O$ and 0.0025 g of $CoCl_2 \cdot 6H_2O$. Water is added to each to bring to 100 mℓ and 10 mℓ of each of these solutions are added to 0.063 g of H_3BO_3. The final volume is brought to 200 mℓ and 20 mℓ are used for 1 ℓ of final medium.

Sodium EDTA-iron sulfate stock solution — Na_2EDTA, 0.373 g and $FeSO_4 \cdot 7H_2O$, 0.278 g are brought to a final volume of 200 mℓ as above and 20 mℓ are used for 1ℓ of final medium.

Inositol stock solution — 1.0 g of *myo*-inositol is brought to a volume of 200 mℓ as above and 20 mℓ are used to make 1 ℓ of final medium.

Vitamin stock solution — The only vitamin used in the LS medium is thiamine·HCl and 0.020 g is brought to 50 mℓ vol. One milliliter of this solution is used for 1 ℓ of final medium. If other vitamins such as pyroxidine·HCl or nicotinic acid are desired they may also be used in this solution.

Sucrose — 30 g are added to 1 ℓ of final medium.

Agar — Although the LS medium calls for 10 g/ℓ of agar, we use only 8 g.

Auxins or cytokinins — If these are desired, they are dissolved in a small amount of 0.1 *N* KOH and stock solutions of each are prepared by diluting to a desired volume with glass-distilled deionized water. The amount of stock solution to be added to the final medium varies, depending on the desired final concentration.

The advantages of the above system for preparing stock solutions are that it minimizes interactions or reactions which might occur between different ingredients (especially precipitates from inorganic compounds) and simplifies preparation of the final medium. Twenty mℓ of all stock solutions, except the vitamins where 1 mℓ is used, are added to make 1ℓ of final medium. If only 250 or 500 mℓ of medium are desired, then only one fourth or one half, respectively, of the stock solutions are needed. The vitamin stock solution is stored in the freezer and all other stock solutions are stored in the refrigerator. The KI stock solution is stored in a brown bottle.

The standard technique for sterilization of culture media and culture apparatus has been by autoclaving at 15 psi for 15 min.[201] The temperature should reach 121°C. Autoclaving creates two major problems; (1) large volumes require longer periods of time to reach the desired temperature than smaller volumes and (2) some of the ingredients, particularly some of the organic constituents, may be chemically modified or broken down. In our laboratory, only the agar is autoclaved; all other ingredients are filter sterilized. Disposable filter funnels of 115 mℓ capacity with a pore size of 0.20 μm work very well. Larger volumes can be sterilized by using large capacity funnels with membrane filters of 0.20 μm or other desired pore size. Various membrane filter sterilization systems are described in scientific supply catalogs.

C. Current Technology for Specific Agronomic Crops

Because of the limited success and few examples of establishment of plants from in vitro culture of many species, it is difficult to generalize procedure with regard to explant, sterilization, medium, light, and temperature conditions, etc. Therefore, the following is an account of specific examples or case histories of successful plant formation from in vitro culture of somatic tissue in agronomic crops. Explant used and a brief account of culture procedure including medium, sterilization, subculture, growth conditions, etc. from the original references are given for one or two examples of most of the crop species listed in Table 1. The procedure is quoted as accurately and as completely as possible from the original references to provide adequate information for culture from time of planting the original explant to the establishment of growing plants. In some cases, however, this is not possible because of the vague account of procedure presented by the original author. The intent is to provide the reader with a starting point for a specific crop species from which modifications may be made to improve or optimize culture conditions and technique. For further examples and references for specific crops the reader may consult those listed in Table 1.

1. Cereals

Although plant formation from callus has been reported for all the major cereals, the frequencies are usually low and further decrease with increasing time of callus maintenance and subculture. The explant often used is either immature or mature embryos. For cloning field established plants, explants must be obtained from tissue other than embryos since these may vary in genetic constitution even when obtained from the same plant. This is especially true in outcrossing species.

a. Oats (Avena sativa L.)

Explants used for callus induction include immature embryos (late milk to early dough stage),[10] apical meristem tissue,[10] or mature embryos.[10,12]

The sterilization and culture procedure for mature embryos by Lorz et al.[12] were as follows: dry seeds were soaked for 16 hr in water and sterilized for 3 min in 0.3% sublimat, rinsed for 5 min in water, 5 min in 70% ethanol, and three changes of water. Two to three days after germination, isolated embryos were sterilized by treatment for 5 min in 0.5% calcium hypochlorite plus 0.05% Tween 80® (Serva), followed by 10 sec in 96% ethanol and three changes of water.

The SH[213] medium with 0.5 mg/ℓ 2,4-D, 2 mg/ℓ p-chlorophenoxyacetic acid (pCPA), 2% sucrose, and 1.2% agar was used to initiate callus. Callus formation at the hypocotyl became visible after 15 to 20 days. This callus together with the adhering hypocotyl was separated from the shoot and root of the seedling after 2 to 3 weeks and subcultured at 4-week intervals. Callus growth during subculture was further stimulated by increasing the 2,4-D concentration to 2 mg/ℓ. After the 8th subculture at 4-

week intervals shoot formation was induced by using 10^{-5} M indole-3-acetic acid (IAA), 3×10^{-5} M NAA, 1.5×10^{-6} M BAP, 4000 lx and 16/8 hr light/dark cycle at 24°C. More than 50% of the calli formed shoots ranging from 1 to 5 shoots per callus. Calli having shoots without roots were transplanted onto hormone-free MS[208] medium where some developed roots. These and calli with both shoots and roots were planted into plotting compost for establishment of plants.

Cummings et al.[10] sterilized immature embryos by first removing the lemma and palea and then submerging the groats for 5 to 10 min in a 2.5% sodium hypochlorite solution containing 0.5% Al Lab detergent (A and L Laboratories, Minneapolis, Minn.) as a wetting agent. The groats were then transferred through three 5 min sterile deionized water rinses. The B5[211] medium containing 20 g sucrose per liter, 7 g agar per liter, and various amounts of 2,4-D ranging from 0.5 to 3 mg/ℓ was used. Callus cultures were incubated at 28 to 30°C in continuous cool-white fluorescent light with an intensity of 3000 lx. After initial callus formation, all cultures were grown on B5 medium containing 1 mg/ℓ of 2,4-D and were subcultured to fresh medium every 4 to 6 weeks. The B5 medium without 2,4-D was used for plantlet regeneration and growth conditions were the same as those for callus initiation. Plantlets were transplanted from culture flasks directly to sterilized potting soil and grown in the greenhouse.

Of 25 genotypes, 23 initiated callus and plants were formed from 16. A total of 133 plants was grown to maturity from 12 genotypes and one cultivar "Lodi" maintained the ability to form plantlets from callus through 13 subcultures for 1½ years.

b. Barley (Hordeum vulgare L.*)*

Seedlings were grown from dehusked seeds which had been previously sterilized for 5 min with 20 fold diluted 5.25% sodium hypochlorite plus 0.2% detergent (Alconox®) and germinated on a solid MS[208] medium with no hormones.[13]

All media were autoclaved for 8 min at 121°C with 1.41 kg/cm² pressure. Callus was induced from apical meristems which were dissected from 1 week-old seedlings and plated on a modified MS medium. These modifications were: 500 mg/ℓ *myo*-inositol, 5 mg/ℓ thiamine·HCl, 8 g/ℓ agar, 10 μM IAA, 15 μM 2,4-D, and 1.5 μM 6-(γ,γ-dimethylallylamino)purine (2iP). Culture conditions for callus induction were in a growth chamber at 25°C and a 16/8 hr light/dark period. The light intensity was 600 fc. After 3 to 5 weeks, the calli were subcultured and maintained by further subculturing every 3 to 4 weeks. The callus maintenance medium contained similar amounts of IAA and 2iP plus one of the following: 5 to 20 μM 2,4-D, 20 to 40 μM NAA, or 20 to 40 μM pCPA. Organ differentiation was obtained by transferring exponentially growing calli to the shoot induction medium which was the same basal medium described above except that no hormones were included. In one cultivar, "Himalaya", 11 of 13 and 6 of 15 calli formed plantlets after 3 and 5 subcultures, respectively. Ability for shoot production was completely lost after 4 to 5 months culture.

In another experiment[14] callus was induced using leaves of young barley seedlings cultured on a modified MS medium with 10 mg/ℓ of 2,4-D and a seedling exctract. However, only one shoot was recovered from one explant.

Dale and Deambrogio[16] tested several explants for callus induction and plantlet formation. Callus production was lowest from the meristem tip, leaf sheath and mesocotyl, and highest from the root, immature and mature embryos. Callus quality also varied with explant. Immature embryos produced firm, nodular, white-yellow callus, whereas all other explants gave soft, watery and generally white callus. The best medium for callus induction was B5[211] with 1 mg/ℓ 2,4-D. Roots were regenerated on callus from most explants, but whole plants were regenerated only from immature embryo callus.

In another study, Orton[17] performed a quantitative analysis of growth and regeneration from tissue cultures of *Hordeum vulgare, H. jubatum* and their interspecific hybrid. Explants included immature ovaries, anthers, stem and leaf sections, shoot and root tips, and crown. These were excised and surface sterilized in 95% ethanol for 30 sec washed twice in sterile distilled water, and placed in contact with either a solid LS[209] medium supplemented with 5 mg/ℓ 2,4-D and 4% sucrose of B5[211] medium supplemented with 4 mg/ℓ 2,4-D and 3% sucrose. All media were solidified with 0.9% Difco Bacto® agar.

After induction, calli were maintained on B5 medium with 3 to 4 mg/ℓ 2,4-D and 2 to 3% sucrose in the dark at 25°C. Callus was successfully induced from root meristems, immature ovaries and peduncles, and 3-day embryos and endosperms. Callus from whole immature ovaries was the most easily obtained and the most vigorous. Uniform callus tissue was subdivided into six subcalli of equal size, plated onto fresh medium and transferred every 2 weeks. Several different media components were tested including different concentrations of various hormones. Petri dishes were randomized with respect to basic and hormone treatments and maintained at 25°C under 16 hr/day of light (2000 to 3250 lx) from GE F96T10-CWX bulbs.

After callus formation, cultures adapted one of five distinct types characterized by differences in tissue consistency, pigmentation, growth rate, and regeneration. Different growth rates were observed among calli of the same callus type. B5 medium supplemented with 1% sucrose, 1% glucose, 0.3 mg/ℓ kinetin and 1 mg/ℓ gibberellic acid (GA_3) maximized regeneration of shoots and roots from hard, yellow, linear, slow growing callus, the only type of callus in which shoot regeneration was observed. The three genotypes exhibited similar patterns of media induced callus growth and regeneration. After shoots were regenerated, roots could easily be induced by transferring to MS medium plus 1 mg/ℓ GA_3. Thus, over 95% of the calli producing shoots could be transformed into whole plants.

c. *Rice (Oryza sativa* L.*)*

The technology for in vitro culture for rice is probably further developed than for any other cereal. This is the first cereal in which haploid production from anther culture was demonstrated.[36] Sustained divisions and callus formation from protoplasts isolated from both callus and mesophyll cells have also been reported.[216] Furthermore, regeneration of roots was observed from callus originating from mesophyll protoplasts.[216] There also have been more reports of plantlet formation from callus in rice than in any other cereal species. Organogenesis has been accomplished from calli derived from embryos,[25,26,29-31] roots,[25,28,31] shoots,[25,31] leaves,[31] immature endosperms,[32] mesocotyls,[34] seedlings,[24,31] and seeds.[35] However, data from earlier experiments concerning the frequency of organogenesis and plantlet formation from callus derived from somatic tissue are sparse.[31] Therefore, this presentation will describe the technology by Henke et al.[31] for successful plantlet formation from root, leaf, and seedling derived calli.

These workers dehusked whole kernels manually and then surface sterilized them by immersing in a solution containing 1.05% sodium hypochlorite and 0.1% liquid detergent (Micro, International Products Corp., Trenton, N.J.) for 30 min followed by a rinse in 0.1 *M* HCl and then several rinses of distilled water. The kernels were then soaked overnight at 5°C in sterile distilled water and germinated on filter sterilized MS medium without added hormones, solidified with 0.8% agar (Difco Bacto®). Germination was for 3 to 4 days at 25°C in the light (1300 lx) from cool-white fluorescent bulbs. The resultant seedlings were used for making explants.

Calli were induced from intact 3- to 4-day-old seedings, primary roots 5 to 10 mm behind the root tip, and leaves 1 to 5 mm from the leaf base. Concentrations of 2,4-D ranging from 0.5 to 10.0 mg/ℓ were added to the MS[208] medium for callus induction. Concentrations of 2.0 mg/ℓ and above promoted best callus growth for roots and seedlings while leaves needed at least 6.0 mg/ℓ. Incubation was at 27°C in the dark for 2 to 3 weeks. Resulting calli and associated explant tissues were transferred to identical fresh medium and incubated for an additional 2 weeks. Subsequent passages were made at 30-day intervals using a 4 to 5 mm^3 portion of each callus as inoculum. At the second passage, care was taken to avoid transferring any tissue of the original explant. The optimum concentrations of 2,4-D used for callus initiation were also used for callus maintenance.

Organogenesis and plantlet formation from calli were initiated by subculturing on MS medium without any hormone or with 0.05 to 5.0 mg/ℓ of 2iP. It has been reported that shoot formation can be initiated by cytokinins except in cereals and grasses.[217] However, in this study results showed that 2iP promoted shoot organogenesis from 16-week-old rice calli (even though the results were variable) with changing concentration. Number of calli exhibiting roots, shoots, or both roots and shoots was determined after 1 month of growth at 27°C in low-irradiance continuous white light. Eight-week-old calli exhibited a higher frequency of organogenesis than 16-week-old calli when treated identically. This result with younger vs. older calli is typical for Gramineae species. After 2 months, intact plantlets were removed from the calli, transferred to peat pots with a potting mix, and grown for several weeks in a mist bed. After this period the young plants were transferred to larger pots and grown to maturity.

An interesting observation in a recent study was that ethylene (C_2H_4) promoted shoot regeneration either alone (10 ppm) or in combination with carbon dioxide.[35] The most favorable effects were obtained for the combination 5 ppm C_2H_4 and 2% CO_2 and for C_2H_4/CO_2 ratios between 2.5×10^{-4} and 5×10^{-4}.

d. Sorghum (Sorghum bicolor L. Moench*)*

Callus has been initiated from the first node of young (5- to 7-day-old) seedlings[57] and mature[56] or immature[55,56] embryos. In one report it was claimed that very large numbers of shoots and embryo-like structures were formed on cultured tissues derived from both immature and mature sexual embryos.[56]

These authors[56] removed caryopses from greenhouse grown plants approximately 10 to 30 days after pollination and then surface sterilized by treating with 0.01% $HgCl_2$ plus one drop of Tween 80®/100 mℓ and washing six times in sterile water. Lengths of time for these procedures were not stated. Immature embryos were isolated and placed with the coleoptile side in contact with a MS[208] agar medium containing 3% sucrose and 2.5 or 5.0 mg/ℓ of 2,4-D. They were incubated under diffuse light of 1000 lx at 27°C (16/8 hr light/dark cycle).

Callus could be observed almost immediately (2 to 3 days) and after 15 days further divisions had given rise to many hundreds of organized structures resembling embryos. After 20 days, the structures had given rise to many hundreds of shoots bearing leaves. Similar structures were also induced on scutellum callus of mature seeds. When plantlets were formed they were transferred to potting compost.

In another report with immature embryos, numerous small leafy shoots also appeared directly on the callus.[55] In this case, zeatin at 10 μM and IAA at 1.0 μM added to MS medium with vitamins, according to Gamborg,[204] greatly enhanced plantlet formation. The medium also contained 30 g/ℓ sucrose and the following compounds in mg/ℓ: glycine, 7.7; L-asparagine, 200; niacinamide 1.3; and calcium pantothenate, 0.25.

Immature seeds were collected from 12-week-old plants and submerged for 60 sec in 70% ethanol, surface sterilized in 20% Javex® (detergent) for 10 min and rinsed with sterile distilled water. The embryos, measuring 1 to 2 mm, were placed with the plumule-radicle axis in contact with the agar medium. The best environmental conditions were a 20 hr/day photoperiod (2000 lx) and 20°C/15°C light/dark temperature. For callus formation, 5 μM 2,4-D and 50 μM zeatin were added to the medium. After 4 to 6 weeks of callus induction, shoot and plantlet formation were enhanced by transfer to the medium with 1.0 μM IAA and 10 μM zeatin. In contrast to results obtained with other cereals and grasses, sorghum (in this case) required cytokinin in the callus inducing medium to ensure morphogenesis. After 8 to 10 weeks, the plantlets produced on the MS-IAA agar were transplanted into vermiculite-sand-peatmoss mixtures in pots, fertilized with Hoagland's solution, and grown in a greenhouse to mature plants. Some of the plants were sterile while others produced normal seeds.

This procedure permitted callus induction and plant formation within 10 weeks. About 20 to 50% of the embryo explants produced a callus from which shoots were obtained.

e. Bread Wheat (Triticum aestivum L.*)*

Various explants have been used to initiate callus in wheat. These include: seed,[65] embryo,[61-63,65] mesocotyl,[34] inflorescence,[61] rachis,[59-62,65] and root.[61,62] O'Hara and Street[62] conducted a recent study using several explant sources and their procedure and results will be presented here.

Explants consisted of mature embryos, coleoptile and seedling root segments, stem and rachis segments, and segments of expanded leaves. The latter failed to yield any callus on any of the media tested. Grains were allowed to imbibe water for 16 hr and then were surface sterilized by treatment with 25% v:v "Domestos" (Lever Brothers, Ltd., London) containing 10% chlorine for 10 min and rinsed with sterile distilled water. Mature embryos were excised and placed with the cut surface in contact with the solid medium. To obtain seedling tissues, grains were germinated on solid culture medium without hormones. When the coleoptiles were about 5-cm long, 0.5 to 2.0 cm segments were cut from the root apices and coleoptiles.

Immature embryos were taken at intervals of 9 to 25 days after anthesis and sterilized by immersion in 25% v:v sodium hypochlorite containing a small amount of wetting agent. They were then washed in four rinses of sterile distilled water. The embryo axis was placed in contact with the medium.

Stem and rachis segments were obtained from mature tillering plants and sterilized by immersion for 20 to 30 min in 0.01% mercuric chloride, with a small amount of wetting agent, followed by washing with sterile water. Stem segments (9 cm length) containing a node were sterilized as above; small end segments were rejected, and the remainder cut into pieces 2- to 10-mm long. Some segments were used directly and others were cut longitudinally (internodes) or obliquely (nodes). The segments were pushed at an oblique angle slightly into the surface of the solid medium. To obtain rachis explants, the spikes were sterilized as above, the spikelets removed and short rachis segments partially imbedded in the solid medium.

The basic MS[208] medium was used containing different amounts of different hormones and solidified with 1% agar. Maximum yield of callus from mature embryos was obtained with 2,4-D at 1.0 or 2.0 mg/ℓ and pCPA at 2.0 or 4.0 mg/ℓ. The callus could be subcultured at any time from 14 days onward. Coleoptile and seedling root segments also readily yielded callus in medium containing 1 mg/ℓ of 2,4-D or 5 mg/ℓ of pCPA. The callus could be subcultured within 4 weeks of initiation. Nodal segments gave callus when 1.0 to 2.0 mg/ℓ of 2,4-D were added to the medium. Callus growth was most prolific in the uppermost node. Callus could be subcultured 3 to 4 weeks after initiation. Internodal segments yielded less callus than nodel segments.

Callus from all sources could be maintained for a prolonged period by serial subculturing on medium containing 1.0 mg/ℓ of 2,4-D. Callus cultures derived from all explants readily formed roots when transferred to medium without 2,4-D and serially propagated subcultures retained the ability to form roots.

The formation of shoots was infrequent and could not be related to any combination of auxin and cytokinin. Old-established callus did not yield shoots and shoot formation was independent of root formation. Where more than one shoot arose in a culture, these were grouped together.

Chin and Scott[61] also observed very low rates of shoot organogenesis from various explants. Cultures of seedling root origin yielded roots, but not shoots and 14% of the clutures of embryo origin yielded roots and 7% yielded shoots.

In another recent study, Gosch-Wackerle et al.[65] utilized bread wheat and several wild and cultivated relatives. In addition, one cultivar of *Triticum aestivum* (Chinese Spring) was represented by one euploid line, 12 lines that were ditelosomic for different chromosomal arms and one diisosomic line.

Explants were obtained from greenhouse plants and included immature rachises, embryos, and seeds. Spikes of various developmental stages, according to the desired explant source, were surface sterilized by transfer through detergent, 70% (v:v) ethanol, approximately 0.5% hypochloride and sterile (tap) water. To obtain rachis callus (R-callus), the spikes were either dissected when the anthers attained the PMC stage or at later stage up to the first microspore mitosis. The spiklets were removed from the rachis and the latter was cut into sections of 0.5 to 1.0 cm. Several sections were placed 5 cm petri dishes containing 5 mℓ of solid medium. Callus from immature embryos (E-callus) was obtained as follows. Spikes were harvested 14 to 18 days after anthesis and sterilized as described above. The immature caryopses were separated, the embryos isolated and each placed in one of 24 wells of a "Linbro" plate (Linbro Scientific Inc., Hamden, Conn.) Each well contained 1 mℓ of solid medium. For seed callus (S-callus) the spikes were harvested somewhat earlier, 10 to 12 days after anthesis, immature seed were separated from the caryopses and placed singly in wells of the Linbro plate with embryos facing the medium.

Three basal media were used: T-medium,[59] B5[211] containing 10 mg/ℓ ascorbic acid and 50 mg/ℓ glutamine, and GP-medium[230] containing 0.2 rather than 1.98 g/ℓ asparagine. All media were solidified with 8.0 g/ℓ Difco Bacto® agar and adjusted to pH 5.7 to 5.8 before autoclaving. Various growth regulators were added. For callus induction and maintenance, the cultures were kept in the dark at 25°C and subcultured every 4 to 5 weeks.

During shoot induction, calli were transferred to continuous light (cool-fluorescent) at approximately 1000 lx. For root regeneration, the cultures having shoots were transferred to media devoid of growth regulators. Resultant plantlets were transplanted to Jiffy® turf-pots and later maintained in the greenhouse.

Frequency of callus induction was highest when immature embryos were used as the explants. More than 90% of the explants produced callus in the species and genotypes tested. Callus was also obtained from R-explants but the percentage was lower fo most genotypes. Callus induction from immature seeds was very poor. The best medium for callus induction was B5 and GP supplemented with 1 and 2 mg/ℓ of 2,4-D, respectively.

R-callus showed a high capacity to regenerate roots and could be retained in *T. aestivum* calli up to nearly 2 years. E- and S-calli showed a high capacity to regenerate shoots. Zeatin at 1 mg/ℓ enhanced the frequency of shoot formation especially when IAA at 1 mg/ℓ was used as the auxin in place of 2,4-D. With this combination, the percentages of E-calli exhibiting shoots were 68, 45, 25, and 33 for the first through fourth subcultures, respectively. An important finding was that the presence of a cytokinin, viz., zeatin enhanced shoot formation. In most other reports of shoot formation from calli in Gramineae, the influence of cytokinins on frequency of shoot formation has been nil or at best variable.

*f. **Durum Wheat (Triticum durum** Dest.)*

Bennici and D'Amato[73] surface sterilized dry kernels by immersion in 95% ethanol for 3 min followed by 1% mercuric chloride for 15 min. After three washes in sterile deionized water, the seeds were sown in tubes (one seed per tube) containing 15 mℓ of sterile water solidified with 8% agar. After 3 days in the dark at 4°C and 3 days at 25°C seedlings deprived of both root and shoot tips were transplanted to flasks containing 30 mℓ of agar solidified Smith's medium supplemented with (mg/ℓ) casein hydrolysate 1000, inositol 100, and NAA 5 (see Reference 34 for medium formulation). The cultures were incubated in a growth room under continuous illumination of 2500 lx with fluorescent lights.

White friable calli were produced from mesocotyl explants. Some rooted plantlets were produced directly from the primary explant. When shoots were produced without roots, root growth was promoted by transferring the callus with shoot to LS medium supplemented with 5 mg/ℓ indole-3-butyric acid (IBA). In general, only one plantlet developed per callus. Young plants could be planted in soil and grown to maturity.

*g. **Corn (Zea mays** L.)*

Immature embryos[75,77] have commonly been used as the explant although mature embryos[76] have also been used. Green et al.[75] used several different genotypes. Ears were removed from field or greenhouse grown plants at 14 to 24 days after pollination. The ears were broken into 5 to 8 cm segments and sterilized for 20 min by submerging in a solution of 1 g of detergent per 100 mℓ of 2.5% sodium hypochlorite. Ears were rinsed three times in sterile deionized distilled water.

Immature embryos were isolated and placed on solid culture medium with the flat plumule-radicle axis side in contact with the medium. Callus was initiated and maintained on MS[208] medium supplemented with 7.7 mg/ℓ glycine, 1.98 g/ℓ L-asparagine, 1.3 mg/ℓ niacin, 0.25 mg/ℓ thiamine·HCl, 0.25 mg/ℓ pyroxidine·HCl, 0.25 mg/ℓ Ca pantothenate, 20 g/ℓ sucrose, and 8 g/ℓ agar. Varying concentrations of hormones were used. Callus was initiated best on medium containing 2 mg/ℓ of 2,4-D. Calli were subcultured every 21 to 28 days on MS medium containing 2 mg/ℓ of 2,4-D. Embryos and subsequent callus cultures were incubated at 28 to 30°C with a 16 hr light per 8 hr dark photoperiod from cool-white fluorescent lights with an intensity of 2000 lx.

Organogenesis was obtained by transferring to MS medium with 0.25 mg/ℓ of 2,4-D. Growth and development after 20 days included many light-green leaves and white compact structures. By 30 days, many curled and wrinkled leaves, additional white scutellar-like structures, and short roots had formed. The differentiation of complete seedlings was accomplished by transferring cultures to 2,4-D free medium. At the two-leaf stage, plants were transplanted to steam sterilized soil and 10 to 15% of the plants became established and grew to maturity. The authors state that more adequate root development would likely lead to a higher frequency of plant establishment. Genotypes A188 and A188 × *R-njR-nj* exhibited differentiation at the highest frequency. Callus of the latter genotype retained the ability to form plantlets for 19 months.

*h. **Millets (Panicum** spp.)*

Callus growth and plantlet formation have been reported in *Panicum miliaceum,*[81] *Paspalum scrobiculatum,*[82] *Eleusine coracana,*[82] and *Pennisetum typhoideum.*[82] Rangan[82] germinated grains in the dark under aseptic conditions. Mesocotyl explants from 5-day-old seedlings were excised and planted on MS[208] medium with one of the auxins, 2,4-D, 2-benzthiozoleacetic acid (BTOA), NAA, IAA, or pCPA. Various cytokinins including kinetin, BAP, 2iP, zeatin, or 6-benzyl-9-tetrahydropyraneadenine (SD 8339) were used at various concentrations for organogenesis. The medium were solidified with 0.8% agar. Undefined media components such as coconut milk, yeast extract, and malt extract were also sometimes added. Cultures were maintained at 25 ± 2°C, 50 to 60% relative humidity, and in diffuse light.

Callus was induced in all species with 10 mg/ℓ of 2,4-D added to the MS medium. Coconut milk (15%) improved callus growth of some species. Shoot buds developed when the calli were transferred to either a no or low concentration auxin medium (0.2 mg/ℓ NAA or IAA). Cytokinins SD 8339 or kinetin at 1 mg/ℓ enhanced shoot development. Complete plantlets could be obtained from culture of excised young shoot buds in a rooting medium.

2. Seed Legumes

Reports of plantlet formation from in vitro culture of legumes grown for their seed are sparse. Organogenesis seems especially difficult to initiate in these species even though callus is easily induced and maintained.

Although adventitious shoot development from hypocotyl sections has been reported in soybean,[9] this species is conspicuously absent from the list of species in which plantlet regeneration from callus has been reported. Using hypocotyls or ovaries as the original explant for callus induction, Beversdorf and Bingham[218] obtained roots and embryo-like structures but no plantlets. Vegetative propagation from cultured stem internodes was also reported for the broad bean (*Vicia faba* L.),[99] but plantlet regeneration from callus has not been reported.

Recently, plant regeneration from apical meristem tips was reported for several seed legumes including, chickpea, lentil, pea, black and green grams.[93] However, the regeneration of plants from subcultured callus tissue was not reported. Plantlet regeneration from callus cultures of pigeon pea resulted if seeds from which explants were finally taken and were exposed to 5 kR of ^{60}Co γ-radiation. Callus from hypocotyls of seeds given no irradiation or higher exposures did not regenerate plantlets.

Two crop species, bean and pea, will be used for examples of plant formation from in vitro culture in this section. The procedure and results from one original reference for bean and two references for pea will be given. Additional information on these two species may be obtained in the chapter on vegetable crops since the genera and species for garden pea and snap bean are the same as those for field pea and dry bean.

***a. Bean (Phaseolus vulgaris** L.)*

Crocomo et al.[92] surface sterilized leaves from 2-week-old plants in 20% v:v hypochlorite for 15 min. Following two rinses in sterile distilled water, the leaves were sectioned into 5 mm squares and inoculated in 20 × 150 mm test tubes containing 10 mℓ of nutrient. Cultures were grown for 8 weeks under constant environmental conditions of 25°C, 12 hr photoperiod, and 200 fc of light from cool-white fluorescent lamps. The medium used contained the salts and vitamins of Veliky and Martin[219] with the addition of auxins, cytokinin, and bean seed extract. The final pH was adjusted to 4.5 and the medium was sterilized for 15 min at 121°C. The medium was solidified with 1% agar. The seed extract was prepared by homogenizing bean seeds in the salts of the above mentioned medium at concentrations ranging from 1/100 bean seed per milliliter to 1 bean seed per milliliter. The extract was filtered through cheese cloth.

Optimal callus proliferation occurred at 4 mg/ℓ kinetin, 1 mg/ℓ IAA, and 1-day-old bean seed extract at a concentration of ¼ bean seed per milliliter. Root induction was best at 2 mg/ℓ kinetin. At a concentration of ¼ bean seed per milliliter, two plantlets were induced in nine cultures. These were the only two plantlets recovered in these experiments representing about 400 cultures. It was not reported as to whether these plantlets were grown to mature plants.

b. Pea (Pisum sativum L.)

In two reports of in vitro culture of pea, shoot apices were used as the explant.[93,94] Seeds were surface sterilized with 1% calcium hypochlorite for 20 min, rinsed with distilled water, and germinated in the dark at 28°C on sterilized filter paper. Apices were excised from 3- to 4-day-old shoots and macerated under sterile conditions. Cell masses were placed on agar medium in erlenmeyer flasks.

Media used were either the B5[211] or MS[208] mineral salt composition with micronutrients and vitamins according to the B5 medium. The medium was solidified with 0.8% agar.

Two environmental conditions were used: One was a temperature controlled room at 26 ± 1°C with an 18 hr light and 6 hr dark cycle with a light intensity of 2000 lx from cool-white fluorescent lamps. The other was a growth cabinet with 18 hr fluorescent light (4000 lx) and 6 hr dark at 25°C/15°C.

Shoots formed on calli within 4 to 6 weeks of culture. The B5 medium was much more effective than the MS medium for shoot production. This may be related to the lower nitrogen concentration (27 m*M* vs. 60 m*M*) and sucrose level (2% vs. 3%) in the B5 vs. the MS medium. Root development occurred on some of the shoots or shoot producing callus. Benzyladenine (BA) at 0.2 to 2.0 m*M* generally improved the frequency of calli forming shoots and the number of shoots per callus. The auxin, NAA, was not necessary and seemed to repress shoot formation. Plants could be transplanted into vermiculite and sand in pots and grown to maturity.

In a recent study, 16 genetic lines of pea were screened for their ability to regenerate whole plants from callus cultures.[95] Malmberg[95] sterilized and germinated seeds by incubating them in 0.5% sodium hypochlorite and 0.01% sodium dodecyl sulfate for 5 min, rinsing in distilled water, and then placing them on basal MS[208] medium with no hormones and solidified with 0.9% agar. The seeds were incubated at 26°C in the dark; after 2 to 5 days, when the radicle had just emerged, the embryos were excised, and slices of epicotyl were placed on callus forming medium (MS with 2 mg/ℓ of NAA and 1 mg/ℓ of BA. Incubation was in the dark at 26°C and calli were transferred monthly to fresh medium and maintained at 26°C in the dark.

For shoot regeneration, callus pieces were placed on solid MS medium with 0.2 mg/ℓ IAA and 5 mg/ℓ BA. Two pieces per Petri plate were incubated at room temperature with a 16/8 hr light/dark period. Light was from cool-white fluorescent bulbs at 60 $\mu E\ m^{-2}s^{-1}$. The greenest most organized regions were subcultured monthly on to fresh regeneration medium until shoots with leaves were obtained. For root induction, well-formed shoots were sliced from the callus just below the lowest node and the leaves from this node were removed. The shoot base was dipped in 1 mg/ℓ NAA for 10 sec and the shoot was planted in solid MS medium. Roots emerging from the node were obtained in more than 75% of the shoots. Rooted plantlets were gently washed with sterile distilled water and planted in sterile soil. They were watered with a half-strength solution of MS salts and kept in a small glass chamber under high humidity. After establishment, plantlets could be transferred to greenhouse or growth chamber conditions.

Regeneration was tested after every other monthly transfer by subculturing on regeneration medium. Nine of the lines, including two commercial cultivars, did not regenerate at all. Six lines regenerated plants from epicotyl slices after 2 months growth as callus. Of these, four still regenerated after 4 months callus growth and two of these after 6 months. Usually two to eight well-developed shoots with leaves occurred on each callus piece. However, regeneration became less easy with time.

3. Forage Grasses

In vitro culture of forage grasses has been attempted much less than in cereal species. Atkin and Barton[220] established tissue cultures of several temperate grass species, but no plantlet formation from callus was reported. There have been reports of plantlet formation from callus in several forage grass species in the past 1 to 3 years; but in most cases there has been only one reference for each species. As with the cereals, callus has been usually obtained from embryonic tissue. The following will account examples and procedures for some forage grasses.

a. Big Bluestem (Andropogon gerardii* Vitman*)

The explant used by Chen et al.[100] was the young unemerged inflorescence taken when the tip of the flag leaf was just emerging. As the authors point out, this explant provides material for cloning field-established individual plants that embryos cannot.

The outer leaves were removed and the inner leaves surface sterilized by swabbing them with 70% ethanol. The young inflorescence was removed and placed in a sterile petri dish. It was cut into 5 to 10 mm pieces and inoculated on a LS[209] medium supplemented with 5 mg/ℓ of 2,4-D and 0 or 0.2 mg/ℓ of kinetin. The medium was adjusted to pH of 5.8 before autoclaving and solidified with 0.8% agar. Twenty mℓ of the medium were dispensed into 25 × 150 mm screw capped culture tubes. The same medium was also used for subculturing. Since no apparent advantage could be detected with kinetin it was omitted from the medium. Callus initiation and maintenance were in the dark at 25°C.

Callus was ready for transfer after an incubation period of 4 weeks. The same medium supplemented with 0, 0.1, or 1 mg of 2,4-D per liter resulted in shoot development from nodules on callus 2 weeks after transfer. When shoots became visible, calli were transferred to medium without hormones and placed in a growth chamber at 27°C and 16 hr fluorescent lighting (2000 lx) and 8 hr dark. Roots developed on the green shoots after 2 weeks and plants at the two-leaf stage were transplanted into vermiculite and watered with one-quarter strength Hoagland's solution. Of 400 transplants, 342 were successfully established in the field or greenhouse.

***b. Indiangrass (Sorghastrum nutans* L.)**

Callus was initiated from segments of young inflorescences explanted on LS[209] medium supplemented with 5 or 10 mg/ℓ of 2,4-D and 0 or 0.2 mg/ℓ of kinetin.[101] The methods of preparing the culture media, disinfecting explants, and inoculation were the same as for big bluestem described above.[100] All callus cultures were placed in the dark at 25°C.

Callus induction occurred more rapidly on medium supplemented with 5 mg/ℓ of 2,4-D. The absence or presence of kinetin (0.2 mg/ℓ) had no influence. Approximately 1 month after explanting, calli were isolated and propagated by subculturing for another month on the callus induction medium. Organ formation was induced when the calli were transferred to medium containing 1 mg/ℓ or less of 2,4-D. Shoots formed faster on LS medium devoid of 2,4-D and this medium was subsequently used for morphogenetic induction. About 300 mg of callus was placed in each flask and incubated in the dark at 25°C for 4 weeks and then in 16 hr cool-white fluorescent light (25 $\mu E/sec^{-1}m^{-2}$/8 hr dark cycles at 27° for another 4 weeks.

Enhancement of root formation on calli with shoots was accomplished by transfer onto the LS medium without hormones with temperature and light conditions as above. When the plantlets were well developed in the culture flasks they were placed in pots or plant bands and transplanted to the field.

c. *Orchardgrass (Dactylis glomerata* L.)

Whole, mature caryopses were dehusked with 50% sulfuric acid:50% water (v:v) for 20 min.[102] They were then rinsed, first with distilled water and then with 70% ethanol. They were surface sterilized for 3 min, with shaking, in 5.25% sodium hypochlorite containing 0.5% (by volume) detergent (Alconox®). The solution was diluted by one half and the sterilization continued for another 12 min. The process was further continued by diluting the sodium hypochlorite to 1% and 15 min shaking. After several rinses with sterile distilled water, the caryopses were hydrated for 3 hr.

Caryopses were placed embryo down on an agar medium in a petri dish. Best results were obtained on a modified SH[213] medium. These modifications were: 27 mg/ℓ $FeSO_4 \cdot 7H_2O$, 37 mg/ℓ Na_2EDTA, and 7 g/ℓ agar.

Growth of orchardgrass in culture involved induction of callus, subculture and maintenance of callus, shoot initiation, and promotion of root growth. The same basal medium was used throughout, but different amounts of growth regulators were used. Fifteen mg/ℓ of 2,4-D and 2.15 mg/ℓ of kinetin were used for callus induction. Calli were large enough (approximately 0.5 to 0.8 cm diam) for subculture 3 weeks after planting when incubated in the dark at 26°C. For the initial subculture, care was taken not to include any callus which contained the original explant. For maintenance of callus during subculture, the 2,4-D concentration was reduced to 5 mg/ℓ. Calli were divided and further subcultured twice at 3 week intervals. Incubation was also in the dark at 26°C. Organogenesis was initiated when the 2,4-D concentration was lowered to 1 mg/ℓ and culture was in the light at 26°C. Root growth, on calli with shoots, was enhanced by transfer to a medium without hormones and containing a one-half concentration of inorganic salts. After sufficient growth, seedlings could easily be transferred to pots and actively growing plants established.

d. *Tall Fescue (Festuca arundinacea* Schreb.)

Embryonic tissue from mature caryopses were also used as the explant, but embryos rather than whole caryopses were plated.[104]

Whole caryopses were dehusked by stirring vigorously for 20 min in 50% sulfuric acid (v:v). They were rinsed with distilled water followed by 70% ethanol and surface sterilized for 20 min with shaking, in 5.25% sodium hypochlorite plus 0.5% (by volume) detergent (Alconox®). After several rinses with sterile distilled water, the embryo was separated from the endosperm just above the scutellum and placed with the flat plumule-radicle axis side in contact with the agar medium in a petri dish. The medium was previously autoclaved for 15 min at 20 psi and the agar (Fisher laboratory grade) was washed several times with deionized water and freeze-dried before use. A modified MS[208] medium consisting of 300 mg/ℓ inositol, 100 mg/ℓ ascorbic acid, 2 g/ℓ casein hydrolysate, 20 g/ℓ sucrose, 8 g/ℓ agar, and 0.5 to 9 mg/ℓ of 2,4-D was used.

Growth of tall fescue in culture involved callus induction, subculture and maintenance of callus, and formation of shoots and roots from callus. The same basal medium was used throughout but with different amounts of 2,4-D. Calli were induced using 9 mg/ℓ of 2,4-D in the medium and culturing in the dark for 28 days at 24°C. After the initial induction, the entire embryo plus existing callus was transferred to a fresh MS medium containing 5 mg/ℓ of 2,4-D and allowed to grow for 28 more days. This was done to allow for further callus growth before serial subculturing began.

At the end of the second 28-day period, the calli were subdivided with one piece plated on a MS (callus maintenance) medium with 5.0 mg/ℓ 2,4-D and the other piece on a MS (organ formation) medium with 0.5 mg/ℓ of 2,4-D. The maintenance cultures were placed in the dark at 24°C and the organogenesis cultures were placed in a 12 hr light and 12 hr dark cycle at 24 C/15°C. Light was from cool-white fluorescent bulbs at 180 μEm^{-2} $sec.^{-1}$ After a third 28-day period, shoot and root formation had occurred on some calli. At this time, the calli in maintenance medium were again divided and tested for organ formation as described above. This procedure was repeated for one additional 28-day period. Calli were scored for frequency of root and shoot formation at 56, 84, and 112 days after the initial plating of embryos.

From a total of more than 800 calli, only about 18% exhibited the ability to form shoots after any one or more of the three successive subcultures. There was a decrease in shoot formation from 9.6% for the first subculture to 3.1% for the third subculture.

Poor root growth on calli where shoots were initiated could be enhanced by transferring to a MS medium without 2,4-D and containing a one-half concentration of inorganic salts. After several weeks, seedlings could be easily transferred to soil in pots and actively growing plants established.

e. *Ryegrass (Lolium* spp.)

Ahloowalia[111] collected immature caryopses from controlled crosses between diploid and tetraploid F_1 hybrids of Italian and perennial ryegrass. These caryopses or those from diploid hybrids or tetraploid perennial ryegrass were used as explant material. Developing caryopses were collected 15 to 20 days after pollination and the lemma and palea were removed. In some cases the whole caryopsis was cultured. Caryopses were soaked in distilled water for 2 to 3 hr, then washed with sterile water, and sterilized by a quick rinse in 7% aqueous solution of sodium hypochlorite. They were rinsed again in sterile distilled water before culturing.

Caryopses were cultured either on Bacto-Orchid agar, 37 g/ℓ (available from Difco Laboratories, Detroit, Mich.) or a modified medium of Niizeki and Oono.[36] The pH was adjusted to 5.8 and all media were autoclaved for 30 min at 15 psi. The caryopses were cultured in 15 mℓ screw-top bottles each with 10 mℓ of medium.

Initiated calli were maintained on modified MS[208] medium by transferring every 30 to 50 days in 100 or 250 mℓ flasks containing 50 or 100 mℓ of medium, respectively. MS medium minus $CoCl_2$ and edamin was supplemented with 6.5 mg/ℓ IAA, 1.5 mg/ℓ 2,4-D, and 2.15 mg/ℓ kinetin. Of the various media tried, both the MS and the modified Niizeki and Oono[36] media with 6.5 mg/ℓ IAA, 1.5 mg/ℓ 2,4-D, and 2.15 mg/ℓ kinetin allowed rapidly growing calli to maintain good growth provided they were transferred to fresh medium every 30 to 40 days.

Root primordia were initiated in callus cultured on either of the above mentioned media containing 0.05 mg/ℓ IAA and 0.15 mg/ℓ 2,4-D. Addition of 1 mℓ/ℓ coconut milk to either medium with or without 2 mg/ℓ IAA and 2 mg/ℓ zeatin stimulated chlorophyll production in the callus and roots and promoted differentiation of shoot primordia.

Calli with root and shoot primordia were transferred to half-strength MS medium with 15 g/ℓ sucrose, 5 g/ℓ agar, 3.75 mg/ℓ IAA, 0,75 mg/ℓ 2,4-D, and 1.075 mg/ℓ kinetin. A number of calli produced numerous plants with shoots and roots within 30 days or longer. Single plants or groups of plantlets were transplanted in soil and grown in a greenhouse. Greenhouse plants could easily be transferred to the field.

f. *Timothy (Phleum pratense* L.)

Dale[103] regenerated plants from meristem tips of timothy, orchardgrass, and different species of ryegrass and fescue. Caryopses were sown in boxes and maintained in a greenhouse at 15 to 20°C. Meristem tips were isolated and cultured 2 to 4 months after sowing. Before removal of the meritstem tips, grass tillers were trimmed and surface sterilized in 50% sodium hypochlorite (5 to 7% w:v available chlorine) for 7 min followed by six washings in sterile water. Meristem tips were dissected aseptically and placed in plastic tubes each containing 10 mℓ of culture medium solidified with 4.4 g/ℓ agar. The tips consisted of the apical dome plus one or several leaf primordia. The MS[208] basal culture medium with 0.01 mg/ℓ 2,4-D and 0.2 mg/ℓ kinetin provided a good general culutre medium for grass meristem tips. Iron was added in the form of Fe·EDTA and the medium was autoclaved at 121°C for 15 min after adjusting the pH to 5.6. Culture was at 25± 1°C and continuous warm-white fluorescent light was provided at an intensity of about 6000 lx.

After 8 to 12 weeks in culture, plants from meristem tips were moved to the soil. During the first week in soil they were covered with plastic funnels or polyethylene sheeting. Plant survival in the soil was 96%.

4. *Forage Legumes*

Generally, the success for in vitro culture of forage legume species has been greater than for the large-seed legumes discussed above. Plant formation from callus has been demonstrated in several species and regeneration of plants from suspension cultures was demonstrated for alfalfa.[120] Successful anther culture was reported for berseem clover.[127]

a. *Alfalfa (Medicago sativa* L.)

Most of the success with in vitro culture experiments in this crop have been reported by Bingham and co-workers.[115-117,120] In an early study,[115] callus was initiated from several explants including immature anthers, immature ovaries, cotyledons, internode sections, and seedling hypocotyls. The account of procedure documented here will be from a more recent paper utilizing seedling hypocotyls.[116]

Scarified seed were surface sterilized in 95% ethanol on a shaker for 15 min followed by 2.5 min in 1.3% sodium hypochlorite, 5 min under vacuum in 1.3% sodium hypochlorite, and 2.5 min in fresh 1.3% sodium hypochlorite. After three 5-min shakes in serile distilled water, seeds were germinated five to a vial on 10 mℓ of semisolid basal medium in 21 × 70 mm screw-top vials with seals removed from the caps. The basal medium was from Blaydes[212] with 100 instead of 1000 mg/ℓ of KNO_3.

Callus was established from 3- to 5-mm long hypocotyl sections cut from germinated seeds and placed on semisolid B II medium. This medium is the same as Blaydes but contains 2 mg/ℓ each of NAA, kinetin, and 2,4-D. After 28 days, calli were subdivided into five replicate pieces and transferred to a semisolid Blaydes medium containing 100 mg/ℓ inositol and 2 g/ℓ Difco Bacto yeast extract.

After 1 month, budding calli were transferred to fresh medium (same as the latter mentioned) to enhance plantlet growth. Two to four weeks after this final transfer, plantlets with and without roots were transferred to Jiffy-7® peat moss pellets, inoculated with *Rhizobium* and kept under clear plastic Olefane to prevent desiccation. About 21 days later most plants were well rooted and were transferred to pots containing a 3:1:1 soil-sand-peat moss mixture. Plants could then be grown to maturity.

b. *White Clover (Trifolium repens* L.)

In one study, a great capacity for regeneration from in vitro culture utilizing cotyledons as the explant was reported.[121] However, the methodology is vague, particularly with regard to media since original references for the formulations are not given. In another report,[122] plantlets were formed from callus derived from germlings (5-mm long) but they could not be established in the soil. Meristem tip culture was utilized in a third study[123] to obtain virus-free plants. These results will be discussed in a later section of this chapter under "Potential Application".

c. *Red Clover (Trifolium pratense* L.)

Radicle meristems, hypocotyl sections, epicotyl sections, cotyledon sections, young primary leaves, and shoot apical meristems of aseptically grown 5- to 17-day-old seedlings provided explant tissues.[124] Scarified seeds were sterilized in 95% ethanol for 5 min and transferred for 20 min to calcium hypochlorite solutions (5 to 15% w:v) using reagent grade powder. The concentration used depended on the extent of seedling contamination. The seeds were then rinsed thoroughly in sterile deionized water and allowed to germinate in the light on moistened, sterile, filter paper. Glassware and other utensils were sterilized in an autoclave for 15 min at 15 psi (120°C). Sterile disposable petri dishes were used as culture vessels.

Phillips and Collins[124] tested more than 60 different media formulations before developing an optimal medium for red clover which also supports growth of other selected legume species. The PC-L2 medium produced nonwatery callus which was friable and nondarkened. The formulation of this medium is given in Tables 2 and 3.

The various seedling explants responded consistently and were about equal in terms of callus growth. Callus was subcultured onto fresh medium at 3-week intervals and maintained at 25°C with continuous, low-flux-density, light from cool-white fluorescent tubes supplemented with incandescent bulbs. Intensity was about 550 $\mu Em^{-2}sec.^{-1}$ Regeneration of plants occurred on the initial medium and during the first, second, and third subcultures. High frequencies of regeneration occurred from meristem-derived callus cultured on medium containing 0.006 mg/ℓ of picloram in combination with BAP at 0.1 to 10.0 mg/ℓ. The auxin NAA at 1.0 to 2.0 mg/ℓ substituted for picloram gave similar results. Regeneration frequencies from nonmeristem-derived callus were much lower. Calli derived from these tissues regenerated more frequently on medium containing 2,4-D (2.0 mg/ℓ) and BAP (2.0 mg/ℓ).

Shoots obtained from regeneration cultures were placed on a rooting medium[124] and 85% of the shoots developed vigorous root systems within 2 to 4 weeks. Approximately 80% survived potting into peat or soil mixtures and could be grown to maturity.

d. *Crimson Clover (Trifolium incarnatum* L.)

Beach and Smith[126] reported plant regeneration from callus cultures of both red and crimson clover. For crimson clover, cultures were established from hypocotyls of germinating seedlings. Seeds were scarified, sterilized for 5 min in 1.3% sodium hypochlorite, then for 5 min in 70% ethanol, then rinsed twice in sterile distilled water and germinated on B5[211] medium without sucrose or hormones. Seeds were germinated for 6 to 10 days under cool-white fluorescent light.

Hypocotyls were sectioned into three pieces approximately 5 mm long and surface sterilized as described above for seeds. Care was taken not to include tissue from root or meristematic regions. All three pieces of hypocotyl from a plant were placed in one bottle. The medium used was B5 with 20 g/ℓ sucrose and 100 mg/ℓ inositol. Various combinations and concentrations of vitamins, auxins, and cytokinins were tried to determine the optimal medium. All media were adjusted to pH 5.8 with NaOH; 10 mℓ was dispensed into 30 mℓ glass bottles with plastic screw caps and autoclaved at 1.1 kg/cm^2 pressure at 120°C for 15 min. Cultures were grown at 25 ± 4°C in 16/8 hr light/dark cycles under cool-white fluorescent light at an intensity of 35 to 45 $\mu E/m^2/sec$.

Best callus growth after 4 weeks was obtained with the above described B5 medium containing 10 mg/ℓ thiamine, 11 μM NAA, 10 μM 2,4-D, and 10 μM kinetin. After 4 weeks, calli were transferred to a second medium for further differentiation. The most effective medium for shoot bud development was B5 with 20 mg/ℓ thiamine, 10 μM NAA, and 15 μM adenine. Roots were also formed on crimson clover on this medium, but for red clover transfer to a third (rooting) medium was usually necessary.[126]

Plants with roots were transferred to a sterile soil-vermiculite mixture and kept in a moist cool (18°C) chamber for 10 to 14 days. They were then transferred to the greenhouse and grown to maturity.

e. Berseem Clover (Trifolium alexandrinum L.)

Mokhtarzadeh and Constantin[127] reported regeneration from both hypocotyl and anther explants of berseem clover. The procedure using hypocotyl explants will be presented here.

The authors surface sterilized seeds by immersing for 1 min in 95% ethanol, rinsing with sterile deionized distilled water, immersing for 3 min in 2.63% sodium hypochlorite solution and then rinsing several times with sterile distilled deionized water. Seeds were germinated on filter-sterilized (0.22 µm pore-size Nalgene filters®) MS[208] basal medium solidified with 0.8% agar contained in petri dishes sealed with parafilm and incubated in continuous darkness at 26°C. Hypocotyls of 6-day-old seedlings were cut 2 mm below the colyledonary node and 2 mm above the hypocotyl-root juncture. Excised hypocotyles were placed immediately onto a thin layer of water. Each hypocotyl was quartered into 0.3 to 0.5 cm segments which were plated for callus induction.

For callus induction, the basal MS medium supplemented with 1.0 mg/ℓ NAA and 1.5 mg/ℓ kinetin produced best results. Callus induction was at 26°C in the dark. At the end of 6 weeks, well-developed calli were excised and divided equally into six replicate pieces. Best callus performance during subculture was observed on 2.0 mg/ℓ NAA and 0.1 mg/ℓ 2iP. Subculture was for 4 weeks at 26°C in the darkness.

Shoots could be produced from either primary or subcultured calli when incubation was at 26°C in continuous flow level light from cool-white fluorescent lamps. The frequency of shoot development was low and variable depending on the auxin and cytokinin combination added to the basal medium. For example, 0.5 mg/ℓ NAA and 0.5 mg/ℓ kinetin gave four of four subcultured calli producing shoots while BAP alone gave one of four primary calli producing shoots. These were the best percentages for both types of calli. Most of the combinations gave 25% or less of the subcultured calli producing shoots and less than 5% of the primary calli producing shoots.

Root development on regenerated shoots was achieved by supplementing the medium with 1 mg/ℓ of IAA plus 0.1 mg/ℓ BAP. Rooted plants were cultured aseptically in peat pots containing a 1:1:1 (by volume) mixture of peat:perlite:sand and later transferred to the greenhouse.

f. Birdsfoot Trefoil (Lotus corniculatus L.)

An in vitro procedure for propagation and maintenance of genotypes has been recently described by Tomes.[128] Plants were maintained in a temperature controlled growth room at approximately 23°C/15°C light/dark temperatures with a 16 hr length day at 65 W/m² illumination. Expanded leaves were removed from the vegetative shoots and the stem was cut into sections at each internode. The stem sections, including the apical section, were surface sterilized in an aqueous solution of 0.6% sodium hypochlorite for 10 min followed by two rinses in sterile distilled water. After surface sterilization, the stem sections were again cut on both sides of the node such that a stem segment approximately 5-mm long containing the node was placed on the culture medium. Shoot tips were also used as explants. These included both the apical meristem and leaf primordia. Culture of both kinds of explants was in small glass vials (22 × 85 mm) containing about 10 mℓ of culture medium.

The inorganic salts and vitamins of the B5[211] medium solidified with 0.8% agar were used for culture. Addition of 0.05 mg/ℓ BA with no auxin produced more shoots per culture than medium with neither auxin nor cytokinin. Root formation, however, was enhanced by culture in medium without BA.

Incubation was at 25°C under a 16 hr photoperiod at low light intensity (8 W/m^2). Shoots were produced directly from the explant and callus was rarely formed. A higher perccentage of the node cultures produced shoots than did the shoot tip cultures making the former the preferred explant.

Small plants were removed from the vials and transplanted into moist Jiffy-7® peat balls. The peat balls were placed in plastic trays, covered with plastic, and placed in a growth room. The cover was removed after 2 weeks. Tomes[128] estimates that it would be possible to produce over 1000 plants from a single explant within 3 months by using two subcultures.

5. Sugar Crops

a. Sugar Beet (Beta vulgaris **L.)**

In vitro culture including the regeneration of plants from callus has been worked on considerably by Soviet workers.[133,134] More recently, Hussey and Hepher[131] in England described a method of clonal propagation of sugar beet plants by tissue culture. Their method utilizes precocious axillary shoot proliferation and the formation of tetraploid plants by colchicine treatment of in vitro tissues.

Monogerm diploid seed was germinated on a basic medium consisting of the MS[208] salt mixture used at one-half strength together with 2% w:v sucrose, 100 mg/ℓ *myo*-inositol and 0.5 mg/ℓ thiamine·HCl solidified with 0.7% agar. The medium (25 mℓ) was dispensed in 90 mℓ polystyrene screw-top jars. Cultures were kept at a constant temperature of 20°C in the light (4000 to 8000 lx) from white fluorescent tubes for 16 hr/day.

When the first pair of leaves was 10- to 15-mm long, the cotyledons and the roots together with all but 4 mm of the hypocotyl were excised and the remainder of each seedling transferred to media containing various concentrations of BAP. Control shoots subcultured to basic medium developed into complete plantlets. With BAP at 0.12 and 0.25 mg/ℓ precocious axillary shoots grew out from the axils of the other leaves, but root formation was inhibited. Axillary branches could be individually separated and recultured for further growth and branching. Thus, very large numbers of shoots could be built up by serial subculture. Shoots transferred to a medium containing 0.03 mg/ℓ or less of BAP eventually formed roots and the resulting plants could be planted in vermiculite.

b. Sugarcane (Saccharum officinarium **L.)**

Sugarcane is one of the few agronomic crops which is propagated vegetatively. In vitro culture has also been extensively studied in this crop although not primarily for the purpose of clonal propagation. Most of the effort has been directed toward recovering genetic variants which are resistant to one or more diseases. Many papers have been published on sugarcane tissue culture and these have been recently reviewed.[142]

Callus is easily induced and shoots and roots are easily obtained. A wide variety of tissues has been used as explants including shoot apices, leaves, and inflorescences.[139] The procedure outlined in an early paper by Heinz and Mee[139] will serve as the example here for in vitro culture and plantlet formation in sugarcane.

Shoot apices were excised from the cane stalk at the 5th or 6th internode from the apical meristem. Young leaves, intact or excised from the shoot apex and inflorescences taken at the premicrosporocyte stage to the mature pollen stage also served as explants. Shoot apices were surface sterilized as follows: ½ to 1 min in 95% ethanol, 4 to 10 min in either 1:400 phenyl mercuric acetate, 1.05% sodium hypochlorite, or 1:400 Wescodyne® (detergent) solution, then finally rinsed in sterile distilled water. Inflorescences were taken when still enclosed in the last leaf. The leaf surface was sterilized by wiping with sterilant and the sections of the inflorescence were removed and explanted.

The basal medium contained the mineral salts of MS[208] with the following organic constituents: 3 mg/ℓ 2,4-D, 1 mg/ℓ thiamine·HC1, 100 mg/ℓ *myo*-inositol, 10% by volume coconut water, 20 g/ℓ sucrose, and 9 g/ℓ agar. All tissue produced callus when 2,4-D was added to the medium. Callus developed from the exposed parenchyma cells of the internodes of excised shoot apices, from the cut surface of leaf sections, and on the cut surfaces, branches, rachis, and glumes of inflorescences. Callus was especially prevalent at the base of the pedicel. The most rapid formation of callus occurred with 3 ppm of 2,4-D and coconut water in the basal medium.

Differentiation of plants from callus was accomplished by transfer to the basal medium without 2,4-D. Greatest numbers of plants were produced from callus which was transferred, at monthly intervals, on 2,4-D medium two or three times and then transferred to medium without 2,4-D. Organ formation occurred on the latter medium within 1 month. Differentiated plants were transferred to water culture for further development of roots and then potted in vermiculite for further growth.

More recently, studies were conducted to improve root initiation and further shoot growth after shoot initiation.[144] Although coconut water stimulated callus growth and shoot differentiation, it sometimes inhibited continued shoot growth. Shoots grown on MS medium in the absence of both coconut water and 2,4-D showed more vigorous growth and developed a darker green color. Addition of 5 mg/ℓ of NAA to the medium improved shoot growth and enhanced root induction.

The method currently used by these workers[144] utilizes MS medium with 10% v:v coconut water and 3 mg/ℓ 2,4-D for callus induction, MS medium with coconut water and without 2,4-D for shoot differentiation, and MS medium with 5 mg/ℓ of NAA for root induction and plant maintenance. The rooted plantlets developed on the latter medium can be transplanted into vermiculite under greenhouse conditions with approximately 90% survival.

6. Oil Seed Crops

a. *Rape (Brassica napus* L.)

The following will outline procedures developed by Kartha et al.[146] to induce shoot and root formation from excised stem segments of *Brassica napus* L.

Plants were raised from seeds under greenhouse conditions in vermiculite. The plants were supplied with additional light from fluorescent lamps to a total intensity of 4.5 mW/cm^2. The temperature and relative humidity were 19 to 21°C and 40 to 45%, respectively. Internodal segments from 6-week-old plants were excised, rinsed in 70% ethanol, sterilized in 3% calcium hypochlorite, and rinsed in sterile distilled water. All subsequent operations were carried out aseptically in a laminar flow cabinet. The internodal segments were cut into 0.5 cm sections and transferred to the culture medium.

The nutrient medium contained the major and minor salts of MS[208] plus 3% sucrose, 0.6% Difco Bacto® agar and vitamins according to the B5[211] formulation. Benzyladenine and NAA or IAA were added depending on the experiment. The pH was adjusted to 5.7 prior to autoclaving. The IAA was filter sterilized. Twenty milliliters aliquots of medium were dispensed into 15 × 2.5 cm glass tubes and autoclaved at 1.46 kg/cm^2 for 20 min. The excised stem segments were transferred to these tubes. Culture was in a growth cabinet at 16 hr light and 8 hr dark cycle at 26°C. Light intensity was 3 mW/cm^2 provided by fluorescent lamps.

Benzyladenine at 5×10^{-6} *M* was optimal for shoot formation and multiple shoots formed on all stem explants. Shoots excised and cultured on MS medium supplemented with NAA or IAA at concentrations ranging from 10^{-6} to 10^{-5} *M* produced well developed roots on agar over a period of 10 to 12 days. Roots could also be produced on shoots transplanted into vermiculite without added hormones. Complete plant formation was obtained when NAA (2×10^{-6}, 5×10^{-6}, or 10^{-5} *M*) was employed in conjunction with BA at 5×10^{-6} *M*. Rooted plants could be transferred to vermiculite in pots and grown to mature plants.

b. Flax (Linum usitatissimum L.)

Plantlet formation has been reported from calli derived from hypocotyls,[157,158] cotyledons,[158-160] and stems.[159] Seeds were first rinsed in 50% ethanol for 1 min, surface sterilized with 1% sodium hypochlorite for 20 min, and rinsed in distilled water.[157] The seeds were germinated on absorbent cotton wetted with 50% Hoagland's solution in 25 × 150 mm screw-cap culture tubes. Five days after germination, the top portion of the seedlings was excised below the cotyledons under sterile conditions and the resulting hypocotyls were further incubated at 26°C, 70% relative humidity, under cool-white fluorescent lighting (about 4000 lx) with a photoperiod of 18 hr.

Gamborg and Shyluk[157] took 4 to 8 mm hypocotyl segments at different times after shoot removal and placed them on a nutrient agar medium. The medium contained the major salts of MS,[208] microelementes and vitamins of B5,[211] 30 g/ℓ sucrose, and 6 g/ℓ agar. The pH was adjusted to 5.8. Generally, two hypocotyls were placed in 4-oz glass jars with screw caps using 25 mℓ of medium. Incubation was in growth chambers under combined fluorescent and incandescent light (about 3000 lx) with a photoperiod of 20 hr. The temperature was 20°C in the light and 15°C in the dark.

After removal of the shoot and cotyledons from the young seedlings, several buds were produced on the hypocotyls and the resulting shoots from these generally continued to grow. When hypocotyl sections (3 to 4 days after removal of the cotyledons and shoot tip) were placed right side up with one end 2 to 3 mm into B5 agar medium at half strength, complete plant development occurred within 3 weeks. Auxins were not essential but 1 μM IAA resulted in development of a more extensive root system. Zeatin at 1 μM plus 0.1 μM IAA or NAA produced best results.

On extended culture of the hypocotyl in nutrient medium with 10 μM BA, a green callus with masses of buds could be produced in 1 month. When 2 to 3 mm^3 sections of the callus were transplanted onto new medium with 1 or 10 μM BA, further callus growth with bud and shoot formation was possible. Transfer of the shoot tips to one-half strength B5 medium with 0.1 μM IAA resulted in complete plant development.

c. Sunflower (Heliantus annus L.)

Stem pith was used as the explant.[156] A stem piece 10 cm in length, starting from the tip, was cut, the leaves were removed, and the stem swabbed with 80% ethanol. Cylinders of pith parenchyma were removed with a sterile No. 2 cork borer and sliced into pieces approximately 2.5 mm thick. These were surface sterilized with a 2% calcium hypochlorite solution. The discs were placed with their flat surface in contact with the medium in culture flasks.

The medium employed was a modified White's[207] medium supplemented with 1 ppm IAA. It consisted of the following in mg/ℓ: NH_4NO_3, 400; KCl, 65; KNO_3, 80;

KH_2PO_4, 12.5; $CaNO_3 \cdot H_2O$, 144; $MgSO_4 \cdot 7H_2O$, 72; Na_2EDTA, 25; H_3BO_3, 1.6; $MnSO_4 \cdot 4H_2O$, 6.5; $ZnSO_4$, 2.7; thiamine, 0.2; nicotinic acid, 0.5; pyroxidine, 0.5; glycine, 2.0; inositol, 100; sucrose, 20,000; and agar, 6000.

Culture was at 30 ± 1.5°C with a low light intensity of 1000 lx for 10 hr/day. After 8 weeks a large (4.57 g), loose, friable callus was obtained. Differentiation occurred after 10 weeks and a number of plantlets developed from a single callus tissue. These plantlets could be transferred to soil and grown to mature plants.

7. Fiber Crops

An account of procedure for flax is given above under oil seed crops. The most important fiber crop species, cotton (*Gossypium hirsutum* L.), has not been regenerated from cell or callus cultures. Recently, the differentiation of somatic embryoids from suspension cultures of a wild relative (*G. klotzschianum* Anderss) was reported.[221] Some of the embryoids developed roots and vestigial leaves, but complete plant development and establishment was not obtained.

8. Drug and Miscellaneous Crops

a. Tobacco (*Nicotiana tabacum* L.)

The genus *Nicotiana* has been one of the most, if not the most, extensively studied in in vitro culture of higher plants. It is considered to be a model system for somatic cell plant genetics.[222] Species in this genus and especially the cultivated species (*N. tabacum* L.), have been and are still being used extensively in basic studies on plant cell and tissue culture to elucidate mechanisms, nutrient requirements, etc. which lead to high frequency plantlet production from somatic tissue, cells, protoplasts, and pollen. The direct formation of haploid plants from anthers was first demonstrated in tobacco[170] and the first report of parasexual hybridization was between two *Nicotiana* species.[183] Obviously, the literature on tobacco culture is very extensive. Even though plants can be regenerated readily from in vitro culture of somatic tissue and even from cells and protoplasts, in vitro techniques are not currently used to propagate this species on a mass scale.

The following will be an account of two reports of procedures and results for plantlet regeneration from somatic tissue. Callus can be easily initiated from almost any tissue, e.g., stems, leaves, roots, inflorescences, etc. in *Nicotiana* species and usually plants can be easily regenerated from callus derived from any of these various tissues. For further information on plantlet regeneration from somatic tissue as well as haploid and protoplast culture the reader may consult the References in Table 1 and those cited in various review papers.

The first account will be that outlined by Walkey and Woolfitt[166] for clonal propagation of *Nicotiana rustica* L. from shoot meristem in culture. These workers aseptically dissected axillary bud tips, 400 to 600 μm in diameter, consisting of the meristem dome plus one or two pairs of primordial leaves. Each tip was placed in a 100 $m\ell$ conical flask containing 25 $m\ell$ of LS[209] medium without the optional constituents, but with 8 mg/ℓ IAA and 2.56 mg/ℓ kinetin. The flasks were rotated in a horizontal position at 8 rpm. The cultures were illuminated, but the intensity was not given. Within 3 to 4 weeks each tip developed into a mass of callus bearing several terminal leaf initials. After another 3 to 4 weeks, protrusions in the callus developed below these initials and each bore several small green leaves. The protrusion became detached and "budded" profusely into a ball of callus units, many of which became detached and continued to proliferate.

Proliferation into callus units continued if portions of the culture were transferred to new flasks and the shaking maintained. If callus units were removed and placed on filter paper bridges in still culture, plantlets developed and grew into normal plants. This method of culture enabled very large numbers of clonal plants to be raised from one apical tip.

The following account by Thorpe and Meier[161] utilized *Nicotiana tabacum* L. "Wisconsin 38". This particular genotype has been used more in plant cell and tissue culture research than any other in any plant species. Stems were cut from 1-m tall tobacco plants grown in the greenhouse. The leaves were removed, and the stems swabbed with 95% ethanol and cut into 5- to 7-cm long cylinders. A 15 cm region of the stem starting 10 cm from the tip was used. Cylinders of pith parenthyma were bored near the center of the stem with a sterile No. 2 cork borer and sliced into discs approximately 2 mm thick. The discs were placed in contact with the MS[208] medium in culture flasks. After a few weeks, firm white callus appeared which could be used directly for experimentation or subdivided and sbucultured for later use.

Callus was maintained on stem pith segments on a three-fourth strength MS medium. For experimental purposes, the tissue was grown on a modified MS medium reported by Thorpe and Murashige,[223] except that the levels of L-tyrosine, adenine sulfate, and $NaH_2PO_4 \cdot H_2O$ were reduced by half. The auxin employed was IAA at 10^{-5} *M*. Different cytokinins were incorporated, but results showed that kinetin at 10^{-5} *M* for shoot formation and 2×10^{-6} *M* for callus proliferation was satisfactory. Sucrose (3% w:v) performed as well or better for shoot formation than other carbon sources tested. Gibberellic acid at 5×10^{-6} *M* was added to the medium before autoclaving.

Cultures were maintained in the darkness or in a 16 hr photoperiod with fluorescent lighting at about 2700 lx. Shoot formation was greatest in the light with sucrose as the carbon source. Erlenmeyer flasks (125 mℓ) containing 50 mℓ of the medium were used as culture vessels. Three or four sections (about $2 \times 4 \times 4$ mm) of stock callus were placed in each flask. Presumably shoots which arose from calli could be further cultured and rooted and actively growing plants established.

V. PROBLEMS AND LIMITATIONS

In addition to the reasons listed earlier in this chapter restricting the need for in vitro propagation of agronomic crops, there are also other problems which exist and act to limit the use of in vitro culture for cloning these crops. These include problems with plant regeneration from callus and suspension cultures and genetic instability in tissue and cell cultures.

A. Plant Regeneration

A basic requirement for in vitro propagation is the demonstration of plantlet formation from callus or regeneration from single cells. Research in this area is not as well developed for most food and feed crops as for fruit, vegetable, ornamental, and tree crops covered in other chapter of this book. Rice et al.[78] have distinguished between *totipotency*, which is an inherent characteristic of most plant cells, and the *competence* of cell type(s) to respond to a given set of conditions used induce organogenesis. The yield of competent cells which develop into cell clones and/or plants for some plant species is very low.[224]

Even though plantlet formation from callus has been demonstrated, and in some cases very large numbers produced in most of the major cereal and several of the forage grass species,[225] the frequency is usually low and decreases with increasing time in callus maintenance and subculture. It is currently considered by some that cellular totipotency has not yet been clearly demonstrated in these species.[226,227]

Recent anatomical and histological studies of culture tissues from maize and other cereals led to the conclusion that the cultures grew not as a callus, but rather by an aberrant root-like mechanism of growth.[11,74] The callus-like structures, even those that had an external appearance of undifferentiated growth, revealed an internal organization different from the typically unorganized and undifferentiated callus growth of tobacco. Organized meristematic centers were present throughout the callus and continued proliferation, which appeared externally as callus, was due to the strong enhancement of branch root initiation.

Therefore, most reports of plant regeneration from cultured cereal tissues are considered by King et al.[227] to be no more then derepression of presumptive shoot primordia which proliferate adventitiously in culture. The same statement may also be made for forage grass species. Plant regeneration from protoplasts in bromegrass was reported.[113] However, a personal communication from the authors stated that even though the results were reproducible they could not claim that cellular totipotency had been demonstrated in cereals and grasses.

The failure to demonstrate cellular totipotency in these species certainly precludes the ability to study and manipulate genetics at the cellular level, e.g., induction and selection of mutants, genetic engineering experiments designed for uptake, incorporation, expression of exogenous DNA, and protoplast fusion experiments. From the standpoint of cloning, however, the multiplication or proliferation of meristems in callus culture which eventually express themselves as individual plantlets still offers a potential for propagation on a larger scale than is possible at the whole plant level. This is particularly true in species which do not possess rhizomes, stolons, or other plant parts readily adaptable for vegetative propagation.

Many tissue culture experiments with cereals and other Gramineae species have utilized embryos (mature or immature) as the original explant. These are not satisfactory tissues for cloning. In outcrossing species embryos on the same plant or even on the same inflorescence probably vary in genotype. This precludes the possibility of cloning genetically identical individuals. Thus, vegetative tissue, leaves, stems, etc. must be used as explants. According to Saalbach and Koblitz,[15] until recently there were no data available concerning callus formation from leaves of grasses. A major exception is sugarcane in which callus has been induced from most plant parts including roots, leaves, and parenchyma of internodal tissue.[142]

Bhojwani et al.[228] spent considerable effort culturing leaf tissue of wheat, maize, and *Sorghum*, but a callus, regardless of the age of the leaf and composition of the nutrient medium, was never obtained. These authors suggest that leaf cells of cereals either lack the potential to divide or have special requirements to elicit this potential. Segments from expanded leaves from 8-day-old wheat plants completely failed to yield callus on any of several media tested.[62] Henke et al.[31] initated callus in rice from leaf segments taken 1 to 5 mm from the base. This callus also formed roots and shoots. The possibility exists, however, that the apical meristem and/or other primordia may have been included in the explant. Using an extract of barley seedlings, Saalbach and Koblitz,[15] claimed success in initiating callus from barley leaves which were detached 3 to 5 mm above the meristem. Furthermore, one explant exhibited plant formation. In rice, calli were initiated from mesophyll protoplasts and roots, but no shoots were obtained.[216] Callus formation was induced from stem protoplasts of maize, but no differentiation was reported.[214]

Environmental factors, including medium composition, which promote shoot formation from callus and maintain this capacity over long periods have not been clearly defined for cereals and other grasses. Most commonly used medium formulations were developed for dicotyledonous species, especially tobacco. Although these media work well in many cases for Gramineae culture, more research is needed to develop an optimum medium for these species. The P2-mod. cereal culture medium[214] developed for maize protoplast culture was not effective for inducing protoplast divisions in that species.[215] Results from most previous studies suggested that cytokinins have little or no influence on shoot formation in Gramineae. However, recent results showed that 2iP promoted shoot organogenesis in rice[31] and that zeatin was required for morphogenesis in sorghum.[55] Zeatin also greatly enhanced the frequency of shoot formation in wheat callus cultures, especially when IAA was used as the auxin in place of 2,4-D.[65] Different kinds and concentrations of auxin may also influence callus induction, growth, and perhaps eventually frequency of shoot formation. For example, it was reported that 2,4,5-T may be superior to 2,4-D for callus induction and growth in ryegrass and orchardgrass and that lower concentrations of the former auxin were needed.[257]

The legumes consisting of both seed and forage species rank second to the grasses as important food and feed crops. These species, especially the seed legumes, are equally or perhaps even more difficult to culture in vitro than grasses.[226] In fact, there has been less success in obtaining plantlet formation from callus. For example, Crocomo et al.[92] obtained only two shoots in 400 primary callus cultures from leaf explants of bean. Higher frequenceis of shoot formation have, however, been obtained with callus cultures of pea.[95]

As stated earlier, there have been no clear reports of successful planlet regeneration in soybeans from callus or cell cultures. Chinese workers[229] (cited by Rice et al.[78]) reported that 21 plantlets were obtained from three of eleven primary soybean hypocotyl calli cultured on MS salts with 2.3 μM kinetin and 2.2 μM BAP. Several of these plantlets survived transfer to the soil. According to Rice et al.[78] the experimental details are not clearly described which makes interpretation of the results difficult. It was not possible to determine the precise frequency of regeneration, whether cotyledonary bud tissue could have been included in the original explant, the age of explant, the frequency of subculture or the number of shoots formed per explant. However, only hypocotyl cultures were capable of shoot formation. Plantlet formation through adventitious bud development from hypocotyl sections of soybean had been reported earlier.[9] Thomas and Wernicke[226] also observed plantlets apparently arising from callus derived from hypocotyl segments. However, closer examination revealed that the shoots arose from minute shoot buds which are present in the axils of cotyledons. This does not constitute plantlet regeneration or totipotency in the opinion of the authors.

Genotypic differences seem to be a major factor limiting both callus induction and plantlet formation from callus. These differences have been observed in many species. In maize, Green et al.[230] observed callus initiation in 70% of 23 single crosses tested and in only 40% of 17 inbreds tested. For plantlet formation from callus, Green and Phillips[75] found one inbred (A188) to be quite superior to four others tested.

The efficiency of plantlet formation from callus of indiangrass varied 100-fold among 10 different genotypes.[101] Shoot formation was quite variable in both tall fescue and orchardgrass where mature embryos were used as the original explant.[102,104] Since these are highly outcrossing species and the genotype may vary from one caryopsis to another, even on the same plant, it is suspected that some of the variability for plantlet formation from callus is due to genotypic differences.

In initial studies with alfalfa, a very low frequency of genotypes (0 to 5%) was found capable of forming plants from callus.[115] This frequency was increased to 67% by two cycles of recurrent selection.[116] These results provide encouragement for improving efficiency of plantlet formation through breeding and selection in other plant species. Therefore, attempts should be made to recognize, select, and manipulate competent cell types to improve plantlet formation from callus in field crops.[78]

B. Genetic Instability

Genetic and cytogenetic instability is a major factor limiting the use of in vitro techniques for cloning. This instability is inherent to plant tissue cultures, especially in callus, and is definitely unwanted when attempting to propagate identical genotypes. Examples of variability are numerous and the reader is referred to various reviews.[231,236] The most common nuclear change is polyploidy including aneuploidy. In agronomic crop species polyploid plants have arisen from diploids in tobacco,[237,238] sugarcane,[139] birdsfoot trefoil,[129] and white clover.[121] Aneuploidy has been shown in tobacco,[188,239,240] sugarcane species hybrids,[141] in both common and emmer wheat,[240] and durum wheat.[73] The degree of aneuploidy increases with callus age. In rice, the frequency was greater in callus derived from roots than in that derived from stem internodes.[241] For a more extensive list of polyploid and aneuploid plants from cell and tissue cultures, the reader is referred to the extensive compilation by D'Amato.[233] Increase in DNA content per cell with increasing maintenance and subculture period has been discussed by D'Amato,[233] Shillito et al.,[242] and Devreux et al.[243]

The use of 2,4-D as an auxin for inducing callus contributed greatly to the ability of researchers to manipulate plants in culture, especially cereals and grasses, as first shown in rye by Carew and Schwarting.[244] This synthetic auxin has been selected because it is not readily destroyed by most plant tissues. However, 2,4-D may not be as favorable as the natural auxin IAA or even NAA for maintaining stability of chromosome number.[245,246] Choice of hormones and other components as well as the concentration for stimulating cell division may help to reduce chromosomal changes.[246] The longer tissue or cells are maintained in subculture, the greater the opportunity for both genetic and cytogenetic abnormalities to occur.

Even though there was a variety of chromosome numbers present in callus cultures of rice[24] and wheat[60] the regenerated plants were found to be diploid. In many other reports of plant regeneration from plant cell and tissue cultures, the nuclear stability of neither the callus nor the regenerated plants was investigated. Further work is needed to confirm the regeneration of only diploid plants in species where they have been reported and chromosome numbers should be studied in regenerated plants from other species. Extensive cytological and genetic analyses should be conducted to define the degree of in vitro stability in the meristem cell line and in callus and suspension cutures of different age since the frequency of genetic aberrant plants increases progressively with each tissue or cell subculture.[247]

VI. POTENTIAL APPLICATION

The most potential for in vitro cloning of agronomic crops might be with species that are propagated vegetatively as a matter of course or with cross-breeding species in which genotypes must be maintained by vegetative propagation during certain stages of a breeding program even though the final planting by the farmer is with seed. Other possible potential uses of in vitro culture in agronomic crops include the use of meristem culture for eradicating viruses in desired clones and the creation of genetic variants to extend the base of germplasm for plant breeding programs. The latter mentioned use is defintely not cloning; however, it does offer a potential tool for plant breeders and therefore will be discussed.

A. Clonal Propagation

There may be less potential for in vitro cloning techniques in species which have plant parts readily available and accessible for easy vegetative propagation. For example, certain cultivars of bermudagrass are propagated vegetatively, but most of these produce abundant rhizomes and/or stolons. The latter can be easily cut and packaged for spreading over large areas to be disced in for easy establishment of a new pasture or turf.

Sugarcane is a good example of a normally vegetatively propagated agronomic species. Tissue culture investigations have been carried out extensively with sugarcane.[142,248-250] The work with sugarcane has not, however, emphasized rapid propagation or mass cloning via in vitro techniques. In fact, Heinz et al.[142] point out that no particular advantage is gained from sugarcane propagation via tissue culture unless selection pressure can be directed toward one particular trait such as resistance to eyespot, Fiji, or downy mildew diseases. Thus, in vitro culture experiments with sugarcane have emphasized the creation of genetic variability or modification of the plant's genome due to passage through the tissue culture system and the use of cultures for the elucidation of biochemical parameters.

As mentioned above, breeding of many types of crop plants depends to an extent on the propagation and maintenance of clones of selected material. Traditionally, cloning has been done by rooting certain plant parts. Recently, the development of in vitro culture techniques has given rise to the prospect of using these methods to maintain selected clones, not only of vegetatively propagated species, but also of species in which the clones are used at some stage in the production of commercial seed. This might be especially true of species in which the genotypes are heterozygous and/or have levels of self-incompatibility. In fact, it could allow the propagation of not only highly self-incompatible parents but also even male sterile parents.[251]

Species which do not produce extensive rhizomes, stolons, or other plant parts readily suitable for vegetative propagation might be especially suitable for in vitro cloning techniques. Certain species of perennial forage crops, including both grasses and legumes, fit into this category. Of course the potential also exists for other crop species in which the breeding procedure is the same or similar, i.e., crop species in which selected plants or clones (usually highly heterozygous) must be maintained by vegetative propagation during various stages of the breeding program.

Bunch grasses (both warm and cool season species) with short or nonexisting rhizomes are not adapted for rapid vegetative propagation. Current methods consist of dividing large clumps into smaller ones or propagation of single tillers. If technology can be developed to the extent that it can be easily applied, in vitro methods might prove an alternative to current methods for rapid clonal propagation of selected genotypes on a large scale. Clones in which only one or a few plants exist, might be especially suitable for in vitro techniques. The main advantage would be an increase in the multiplication rate. Theoretically, the potential exists for a thousand- or even a millionfold increase in the rate of clonal propagation over conventional methods. Millionfold increases may require the development of techniques to regenerate plants from cells.

If plants could be regenerated directly from meristematic tissue this might be advantageous over the technique of inducing callus since, as discussed above, plants regenerated from callus are liable to exhibit genetic aberrations. Also, the techniques of inducing callus and regeneration of plants from callus are complex and demand basic research which the plant breeder may not be able to justify. Plants regenerated from meristems display no more genetic abnormalities than do those propagated by conventional methods of clonal propagation.[251]

The two main problems which may exist are that appropriate meristems may not be available at the ontogenetic stage from which the plant breeder may wish to multiply the plant clonally and different crop plants differ widely in their potential multiplication rates. Cultured meristems from some plant species may give rise to only a single plant while those from other species may give rise to several plants. The inflorescence of many grass species consists of many florets and since there are several tillers on each plant the potential exists for regeneration of many plantlets from a single mother plant. Plants have been produced directly from explants without going through a callus state in birdsfoot trefoil,[128] flax,[157] and sugar beet[131] to name a few.

B. Meristem Tip Culture

The extensive use of meristem tip culture for the production of virus-free plants has resulted in it being the most economically important of all tissue culture techniques. This is especially true for ornamental and vegetable crops.[252-254] In most agronomic crops, virus transmission from generation to generation is not a serious problem because commercial propagation and distribution is by seed and the important viruses are not seed-transmitted.[109] Eradication of viruses, however, may be important in some outcrossing species where it is desirable to maintain genotypes in a healthy and vigorous condition and increase them by clonal propagation. The following will describe cases of virus elimination by meristem tip culture in *Nicotiana rustica*,[167] white clover,[123] red clover,[125] and ryegrass.[109]

1. Nicotiana rustica L.

Meristem tip culture was used to eradicate or greatly diminish the concentrations of cucumber mosaic (CMV) and alfalfa mosiac (ALFMV) viruses.[167] Less success was obtained in eradication of tobacco mosaic virus (TMV).

Meristem tips consisting of the dome surrounded by two or more pairs of leaf primordia were excised and placed in 100 mℓ flasks containing 25 mℓ of LS[209] medium without the optional constituents and with 8 mg/ℓ IAA and 2.56 mg/ℓ kinetin.[166] Culture was at constant temperature either in static tubes or on filter-paper bridges in moving flasks. Tissues were assayed for virus by grinding tips, whole cultures, proliferating tissue units, or 0.5 mm leaf discs in 0.1 *M* potassium phosphate buffer at pH 7.5 (1 g tissue per 1 mℓ buffer) and inoculating the extract to the following plants: *Chenopodium quinoa* (CMV and ALFMV), *N. tabacum* (ALFMV), and *N. glutinosa* (TMV).

At 34°C, CMV was almost completely eradicated after 46 days, but most cultures grown at 14 or 22°C remained infected. In static culture at 22°C, the concentration of ALFMV remained high, but was lower at 32°C. In shake culture, the concentration also remained high at 22°C, but at 32°C the virus was greatly reduced after 98 days and almost completely lost after 120 days in culture. The eradication of TMV was also attempted, but was not successful at either temperature.

2. White Clover (Trifolium repens L.)

Plants were grown at 10°C in controlled environment chambers and the apical 2 to 3 cm tip of stolons were used as explants.[123] The leaves were removed under aseptic conditions. The stolen tips were washed in water, surface sterilized in 70% ethanol for about 2 min, and washed again in sterile distilled water. The meristem tips, consisting of the meristem dome plus one or two leaf primordia, was excised and placed on the surface of the medium. The basic medium was MS[208] with 50 g/ℓ sucrose, 5 g/ℓ agar, and 10^{-5} mg/ℓ IAA. The culture vessels were flat-bottom glass vials, 24 × 72 mm, capped with inverted flat-bottom glass vials, 29 × 65 mm.

The presence or absence of virus was determined by inoculation from each plant that developed from a meristem tip to the following plants: *Chenopodium amaranticolor, C. quinoa, Nicotiana clevelandii, N. tabacum* "X-73", *Phaseolus vulgaris* "Bountiful", *Pisum sativum* "Alaska", and *Vigna unguiculata* subsp. *unguiculata.*

Each plant in which no virus was detected was assayed at least three times during 18 months. Assays were conducted for white clover mosaic virus (WCMV), clover yellow vein virus (CYVV), alfalfa virus (AMV), and peanut stunt virus (PSV).

About 9% of the tips survived, grew, and were transplanted into sand. Of the plants derived from meristem tip culture from each of the original virus-infected clones, 57% were free of all detectable viruses. All of the plants from clones infected with CYVV and PSV were free of the viruses and 83 and 60% of the plants infected with AMV and WCMV, respectively, were freed by meristem tip culture.

3. Red Clover (Trifolium pratense L.)

Meristem culture techniques were used to eliminate viruses from red clover.[125] Major viruses included bean yellow mosaic, peanut stunt, clover mosaic, and alfalfa mosaic viruses. The authors obtained meristem explants from crown tissue of nonflowering plants and from axillary meristems from flowering stems of plants with inflorescences. Explants measuring 2 to 5 cm in length and containing crown or axillary shoot meristems were surface sterilized by placing them in 70% ethanol for 5 min and then in 1% sodium hypochlorite for 8 min. They were rinsed in sterile deionized water for at least 5 min. Excised meristems containing the meristematic dome and one foliar primordium were compared with meristems containing the dome, second foliar primordium and primary leaf. The medium used was the PC-L2 medium previously described by the authors.[124] The composition of this medium is listed in Tables 2 and 3. Best results for shoot initiation was obtained with this medium plus 4.0 μg/ℓ picloram and 1.0 mg/ℓ BAP.

When shoots were 2.0- to 2.5-cm tall, they were transferred to a rotting medium which is half strength PC-L2 basal medium without picloram and BAP, but with the full complement of ferrous sulfate and potassium phosphate. Sucrose were reduced from 25 to 15 g/ℓ and 1 mg/ℓ nicotinic acid, 2.5 mg/ℓ 3-aminopyridine, and 0.02 mg/ℓ IAA were added. Light intensity during rooting was at about 800 $\mu Em^{-2}sec.^{-1}$ Rooted shoots were transferred to soil mixed with vermiculite and sand (4:2:1) and maintained under a plastic humidity tent. Plants were subsequently potted in soil and grown to maturity in the greenhouse.

For the bioassay, leaves were ground in a mortar with carborundum (2:1 on a w:w basis). One gram of tissue was mixed with 10 mℓ of a 2.0 g/ℓ solution of sodium diethyldithiocarbamate. Leaves of *Chenopodium amaranticolor* Coste and Reyn. were rubbed with the tissue homogenate. Virus symptoms were checked after 1 and 2 weeks.

Best results were obtained with plants derived from crown meristems of nonflowering plants and it was concluded that only crown-derived meristems should be used as explants in shoot tip cultures. A lower percentage (65 vs. 80%) of the cultured excised shoot tips containing the meristematic dome and first foliar primordium grew as compared to tips containing the dome, primary leaf, and second foliar primordium. However, the smaller explants yielded a higher percentage of symptomless plants (85 vs. 50%).

4. Ryegrass (Lolium spp.)

Plants were established from seed in boxes containing potting compost and grown during the winter months in a greenhouse heated to 15 to 20°C.[109] Half of the plants were used as uninoculated controls and the other half were inoculated with ryegrass mosaic virus (RMV) at about the three tiller stage. Inoculum was prepared by homogenizing leaves showing the disease symptoms with 0.15 *M* phosphate buffer (pH 7). Celite was added to the extract as an abrasive and the mixture was rubbed onto the leaves. Meristem tips from inoculated material showing RMV symptoms were cultured 9 weeks after inoculation.

Various sizes of the tips (0.2 to 1.1 m) were cultured consisting of the meristem dome plus one or several pairs of leaf primordia. It was found that size of tip did not influence effectiveness of virus eradication, but longer tips cultured more easily and a higher percentage gave rise to plants. Before dissection, tillers were trimmed and surface sterilized in a solution of sodium hypochlorite (6% available chlorine) for 7 min and washed six times with sterile water.

The MS[208] basal culture medium was used except for the addition of iron in the form of Fe-EDTA and the omission of $FeSO_4$ and Na_2-EDTA. The medium was solidified with 4.4 g/ℓ of agar. All components of the medium were autoclaved at 121°C for 15 min after adjusting to pH 5.6. Culture was in plastic tubes (10 mℓ per tube) at 25 ± 1°C and illuminated continuously at 6000 lx with warm-white fluorescent tubes. Plantlets produced from infected and noninfected material were recorded 9 weeks after culturing. At this time they were transferred to the soil. After 6 weeks in the soil, symptoms were not observed in any of the plants. The plants were subsequently checked for a period of 10 months and no symptoms appeared over that time.

C. Creation of Genetic Variability

The problem of genetic variability and its undesirable qualities for cloning were mentioned above. However, from a positive standpoint, it represents a potentially powerful tool for creating genetic variation for plant breeding programs. As pointed out by Meredith and Carlson[231] a plant cell culture should not be considered static. Its genetic character is not fixed but is quite changeable, a property which makes cultured plant cells amenable to manipulation for a variety of purposes. In studies with sugarcane, Liu and Chen[137] stated that callus culture per se is an effective mutagen because callus cells maintained in vitro over a long period are usually cytologically unstable and give rise to plantlets which are often characterized by genetic variability.

Selection of mutants in tissue cultures has been very successful in sugarcane.[142,250] Inherent variation which occurs in tissue culture has been utilized in addition to treating the cultures with both chemical and physical mutagens. Resistance to eyespot disease was obtained by regenerating plants from suspension cultures which were previously treated with methyl methanesulfonate or ionizing radiation. Resistance to other diseases such as downy mildew and Fiji disease was obtained by screening subclones derived directly from tissue cultures.

Sugarcane is a vegetatively propagated crop and the creation of genetic variability through sexual hybridization is not a readily available tool. However, even in sexually propagated species tissue cultures have been used to create genetic variability. In rice, from 75 seed calli of a genetically pure line, 1121 plants (D_1) were regenerated and 83 plants among them were albina.[33] In the D_2 generation, 6500 plants were examined for various morphological and chlorophyll mutants. Only 28.1% of the plants were normal for all characters examined. Cell lines of *Zea mays* resistant to *Helminthosporium maydis* toxin were selected from callus cultures originating from an originally susceptible CMS-T genotype. All plants regenerated from the fifth and later selection cycles were toxin resistant.[255] This represents a significant finding even though the character is cytoplasmically inherited. The feasibility of exploiting cytogenetic and genetic variation in tissue cultures of durum wheat has also been explored.[73] Chromosome counts in shoot and root apices of plantlets formed from callus showed chromosome number mosaicism (one or more aneuploid chromosome numbers in addition to the diploid number) in all apical meristems analyzed. There was a high frequency of reduced chromosome numbers (hypohaploid, haploid, and hypodiploid).

Polyploid and/or self-incompatible plant species are not as amenable to "conventional" mutation breeding techniques as diploid, self-fertilizing species.[256] More than

2000 ryegrass plants were regenerated by continuous culture over 5 years from the callus of an aborted triploid embryo, obtained from the diploid (*Lolium perenne* × *L. multiflorum*) × tetraploid *L. perenne*.[112] The regenerated plants from the initial callus were triploid, but those differentiated from successive subcultures showed variations of chromosome number and structure. The plants showed a wide variation in leaf-shape, size, floral development, growth vigor, survival, and perenniality. Some of the novel recombinants have not been produced via sexual hybridization. The seed progeny of the plants from tissue culture showed a similar wide variation of phenotypes which combined valuable agronomic characters of the parental taxa. Ahloowalia[112] concluded that tissue culture is a valuable technique for generating new variants of ryegrass.

All plants regenerated from young callus of annual ryegrass × tall fescue had 28 chromosomes and were similar to the source plants.[106] However, those regenerated from aged callus included several phenotypic variants and several plants with a doubled number of chromosomes. More than 1200 regenerated plants were evaluated under field conditions for useful agronomic traits. In another study with tall fescue, about 12% of the recovered plants from callus exhibited various degrees of albinism.[104]

Inducing mutations in somatic tissue creates the problem of chimeras. However, new cultivars of highly heterozygous, outcrossing, self-incompatible species, such as tall fescue, are produced by combining several different clones. The performance of the progeny at least four generations removed from the polycross of the parent clones is the important factor and a high amount of heterozygosity is tolerated within cultivars and even on the same plant. Hopefully, the combination of outstanding clones, even though they may be heterogeneous, will result in improved new cultivars.

VII. SUMMARY AND CONCLUSIONS

Rapid cloning or mass propagation by in vitro techniques are currently being used very little, if at all, commercially in agronomic crops. Most agronomic crops are propagated by seed, thus the need for propagation by cell and tissue culture methods is not as great as in other classes of agricultural plants. With the exception of sugarcane, species in the two major families which comprise most of the important agronomic crops, i.e., the Gramineae and Leguminosae, are difficult to culture, the frequency of plantlet formation from callus is low, and it further decreases with increasing time of callus maintenance and subculture. Also, the maintenance of genetic and cytogenetic stability in cell and tissue cultures is a major problem limiting the use of in vitro techniques for cloning since the purpose is to maintain and propagate specific genotypes.

The preceding has documented technology, both in a general sense and for a number of specific crop species from the original literature sources. Several media compositions, including both inorganic and organic constituents have also been given. Agronomic species which might have the most potential for utilizing in vitro techniques for propagation include those which are normally propagated vegetatively and outcrossing species in which clones must be maintained by vegetative propagation during certain stages of a breeding program for the ultimate production of commercial seed. The use of meristem culture for virus eradication might also be most applicable in species fitting either of these two categories. The potential also exists for using in vitro cloning methods for maintaining any specific genotype of any crop species including heterozygotes, homozygotes, and self-incompatible species maintained by seed.

The regeneration of many plants directly from meristems without passing through the callus state might have advantages for maintaining genetic stability. However, the utilization of genetic and cytogenetic instability for creating genetic variation for plant breeding programs may be a useful tool in the improvement of certain species.

ACKNOWLEDGMENTS

The author extends his sincere appreciation to Professor L. N. Skold and Dr. L. F. Seatz for their critical reading of the manuscript and to Miss Alma Ramsey for her patience and diligence in typing the manuscript.

ADDENDUM

During the interim period of submission and publication of this manuscript there have been several additional reports of plantlet regeneration from tissue cultures of various agronomic crop species. Most important however, are reports of plantlet regeneration and recovery of whole plants from protoplasts of pearl millet [*Pennisetum americanum* (L.) K. Schum.],[260] white clover (*Trifolium repens* L.),[261] and alfalfa *Medicago sativa* L.).[262,263] These are the first reports of plantlet regeneration from protoplasts in the two most economically important agronomic families, i.e., the Gramineae and Leguminosae. The results with alfalfa are especially significant since the reports from both laboratories deal with plantlet regeneration from leaf mesophyll protoplasts. These reports provide encouragement for the demonstration of cellular totipotency in cereal and legume species. Hopefully, similar results with other important species in these two families are forthcoming so that routine manipulation, experimentation, and even improvement of agronomic crops at the cellular level may not be too far in the distant future.

REFERENCES

1. **Leonard, W. H. and Martin, J. H.,** *Cereal Crops,* Macmillan, New York, 1963, chap. 1.
2. **Janick, J., Schery, R. W., Woods, F. W., and Ruttan, V. W.,** *Plant Science — An Introduction to World Crops,* 2nd ed., W. H. Freeman, San Francisco, 1974, chap. 20.
3. **Martin, J. H., Leonard, W. H., and Stamp, D. L.,** *Principles of Field Crop Production,* 3rd ed., Macmillan, New York, 1976, chap. 1.
4. **Hodgson, H. J.,** Forage crops, *Sci. Am.,* 234, 61, 1976.
5. **Murashige, T.,** Plant propagation through tissue cultures, *Annu. Rev. Plant Physiol.,* 25, 135, 1974.
6. **Murashige, T.,** The impact of plant tissue culture on agriculture, in *Frontiers of Plant Tissue Culture 1978,* Thorpe, T. A., Ed., University of Calgary Printing Service, Canada, 1978, 15.
7. **Conger, B. V.,** Problems and potentials of cloning agronomic crops via, *in vitro* techniques, in Propagation of Higher Plants through Tissue Culture, Hughes, K. W., Henke, R. R., and Constantin, M. J., Eds., Technical Information Center, U.S. Department of Energy, Springfield, Va., 1978, 62.
8. **Murashige, T.,** Principles of rapid propagation, in Propagation of Higher Plants through Tissue Culture, Hughes, K. W., Henke, R. R., and Constantin, M. J., Eds., Technical Information Center, U.S. Department of Energy, Springfield, Va., 1978, 14.
9. **Kimball, S. L. and Bingham, E. T.,** Adventitious bud development of soybean hypocotyl sections in culture, *Crop Sci.,* 13, 758, 1973.
10. **Cummings, D. P., Green, C. E., and Stuthman, D. D.,** Callus induction and plant regeneration in oats, *Crop Sci.,* 16, 465, 1976.
11. **Cure, W. W. and Mott, R. L.,** A comparative anatomical study of organogenesis in cultured tissues of maize, wheat and oats, *Physiol. Plant.,* 42, 91, 1978.
12. **Lorz, H., Harms, C. T., and Potrykus, I.,** Regeneration of plants from callus in *Avena sativa* L., *Z. Pflanzenzuecht.,* 77, 257, 1976.
13. **Cheng, T. Y. and Smith, H. H.,** Organogenesis from callus culture of *Hordeum vulgare, Planta,* 123, 307, 1975.
14. **Koblitz, H. and Saalbach, G.,** Callus cultures from apical meristems of barley *(Hordeum vulgare), Biochem. Physiol. Pflanz.,* 170, 97, 1976.

15. **Saalbach, G. and Koblitz, H.**, Attempts to initiate callus formation from barley leaves, *Plant Sci. Lett.*, 13, 165, 1978.
16. **Dale, P. J. and Deambrogio, E**,.A comparison of callus induction and plant regeneration from different explants of *Hordeum vulgare, Z. Pflanzenphysiol.*, 94, 65, 1979.
17. **Orton, T. J.**, A quantitative analysis of growth and regeneration from tissue cultures of *Hordeum vulgare, H. jubatum* and their interspecific hybrid., *Environ. Exp. Bot.*, 19, 319, 1979.
18. **Clapham, D.**, Haploid *Hordeum* plants from anthers, *in vitro, Z. Pflanzenzuecht.*, 69, 142, 1973.
19. **Foroughi-Wehr, B., Mix, G., Gaul, H., and Wilson, H. M.**, Plant production from cultured anthers of *Hordeum vulgare* L., *Z. Pflanzenzuecht.*, 77, 198, 1976.
20. **Mix, G., Wilson, H. M., and Foroughi-Wehr, B.** the cytological status of plants of *Hordeum vulgare* L. regenerated from microspore callus, *Z. Pflanzenzuecht.*, 80, 89, 1978.
21. **Malepszy, S. and Grunewald, G.**, Ein Beitrag zur Erzeugung von Haploiden bei *Hordeum vulgare, Z. Pflanzenzuecht*, 72, 206, 1974.
22. **Gonzalez-Medina, M. and Bouharmont, J.**, Experiments on anther culture in barley. Influence of culture methods on cell proliferation and organ differentiation, *Euphytica*, 27, 553, 1978.
23. **Foroughi-Wehr, B. and Mix, G.**, *In vitro* response of *Hordeum vulgare* L. anthers cultured from plants grown under different environments, *Environ. Exp. Bot.*, 19, 303, 1979.
24. **Nishi, T., Yamada, Y., and Takahashi, E.**, Organ redifferentiation and plant restoration in rice callus, *Nature (London)*, 219, 508, 1968.
25. **Nishi, T., Yamada, Y., and Takahashi, E.**, The role of auxins in differentiation of rice tissue cultures *in vitro, Bot. Mag.*, 86, 183, 1973.
26. **Tamura, S.**, Shoot formation in calli originated from rice embryo, *Proc. Jpn. Acad.*, 44, 544, 1968.
27. **Lai, K. L. and Liu, D. J.**, Morphogenetical studies on the callus originated from excised roots of rice plants. Organ induction and the successful rising of green plants, *Mem. Coll. Agric. Natl. Taiwan Univ.*, 11, 106, 1971.
28. **Kawata, S. and Ishihara, A.** The regeneration of rice plant, *Oryza sativa* L., in the callus derived from the seminal root, *Proc. Jpn. Acad.*, 44, 549, 1968.
29. **Maeda, E.**, Subculture and organ formation in the callus derived from rice embryos *in vitro, Proc. Crop. Sci. Soc. Jpn.*, 37, 51, 1968.
30. **Saka, H. and Maeda, E.**, Effect of kinetin on organ formation in callus tissue derived from rice embryos, *Proc. Crop Sci. Soc. Jpn.*, 38, 668, 1969.
31. **Henke, R. R., Mansur, M. A., and Constantin, M. J.**, Organogenesis and plantlet formation from organ-and seedling-derived calli of rice *(Oryza sativa), Physiol. Plant.*, 44, 11, 1978.
32. **Nakano, H., Tashiro, T., and Maeda, E.**, Plant differentiation in callus tissue induced from immature endosperm of *Oryza sativa* L., *Z. Pflanzenphysiol.*, 76, 444, 1975.
33. **Oono, K.**, Test tube breeding of rice by tissue culture, *Trop. Agric. Res. Ser.*, 11, 109, 1978.
34. **Mascarenhas, A. F., Pathak, M., Hendre, R. R., Ghugale, D. D., and Jagannathan, V.**, Tissue culture of maize, wheat, rice, and *Sorghum.* IV. Studies of organ differentiation in tissue cultures of maize, wheat, and rice, *Indian J. Exp. Biol.*, 13, 116, 1975.
35. **Cornejo-Martin, M. J., Mingo-Castel, A. M., and Primo-Millo, E.**, Organ redifferentiation in rice callus: effects of $C_2H_4CO_2$ and cytokinins, *Z. Pflanzenphysiol.*, 94, 117, 1979.
36. **Niizeki, H. and Oono, K.**, Induction of haploid rice plant from another culture, *Proc. Jpn. Acad.*, 44, 554, 1968.
37. **Woo, S. C. and Su, H. Y.**, Doubled haploid rice from Indica and Japonica hybrids through anther culture, *Bot. Bull. Acad. Sin.*, 16, 19, 1975.
38. **Guha, S., Iyer, R. D., Gupta, N., and Swaminathan, M. S.**, Totipotency of gametic cells and the production of haploids in rice, *Curr. Sci.*, 8, 174, 1970.
39. **Iyer, R. D. and Raina, S. K.**, The early ontogeny of embryoids and callus from pollen and subsequent organogenesis in anther culture of *Datura metel* and rice, *Planta*, 104, 146, 1972.
40. **Nishi, T. and Mitsuoka, S.**, Occurrence of various ploidy plants from anther and ovary culture of rice plant, *Jpn. J. Genet.*, 44, 341, 1969.
41. **Academia Sinica**, Studies on anther culture *in vitro* of *Oryza sativa* subsp. Shien I., *Acta Genet. Sin.*, 2, 81, 1975.
42. **Guha-Muherjee, S.**, Genotypic differences in the *in vitro* formation of embryoids from rice pollen, *J. Exp. Bot.*, 24, 139, 1973.
43. **Woo, S. C. and Tung, I. J.**, Induction of rice plants from hybrid anthers of Indica and Japonica cross, *Bot. Bull. Acad. Sin.*, 14, 61, 1973.
44. **Niizeki, H. and Oono, K.**, Induction of haploid rice plant from anther culture, *Proc. Jpn. Acad.*, 44, 554, 1968.
45. **Niizeki, H. and Oono, K.**, Rice plants obtained by anther culture, *Colloq. Int. C. N. R. S.*, 193, 251, 1971.
46. **Chen, C. C.**, *In vitro* development of plants from microspores of rice, *In Vitro*, 13, 484, 1977.

47. **Chen. C. C. and Lin, M. H.**, Induction of rice plantlets from anther culture, *Bot. Bull. Acad. Sin*, 17, 18, 1976.
48. **Academia Sinica**, Investigation on the induction and genetic expression of rice pollen plants, *Sci. Sin.*, 17, 209, 1974.
49. **Genovesi, A. D. and Magill, C. W.**, Improved rate of callus and green plant production from rice anther culture, *Crop. Sci.*, 19, 662, 1979.
50. **Wenzel, G.**, Production of haploids of rape and rye, in *Production of Natural Compounds by Cell Culture Methods*, BPT Rep. 1/78, Alfermann, A. W. and Reinhard, E., Eds., Gesellschaft fur Strahlen und Umweltforschung mbH, Munich, 1978, 312.
51. **Thomas, E., Hoffman, F., and Wenzel, G.**, Haploid plantlets from microspores of rye, *Z. Pflanzenzuecht.*, 75, 106, 1975.
52. **Thomas, E. and Wenzel, G.**, Embryogenesis from microspores of rye, *Naturwissenschaften*, 62, 40, 1975.
53. **Wenzel, G. and Thomas, E.**, Observations on the growth in culture of anthers of *Secale cereale*, *Z. Pflanzenzuecht.*, 72, 89, 1974.
54. **Hoffmann, F.**, Mutation and selection of haploid cell culture systems of rape and rye, in *Production of Natural Compounds by Cell Culture Methods*, BPT Rep. 1/78, Alfermann, A. W. and Reinhard, E., Eds., Gesellschaft fur Strahlen- und Umweltforschung mbH, Munich, 1978, 319.
55. **Gamborg, O. L., Shyluk, J. P., Brar, D. S., and Constabel, F.**, Morphogenesis and plant regeneration from callus of immature embryos of *Sorghum*, *Plant Sci. Lett.*, 10, 67, 1977.
56. **Thomas, E., King, P. J., and Potrykus, I.**, Shoot and embryo-like structure formation from cultured tissues of *Sorghum bicolor*, *Naturwissenschaften*, 64, 587, 1977.
57. **Masteller, V. J. and Holden, D. J.**, The growth of and organ formation from callus tissue of *Sorghum*, *Plant Physiol.*, 45, 362, 1970.
58. **Dunstan, D. I., Short, K. C., and Thomas, E.**, The anatomy of secondary morphogenesis in cultured scutellum tissues of *Sorghum bicolor*, *Protoplasma*, 97, 251, 1978.
59. **Dudits, D., Nemet, G., and Haydu, Z.**, Study of callus growth and organ formation in wheat *(Triticum aestivum)* tissue cultures, *Can. J. Bot.*, 53, 957, 1975.
60. **Shimada, T., Sasakuma, T., and Tsunewaki, K.**, *In vitro* culture of wheat tissues. I. Callus formation, organ redifferentiation and single cell culture, *Can. J. Genet. Cytol.*, 11, 294, 1969.
61. **Chin, J. C. and Scott, K. J.**, Studies on the formation of roots and shoots in wheat callus cultures, *Ann. Bot., (London)*, 41, 473, 1977.
62. **O'Hara, J. F. and Street, H. E.**, Wheat callus culture: the initiation, growth and organogenesis of callus derived from various explant sources, *Ann. Bot. (London)*, 42, 1029, 1978.
63. **Shimada, T.**, Plant regeneration from the callus induced from wheat embryo, *Jpn. J. Genet.*, 53, 371, 1978.
64. **Bhojwani, S. S. and Hayward, C.**, Some observations and comments on tissue culture of wheat, *Z. Pflanzenphysiol.*, 85, 341, 1977.
65. **Gosch-Wackerle, G., Avivi, L., and Galun, E.**, Induction, culture and differentiation of callus from immature rachises, seeds and embryos of *Triticum*, *Z. Pflanzenphysiol.*, 91, 267, 1979.
66. **Ouyang, T. W., Hu, H., Chuang, C. C., and Tseng, C. C.**, Induction of pollen plants from anthers of *Triticum aestivum* L. cultured *in vitro*, *Sci. Sin.*, 16, 79, 1973.
67. **Picard, E., and deBuyser, J.**, Obtaining haploid plants of *Triticum aestivum* L. by the culture of anthers *in vitro*, *C. R. Acad. Sci. Ser. D*, 277, 1463, 1973.
68. **Wang, C. C., Chu, C. C., Sun, C. S., Wu, S. H., Yin, K. C., and Hsu, C.**, The androgenesis in wheat (*Triticum aestivum)* anthers cultured *in vitro*, *Sci. Sin.*, 16, 218, 1973.
69. **Heszky, L. and Mesch, J.**, Anther culture investigations in cereal gene bank collections, *Z. Pflanzenzuecht.*, 77, 187, 1976.
70. **Shimada, T. and Makino, T.**, *In vitro* culture of wheat. III. Anther culture of the A genome aneuploids in common wheat, *Theor. Appl. Genet.*, 46, 407, 1975.
71. **DeBuyser, J. and Henry, Y.**, Androgenèse sur des blés tendres en cours de sélection l. L'Obtention des plantes, *in vitro*, *Z. Pflanzenzuecht.*, 83, 49, 1979.
72. **Schaeffer, G. W., Baenziger, P. S., and Worley, J.**, Haploid plant development from anthers and *in vitro* embryo culture of wheat, *Crop Sci.*, 19, 697, 1979.
73. **Bennici, A. and D'Amato, F.**, *In vitro* regeneration of durum wheat plants. I. Chromosome numbers of regenerated plantlets, *Z. Planzenzuecht.*, 81, 305, 1978.
74. **Mott, R. L. and Cure, W. W.**, Anatomy of maize tissue cultures, *Physiol. Plant.*, 42, 139, 1978.
75. **Green, C. E. and Phillips, R. L.**, Plant regeneration from tissue cultures of maize, *Crop Sci.*, 15, 417, 1975.
76. **Harms, C. T., Lorz, H., and Potrykus, I.**, Regeneration of plantlets from callus cultures of *Zea mays* L., *Z. Pflanzenzuecht.*, 77, 347, 1976.
77. **Freeling, M., Woodman, J. C., and Cheng, D. S. K.**, Developmental potentials of maize tissue cultures, *Maydica*, 21, 97, 1976.

78. **Rice, T. B., Reid, R. K., and Gordon, P. N.**, Morphogenesis in field crops, in Propagation of Higher Plants through Tissue Culture, Hughes, K. W., Henke, R. R., and Constantin, M. J., Eds., Technical Information Center, U.S. Departmental of Energy, Springfield, Va., 1978, 262.
79. **Academia Sinica**, People's Republic of China develops maize haploid, *New China News Agency*, Hong Kong, Oct. 27, 1977.
80. Institute of Genetics, 401 Research Group, Academia Sinica, Peking, Primary study on induction of pollen plants of *Zea mays*, *Acta Genet. Sin.*, 2, 138, 1975.
81. **Rangan, T. S.**, Morphogenic investigations on tissue cultures of *Panicum miliaceum*, *Z. Pflanzenphysiol.*, 72, 456, 1974.
82. **Rangan, T. S.**, Growth and plantlet regeneration in tissue cultures of some Indian millets: *Paspalum scrobiculatum* L., *Eleusine coracana* Gaertn. and *Pennisetum typhoideum* Pers., *Z. Pflanzenphysiol.*, 78, 208, 1976.
83. **Ban, Y., Kokubu, T., and Miyaji, Y.**, Production of haploid plants by anther culture of *Setaria italicà*, *Bull. Fac. Agric. Kagoshima Univ.*, 21, 77, 1971.
84. **Kimato, M. and Sakamoto, S.**, Production of haploid albino plants of *Aegilops* by anther culture, *Jpn. J. Genet.*, 47, 61, 1972.
85. **Ono, H. and Larter, E. N.**, Anther culture of triticale, *Crop Sci.*, 16, 120, 1976.
86. **Sun, C. S., Wang, C. C., and Chu, C. C.**, Cell divisions and differentiation of pollen grains in *Triticale* anthers cultured *in vitro*, *Sci. Sin.*, 17, 47, 1974.
87. **Bernard, S.**, Étude de quelques facteurs contribuant à la réussite de l'androgénèse par culture d'anthères *in vitro* chez le triticale hexaploide, *Ann. Amelior. Plant. (Paris)*, 27, 639, 1977.
88. **Wang, Y. Y., Sun, C. C., Wang, C. C., and Chien, N. F.**, The induction of the pollen plantlets of *Triticale* and *Capsicum annuum* from anther culture, *Sci. Sin.*, 16, 147, 1973.
89. **Wang, C. C., Chu, Z. C., Sun, C. S., Hsu, C., Yin, K. C., and Bi, F. Y.**, Induction of pollen plants from the anther culture of *Triticum vulgare* — *Agropyron glaucum* hybrid, *Acta Genet. Sin.*, 2, 72, 1975.
90. **Cooper, K. V., Dale, J. E., Dyer, A. F., Lyne, R. L., and Walker, J. T.**, Hybrid plants from the barley × rye cross, *Plant Sci. Lett.*, 12, 293, 1978.
91. **Sangwan, R. S. and Gorenflot, R.**, *In vitro* culture of *Phragmites* tissues. Callus formation, organ differentiation and cell suspension culture, *Z. Pflanzenphysiol.*, 75, 256, 1975.
92. **Crocomo, O. J., Sharp, W. R., and Peters, J. E.**, Plantlet morphogenesis and the control of callus growth and root induction of *Phaseolus vulgaris* with the addition of a bean seed extract, *Z. Pflanzenphysiol.*, 78, 456, 1976.
93. **Bajaj, Y. P. S. and Dhanju, M. S.**, Regeneration of plants from apical meristem tips of some legumes, *Curr. Sci.*, 48, 906, 1979.
94. **Kartha, K. K., Gamborg, O. L., and Constabel, F.**, Regeneration of pea (*Pisum sativum* L.) plants from shoot apical meristems, *Z. Pflanzenphysiol.*, 72, 172, 1974.
95. **Malmberg, R. L.**, Regeneration of whole plants from callus culture of diverse genetic lines of *Pisum sativum* L., *Planta*, 146, 243, 1979.
96. **Kartha, K. K., Leung, N. L., and Gamborg, O. L.**, Freeze-preservation of pea meristems in liquid nitrogen and subsequent plant regeneration, *Plant Sci. Lett.*, 15, 7, 1979.
97. **Gamborg, O. L., Constabel, F., and Shyluk, J. P.**, Organogenesis in callus from shoot apices of *Pisum sativum*, *Physiol. Plant.*, 30, 125, 1974.
98. **Shama Rao, H. K. and Narayanaswamy, S.**, Effect of gamma irradiation on cell proliferation and regeneration in explanted tissues of pigeon pea, *Cajanus cajan* (L.) Mills P., *Radiat. Bot.*, 15, 301, 1975.
99. **Aubrey, A. M., Dutuit, P., Thiellement, H., and Bervillé, A.**, Propagation végétative de la féverole (*Vicia faba*) a patir de fragments de tiges, *Ann. Amelior. Plant. (Paris)*, 25, 225, 1975.
100. **Chen, C. H., Stenberg, N. E., and Ross, J. G.**, Clonal propagation of big bluestem by tissue culture, *Crop Sci.*, 17, 847, 1977.
101. **Chen, C. H., Lo, P. F., and Ross, J. G.**, Regeneration of plantlets from callus cultures of indiangrass, *Crop Sci.*, 19, 117, 1979.
102. **Conger, B. V. and Carabia, J. V.**, Callus induction and plantlet regeneration in orchardgrass, *Crop. Sci.*, 18, 157, 1978.
103. **Dale, P. J.**, Meristem tip culture in *Lolium, Festuca, Phleum,* and *Dactylis*, *Plant Sci. Lett.*, 9, 333, 1977.
104. **Lowe, K. W. and Conger, B. V.**, Root and shoot formation from callus cultures of tall fescue, *Crop Sci.*, 19, 397, 1979.
105. **Kasperbauer, M. J., Buckner, R. C., and Springer, W. D.**, Haploid plants by anther-panicle culture of tall fescue, *Crop Sci.*, 19, 103, 1980.
106. **Kasperbauer M. J., Buckner, R. C., and Bush, L. P.**, Tissue culture of annual ryegrass × tall fescue F_1 hybrids: callus establishment and plant regeneration, *Crop Sci.*, 19, 457, 1979.

107. **Nitzsche, W.**, Herstellung haploider Pflanzen aus *Festuca — Lolium* Bastarden, *Naturwissenschaften,* 57, 199, 1970.
108. **Dale, P. J.**, Meristem tip culture in *Lolium multiflorum, J. Exp. Bot.,* 26, 731, 1975.
109. **Dale, P. J.**, The elimination of ryegrass mosaic virus from *Lolium multiflorum* by meristem tip culture, *Ann. Appl. Biol.,* 85, 93, 1977.
110. **Clapham, D.**, *In vitro* development of callus from the pollen of *Lolium* and *Hordeum, Z. Pflanzenzuecht.,* 65, 285, 1971.
111. **Ahloowalia, B. S.**, Regeneration of ryegrass plants in tissue culture, *Crop Sci.,* 15, 449, 1975.
112. **Ahloowalia, B. S.**, Chromosomal changes in parasexually produced ryegrass, in *Current Chromosome Research,* Jones, K. and Brandham, P. E., Eds., Elsevier/North-Holland Biomedical Press, Amsterdam, 1976, 115.
113. **Gamborg, O. L., Constabel, F., and Miller, R. A.**, Embryogenesis and production of albino plants from cell cultures of *Bromus inermis, Planta,* 95, 355, 1970.
114. **Kao, K. N., Gamborg, O. L., Michayluk, M. R., and Keller, W. A.**, Effect of sugars and inorganic salts on cell regeneration and sustained division in plant protoplasts, *Colloq. Int. C. N. R. S.,* 212, 207, 1973.
115. **Saunders, J. W. and Bingham, E. T.**, Production of alfalfa plants from callus tissue, *Crop Sci.,* 12, 804, 1972.
116. **Bingham, E. T., Hurley, L. V., Kaatz, D. M., and Saunders, J. W.**, Breeding alfalfa which regenerates from callus tissue in culture, *Crop Sci.,* 15, 719, 1975.
117. **Saunders, J. W. and Bingham, E. T.**, Growth regulator effects on bud initiation in callus cultures of *Medicago sativa, Am. J. Bot.,* 62, 850, 1975.
118. **Walker, K. A., Yu, P. C., Sato, S. J., and Jaworski, E. G.**, The hormonal control of organ formation in callus of *Medicago sativa* L. cultured *in vitro, Am. J. Bot.,* 65, 654, 1978.
119. **Walker, K. A., Wendeln, M. L., and Jaworski, E. G.**, Organogenesis in callus tissue of *Medicago sativa.* The temporal separation of induction processes from differentiation processes, *Plant Sci. Lett.,* 16, 23, 1979.
120. **McCoy, T. J. and Bingham, E. T.**, Regeneration of diploid alfalfa plants from cells grown in suspension culture, *Plant Sci. Lett.,* 10, 59, 1977.
121. **Pelletier, G. and Pelletier, A.**, Culture *in vitro* de tissus de trèfle blanc *(Trifolium repens)*; variabilité des plantes régénérées, *Ann. Amelior. Plant. (Paris),* 21, 221, 1971.
122. **Oswald, T. H., Smith, A. E., and Phillips, D. V.**, Callus and plantlet regeneration from cell cultures of ladino clover and soybeans, *Physiol. Plant.,* 39, 129, 1977.
123. **Barnett, O. W., Gibson, P. B., and Seo, A.**, A comparison of heat treatment, cold treatment and meristem-tip culture for obtaining virus-free plants of *Trifolium repens, Plant Dis. Rep.,* 59, 834, 1975.
124. **Phillips, G. C. and Collins, G. B.**, *In vitro* tissue culture of selected legumes and plant regeneration from callus cultures of red clover, *Crop Sci.,* 19, 59, 1979.
125. **Phillips, G. C. and Collins, G. B.**, Virus symptom-free plants of red clover using meristem culture, *Crop. Sci.,* 19, 213, 1979.
126. **Beach, K. H. and Smith, R. R.**, Plant regeneration from callus of red and crimson clover, *Plant Sci. Lett.,* 16, 231, 1979.
127. **Mokhtarzadeh, A. and Constantin, M. J.**, Plant regeneration from hypocotyl- and anther-derived callus of bersemm clover, *Crop Sci.,* 18, 567, 1978.
128. **Tomes, D. T.**, A tissue culture procedure for propagation and maintenance of *Lotus corniculatus* genotypes, *Can. J. Bot.,* 57, 137, 1979.
129. **Niizeki, M. and Grant, W. F.**, Callus, plantlet formation, and polyploidy from cultured anthers of *Lotus* and *Nicotiana, Can. J. Bot.,* 49, 2041, 1971.
130. **Scowcroft, W. R. and Adamson, J. A.**, Organogenesis from callus cultures of the legume, *Stylosanthes hamata, Plant Sci. Lett.,* 7, 39, 1976.
131. **Hussey, G. and Hepher, A.**, Clonal propagation of sugar beet plants and the formation of polyploids by tissue culture, *Ann. Bot. (London),* 42, 477, 1978.
132. **Margara, J.**, Néoformation de bourgeons *in vitro* chez la betterave sucrière, *C. R. Acad. Sci. Ser. D,* 270, 698, 1970.
133. **Atanassov, A. and Kikindonov, T.**, Organogenesis of sugarbeet cultures *in vitro* compared with other species, *Genet. Sel.,* 5, 399, 1972.
134. **Butenko, R. G., Atanassov, A., and Urmantseva, W.**, Some features of sugarbeet tissue culture, *Phytomorphology,* 22, 140, 1972.
135. **Rogozinska, J. H., Goska, M., Kuzdowicz, A.**, Induction of plants from anthers of *Beta vulgaris* cultured *in vitro, Acta Soc. Bot. Pol.,* 46, 471, 1977.
136. **Liu, M. C., Huang, Y. J., and Shih, S. C.**, The *in vitro* production of plants from several tissues of *Saccarhum* species, *J. Agric. Assoc. China,* 77, 52, 1972.

137. **Liu, M. C. and Chen, W. H.**, Tissue and cell culture as aids to sugarcane breeding. I. Creation of genetic variation through callus culture, *Euphytica*, 25, 393, 1976.
138. **Liu, M. C. and Chen, W. H.**, Tissue and cell culture as aids to sugarcane breeding. II. Performance and yield potential of callus derived lines, *Euphytica*, 27, 273, 1978.
139. **Heinz, D. J. and Mee, G. W. P.**, Plant differentiation from callus tissue of *Saccharum* sp., *Crop Sci.*, 9, 346, 1969.
140. **Heinz, D. J. and Mee, G. W. P.**, Colchicine-induced polyploids from cell suspension cultures of sugar cane, *Crop Sci.*, 10, 696, 1970.
141. **Heinz, D. J. and Mee, G. W. P.**, Morphologic, cytogenetic and enzymatic variation in *Saccharum* species hybrid clones derived from callus tissue, *Am. J. Bot.*, 58, 257, 1971.
142. **Heinz, D. J., Krishnamurthi, M., Nickell, L. G., and Maretzki, A.**, Cell, tissue and organ culture in sugarcane improvmeent, in *Plant Cell, Tissue, and Organ Culture*, Reinert, J. and Bajaj, Y. P. S., Eds., Springer-Verlag, Berlin, 1977, 3.
143. **Barba, R. and Nickell, L. G.**, Nutrition and organ differentiation in tissue cultures of sugarcane, a monocotyledon, *Planta*, 89, 299, 1969.
144. **Nadar, H. M. and Heinz, D. J.**, Root and shoot development from sugarcane callus tissue, *Crop Sci.*, 17, 814, 1977.
145. **Heinz, D. J., Mee, G. W. P., and Nickell, L. G.**, Chromosome numbers of some *Saccharum* species hybrids and their cell suspension cultures, *Am. J. Bot.*, 56, 450, 1969.
146. **Kartha, K. K., Gamborg, O. L., and Constabel, F.**, *In vitro* plant formation from stem explants of rape (*Brassica napus* cv. Zephyr.), *Physiol. Plant*, 31, 217, 1974.
147. **Kameya, T. and Hinata, K.**, Induction of haploid plants from pollen grains of *Brassica*, *Jpn. J. Breed.*, 20, 82, 1970.
148. **Thomas, E. and Wenzel, G.**, Embryogenesis from microspores of *Brassica napus*, *Z. Pflanzenzuecht.*, 74, 77, 1975.
149. **Thomas, E., Hoffmann, F., Potrykus, I., and Wenzel, G.**, Protoplast regeneration and stem embryogenesis of haploid androgenetic rape, *Mol. Gen. Genet.*, 145, 245, 1976.
150. **Keller, W. A. and Armstrong, K. C.**, Embryogenesis and plant regeneration in *Brassica napus* anther cultures, *Can. J. Bot.*, 55, 1383, 1977.
151. **Stringham, G. R.**, Regeneration in leaf-callus cultures of haploid rapeseed (*Brassica napus* L.), *Z. Pflanzenphysiol.*, 92, 459, 1979.
152. **Kartha, K. K., Michayluk, M. R., Kao, K. N., Gamborg, O. L., and Constabel, F.**, Callus formation and plant regeneration from mesophyll protoplasts of rape plants (*Brassica napus* L. cv. Zephyr), *Plant Sci. Lett.*, 3, 265, 1974.
153. **Keller, W. A., Rajhathy, A., and Lacarpa, J.**, *In vitro* production of plants from pollen in *Brassica campestris*, *Can. J. Genet. Cytol.*, 17, 655, 1975.
154. **Laboratory of Genetics, Kwangtung Institute of Botany**, Haploid plantlet of *Brassica campestris* from anther culture *in vitro*, *Acta Bot. Sin.*, 17, 167, 1975.
155. **Bowes, B. G.**, *In vitro* morphogenesis of *Crambe maritima* L., *Protoplasma*, 89, 185, 1976.
156. **Sadhu, M. K.**, Effect of different auxins on growth and differentiation in callus tissue derived from sunflower stem pith, *Indian J. Exp. Biol.*, 12, 110, 1974.
157. **Gamborg, O. L. and Shyluk, J. P.**, Tissue culture, protoplasts, and morphogenesis in flax, *Bot. Gaz. (Chicago)*, 137, 301, 1976.
158. **Mathews, H. V. and Narayanaswamy, S.**, Phytohormone control of regeneration in cultured tissues of flax, *Z. Pflazenphysiol.*, 80, 436, 1976.
159. **Murray, B. E., Handyside, R. J., and Keller, W. A.**, *In vitro* regeneration of shoots on stem explants of haploid and diploid flax *(Linum usitatissimum)*, *Can. J. Genet. Cytol.*, 19, 177, 1977.
160. **Rybczynski, J. J.**, Callus formation and organogenesis of mature cotyledons of *Linum usitatissimum* L. var. Szokijskij *in vitro* culture, *Genet. Pol.*, 16, 161, 1975.
161. **Thorpe, T. A. and Meier, D. D.**, Starch metabolism, respiration and shoot formation in tobacco callus cultures, *Physiol. Plant.*, 27, 365, 1972.
162. **Ross, M. K. and Thrope, T. A.**, Physiological gradients and shoot initiation in tobacco callus cultures, *Plant Cell Physiol.*, 14, 473, 1973.
163. **Noriko, S.**, Progressive change in organ-forming capacity of tobacco callus during single subculture period, *Jpn. J. Genet.*, 47, 53, 1972.
164. **Gupta, G. R. P., Guha, S., and Maheshwari, S. C.**, Differentiation of buds from leaves of *Nicotiana tabacum* Linn. in sterile culture, *Phytomorphology*, 16, 175, 1966.
165. **Engelke, A. L., Hamzi, H. Q., and Skoog, F.**, Cytokinin-gibberellin regulation of shoot development and leaf form in tobacco plantlets, *Am. J. Bot.*, 60, 491, 1973.
166. **Walkey, D. G. A. and Woolfitt, J. M. G.**, Clonal multiplication of *Nicotiana rustica* L. from shoot meristems in culture, *Nature (London)*, 220, 1346, 1968.
167. **Walkey, D. G. A. and Cooper, V. C.**, Effect of temperature on virus eradication and growth of infested tissue cultures, *Ann. Appl. Biol.*, 80, 185, 1975.

168. **Collins, G. B., Legg, P. D., and Kasperbauer, M. J.**, Chromosome numbers in anther-derived haploids of two *Nicotiana* species, *N. tabacum* L. and *N. otophora* Gris., *J. Hered.*, 63, 113, 1972.
169. **Bourgin, J. P. and Nitsch, J. P.**, Obtention de *Nicotiana* haploides a partir d'étamines cultivées, *in vitro, Ann. Physiol. Veg. (Paris)*, 9, 377, 1967.
170. **Nitsch, J. P. and Nitsch, C.**, Haploid plants from pollen grains, *Science*, 163, 85, 1969.
171. **Kochar, T., Sabharwal, P., and Engelberg, J.**, Production of homozygous diploid plants by tissue culture technique, *J. Hered.*, 62, 59, 1971.
172. **Nakata, K. and Tanaka M,.** Differentiation of embryoids from developing germ cells in anther culture of tobacco, *Jpn. J. Genet.*, 43, 65, 1968.
173. **Sunderland, N. and Wicks, F. M.**, Cultivation of haploid plants from tobacco pollen, *Nature (London)*, 224, 1227, 1969.
174. **Kasperbauer, M. J. and Collins, G. B.**, Reconstitution of diploids from leaf tissue of anther-derived haploids in tobacco, *Crop Sci.*, 12, 98, 1972.
175. **Sunderland, N.**, Strategies in the improvement of yields in anther culture, in *Proc. Symp. on Plant Tissue Culture*, Sciences Press, Peking, 1978, 65.
176. **Nitsch, J. P. and Ohyama, K.**, Obtention de plantes à partir de protoplastes haploides cultivées, *in vitro, C. R. Acad. Sci. Ser. D*, 273, 801, 1971.
177. **Bajaj, Y. P. S.**, Protoplast culture and regeneration of haploid tobacco plants, *Am. J. Bot.*, 59, 647, 1972.
178. **Hlasnikova, A.**, Androgenesis *in vitro* evaluated from the aspects of genetics, *Z. Pflanzenzuecht.*, 78, 44, 1977.
179. **Ohyama, K. and Nitsch, J. P.**, Flowering haploid plants obtained from protoplasts of tobacco leaves, *Plant Cell Physiol.*, 13, 229, 1972.
180. **Nitsch, J. P.**, Haploid plants from pollen, *Z. Pflanzenzuecht.*, 67, 3, 1972.
181. **Takebe, I., Labib, G., and Melchers, G.**, Regeneration of whole plants from isolated mesophyll protoplasts of tobacco, *Naturwissenschaften*, 58, 318, 1971.
182. **Aviv, D. and Galun, E.**, Isolation of tobacco protoplasts in the presence of isopropyl n-phenylcarbamate and their culture and regeneration into plants, *Z. Pflanzenphysiol.*, 83, 267, 1977.
183. **Carlson, P. S., Smith, H. H., and Dearing, R. D.**, Parasexual interspecific plant hybridization, *Proc. Natl. Acad. Sci. U.S.A.*, 69, 2292, 1972.
184. **Davey, M. R., Frearson, E. M., Withers, L. A., and Powers, J. B.**, Observations on the morphology, ultrastructure and regeneration of tobacco epidermal protoplasts, *Plant Sci. Lett.*, 2, 23, 1974.
185. **Gleba, Y. Y., Shvydkaya, L. G., Butenko, R. G., and Sytnik, K. M.**, Cultivation of isolated protoplasts, *Sov. Plant Physiol.*, 21, 486, 1974.
186. **Raveh, D. and Galun, E.**, Rapid regeneration of plants from tobacco protoplasts plated at low densities, *Z. Pflanzenphysiol.*, 76, 76, 1975.
187. **Melchers, G. and Labib, G.**, Somatic hybridization of plants by fusion of protoplasts. I. Selection of light resistant hybrid of "haploid" light sensitive varieties of tobacco, *Mol. Gen. Genet.*, 135, 277, 1974.
188. **Smith, H. H., Kao, K. N., and Combatti, N. C.**, Interspecific hybridization by protoplast fusion in *Nicotiana:* confirmation and extension, *J. Hered.*, 67, 123, 1976.
189. **Nagata, T. and Takebe, I.**, Plating of isolated tobacco mesophyll protoplasts on agar medium, *Planta*, 99, 12, 1971.
190. **Bourgin, J. P., Chupeau, Y., and Missonier, C.**, Plant regeneration from mesophyll protoplasts of several *Nicotiana* species, *Physiol. Plant.*, 45, 288, 1979.
191. **Clapham, D. H.**, Haploid induction in cereals, in *Applied and Fundamental Aspects of Plant Cell, Tissue, and Organ Culture*, Reinert, J. and Bajaj, Y. P. S., Eds., Springer-Verlag, Berlin, 1977, 279.
192. **Vasil, I. K. and Nitsch, C.**, Experimental production of pollen haploids and their uses, *Z. Pflanzenphysiol.*, 76, 191, 1975.
193. **Vasil, I. K.**, The progress, problems, and prospects of plant protoplast research, *Adv. Agron.*, 28, 119, 1976.
194. **Scowcroft, W. R.**, Somatic cell genetics and plant improvement, *Adv. Agron.*, 29, 39, 1977.
195. **Bajaj, Y. P. S.**, Protoplast isolation, culture and somatic hybridization, in *Applied and Fundamental Aspects of Plant Cell, Tissue, and Organ Culture*, Reinert, J. and Bajaj, Y. P. S., Eds., Springer-Verlag, Berlin, 1977, 467.
196. **Bajaj, Y. P. S.**, Potentials of protoplast culture work in agriculture, *Euphytica*, 23, 633, 1974.
197. **Bhojwani, S. S., Evans, P. K., and Cocking, E. C.**, Protoplast technology in relation to crop plants: progress and problems, *Euphytica*, 26, 343, 1977.
198. **Bottino, P. J.**, The potential of genetic manipulation in plant cell cultures for plant breeding, *Radiat. Bot.*, 15, 1, 1975.
199. **Hu, H., Hsi, T. Y., Tseng, C. C., Ouyang, T. W., and Ching, C. K.**, Application of anther culture to crop plants, in *Frontiers of Plant Tissue Culture 1978*, Thorpe, T. A., Ed., University of Calgary Printing Service, Canada, 1978, 123.

200. **Vasil, I. K., Ahuja, M. R., and Vasil, V.,** Plant tissue cultures in genetics and plant breeding, *Adv. Genet.*, 20, 127, 1979.
201. **Street, H. E.,** Laboratory organization, in *Plant Tissue and Cell Culture*, 2nd ed., Street, H. E., Ed., University of California Press, Berkeley, 1977, 11.
202. **Yeoman, M. M. and MacLeod, A. J.,** Tissue (callus) cultures- techniques, in *Plant Tissue and Cell Culture*, 2nd ed., Street, H. E., Ed., University of California Press, Berkeley, 1977, 31.
203. **Hussey, G.,** The application of tissue culture to the vegetative propagation of plants, *Sci. Prog. (London)*, 65, 185, 1978.
204. **Gamborg, O. L.,** Callus and cell culture, in *Plant Tissue Culture Methods*, Gamborg, O. L. and Wetter, L. R., Eds., National Research Council of Canada, Saskatoon, 1975, 1.
205. **Fossard, R. A. de,** *Tissue Culture for Plant Propagators*, University of New England Printing, Armidale, 1976, chap. 4.
206. **Gamborg, O. L., Murashige, T., Thorpe, T. A., and Vasil, I. K.,** Plant tissue culture media, *In Vitro*, 12, 473, 1976.
207. **White, P. R.,** *The Cultivation of Animal and Plant Cells*, 2nd ed., Ronald Press, New York, 1963, chap. 4.
208. **Murashige, T. and Skoog, F.,** A revised medium for rapid growth and bioassays with tobacco tissue cultures, *Physiol. Plant.*, 15, 473, 1962.
209. **Linsmaier, E. M. and Skoog, F.,** Organic growth factor requirements of tobacco tissue cultures, *Physiol. Plant.*, 18, 100, 1965.
210. **Eriksson, T.,** Studies on the growth requirements and growth measurements of cell cultures of *Haplopappus gracilis*, *Physiol. Plant.*, 18, 976, 1965.
211. **Gamborg, O. L., Miller, R. A., and Ojima, K.,** Nutrient requirements of suspension cultures of soybean root cells, *Exp. Cell Res.*, 50, 148, 1968.
212. **Blaydes, D. F.,** Interaction of kinetin and various inhibitors in the growth of soybean tissue, *Physiol. Plant.*, 19, 748, 1966.
213. **Schenk, R. U. and Hildebrandt, A. C.,** Medium and techniques for induction and growth of monocotyledonous and dicotyledonous plant cell cultures, *Can. J. Bot.*, 50, 199, 1972.
214. **Potrykus, I., Harms, C. T., Lorz, H., and Thomas, E.,** Callus formation from stem protoplasts of corn (*Zea mays* L.), *Mol. Gen. Genet.*, 156, 347, 1977.
215. **Potrykus, I., Harms, C. T., and Lorz, H.,** Callus formation from cell culture protoplasts of corn (*Zea mays* L.), *Theor. Appl. Genet.*, 54, 209, 1979.
216. **Deka, P. C. and Sen, S. K.,** Differentiation of calli originated from isolated protoplasts of rice (*Oryza sativa* L.) through plating techniques, *Mol. Gen. Genet.*, 145, 239, 1976.
217. **Gamborg, O. L., Constabel, F., Fowke, L., Kao, K. N., Ohyama, K., Kartha, K., and Pelcher, L.,** Protoplast and cell culture methods in somatic hybridization in higher plants, *Can. J. Genet. Cytol.*, 16, 737, 1974.
218. **Beversdorf, W. D. and Bingham, E. T.,** Degrees of differentiation obtained in tissue cultures of *Glycine* species, *Crop Sci.*, 17, 307, 1977.
219. **Veliky, I. A. and Martin, S. M.,** A fermenter for plant cell suspension cultures, *Can. J. Microbiol.*, 16, 223, 1970.
220. **Atkin, R. K. and Barton, G. E.,** The establishment of tissue cultures of temperate grasses, *J. Exp. Bot.*, 24, 689, 1973.
221. **Price, H. J. and Smith, R. H.,** Somatic embryogenesis in suspension cultures of *Gossypium klotzschianum* Anderss, *Planta*, 145, 305, 1979.
222. **Smith, H. H.,** Model systems for somatic cell plant genetics, *BioScience*, 24, 269, 1974.
223. **Thorpe, T. A. and Murashige, T.,** Some histochemical changes underlying shoot initiation in tobacco callus cultures, *Can. J. Bot.*, 48, 277, 1970.
224. **Thomas, E., King, P. J., and Potrykus, I.,** Improvement of crop plants via single cells *in vitro* — An assessment, *Z. Pflanzenzuecht*, 82, 1, 1979.
225. **Green, C. E.,** *In vitro* plant regeneration in cereals and grasses, in *Frontiers of Plant Tissue Culture 1978*, Thorpe, T. A., Ed., University of Calgary Printing Service, Canada, 1978, 411.
226. **Thomas, E. and Wernicke, W.,** Morphogenesis in herbaceous crop plants, in *Frontiers of Plant Tissue Culture 1978*, Thorpe, T. A., Ed., University of Calgary Printing Service, Canada, 1978, 403.
227. **King, P. J., Potrykus, I., and Thomas, E.,** *In vitro* genetics of cereals: problems and perspectives, *Physiol. Veg.*, 16, 381, 1978.
228. **Bhojwani, S. S., Evans, P. K., and Cocking, E. C.,** Protoplast technology in relation to crop plants: progress and problems, *Euphytica*, 26, 343, 1977.
229. **Research Group of Soya Bean Tissue Culture of the Kirin Academy of Agriculture,** Successful induction of the plantlets from the callus culture of soy hypocotyl, *Acta Bot. Sin.*, 18, 258, 1976.
230. **Green, C. E., Phillips, R. L., and Kleese, R. A.,** Tissue cultures of maize (*Zea mays* L.): initiation, maintenance, and organic growth factors, *Crop Sci.*, 14, 54, 1974.

231. **Meredith, C. P. and Carlson, P. S.,** Genetic variation in cultured plant cells, in Propagation of Higher Plants through Tissue Culture, Hughes, K. W., Henke, R. R., and Constantin, M. J., Eds., Technical Information Center, U.S. Department of Energy, Springfield, Va., 1978, 166.
232. **D'Amato, F.,** The problem of genetic stability in plant tissue and cell cultures, in *Crop Genetic Resources for Today and Tomorrow,* Frankel, O. H. and Hawkes, J. G., Eds., Cambridge University Press, London, 1975, 333.
233. **D'Amato, F.,** Cytogenetics of differentiation in tissue and cell cultures, in *Plant Cell, Tissue, and Organ Culture,* Reinert, J. and Bajaj, Y. P. S., Eds., Springer-Verlag, Berlin, 1977, 343.
234. **Dulieu, H.,** The combination of cell and tissue culture with mutagenesis for the induction and isolation of morophological or developmental mutants, *Phytomorphology,* 22, 283, 1972.
235. **Nabors, M. W.,** Using spontaneously occurring and induced mutations to obtain agriculturally useful plants, *BioScience,* 26, 761, 1976.
236. **Skirvin, R. M.,** Natural and induced variation in tissue culture, *Euphytica,* 27, 241, 1978.
237. **Murashige, T. and Nakano, R.,** Tissue culture as a potential tool in obtaining polyploid plants, *J. Hered.,* 57, 115, 1966.
238. **Sacristan, M. D. and Melchers, G.,** The caryological analysis of plants regenerated from tumorous and other callus cultures of tobacco, *Mol. Gen. Genet.,* 105, 317, 1969.
239. **Murashige, T. and Nakano, R.,** Chromosome complement as a determinant of the morphogenic potential of tobacco cells, *Am. J. Bot.,* 54, 963, 1967.
240. **Shimada, T.,** Chromosome constitution of tobacco and wheat callus cells, *Jpn. J. Genet.,* 46, 235, 1971.
241. **Yamada, Y., Tanaka, K., and Takahashi, E.,** Callus induction in rice. *Oryza sativa* L., *Proc. Jpn. Acad.,* 43, 156, 1967.
242. **Shillito, R., Robinson, N. E., and Street, H. E.,** Isolation and characterization of mutant cell lines via cell cultures, in *Experimental Mutagenesis in Plants,* Bulgarian Academy of Science, Sofia, 1978, 275.
243. **Devreux, M., Saccardo, F., and Brunori, A.,** Plantes haploides et lignes isogéniques de *Nicotiana tabacum* obtenues par culture d'anthères et de tiges *in vitro, Caryologia,* 24, 141, 1971.
244. **Carew, D. P. and Schwarting, A. E.,** Production of rye embryo callus, *Bot. Gaz. (Chicago),* 119, 237, 1958.
245. **Nickell, L. G. and Torrey, J. G.,** Crop improvement through plant cell and tissue culture, *Science,* 166, 1068, 1969.
246. **Torrey, J. G.,** Cytodifferentiation in cultured cells and tissues, *HortScience,* 12, 138, 1977.
247. **Murashige, T.,** Current status of plant cell and organ cultures, *HortScience,* 12, 127, 1977.
248. **Heinz, D. J.,** Sugarcane improvement through induced mutations using vegetative propagules and cell culture techniques, in *Induced Mutations in Vegetatively Propagated Plants,* International Atomic Energy Agency, Vienna, 1973, 53.
249. **Nickell, L. G. and Heinz, D. J.,** Potential of cell and tissue culture techniques as aids in economic plant improvement, in *Genes, Enzymes and Populations,* Srb, A. M., Ed., Plenum Press, New York, 1973, 109.
250. **Nickell, L. G.,** Crop improvement in sugarcane: studies using *in vitro* methods, *Crop Sci.,* 17, 717, 1977.
251. **Crisp, P. and Walkey, D. G. A.,** The use of aseptic meristem culture in cauliflower breeding, *Euphytica,* 23, 305, 1974.
252. **Walkey, D. G. A.,** *In vitro* methods for virus elimination, in *Frontiers of Plant Tissue Culture 1978,* Thorpe, T. A., Ed., University of Calgary Printing Service, Canada, 1978, 245.
253. **Langhans, R. W., Horst, R. K., and Earle, E. D.,** Disease-free plants via tissue culture propagation, *HortScience,* 12, 149, 1977.
254. **Quak, F.,** Meristem culture and virus-free plants, in *Plant Cell, Tissue, and Organ Culture,* Reinert, J. and Bajaj, Y. P. S., Eds., Springer-Verlag, Berlin, 1977, 598.
255. **Gengenbach, B. G., Green, C. E., and Donovan, C. M.,** Inheritance of selected pathotoxin resistance in maize plants regenerated from cell cultures, *Proc. Natl. Acad. Sci. U.S.A.,* 74, 5113, 1977.
256. **Conger, B. V.,** Experimental mutagenesis in plants: an introduction, in *Radiation Research: Proc. 6th Int. Congr. Radiat. Res.,* Okada, S., Imamura, M., Terasima, T., and Yamaguchi, H., Eds., Toppan Printing, Tokyo, 1979, 562.
257. **Conger, B. V., Carabia, J. V., and Lowe, K. W.,** Comparison of 2,4-D and 2,4,5-T on callus induction and growth in three Gramineae species, *Environ. Exp. Bot.,* 18, 163, 1978.
258. **Shimada, T. and Yamada, Y.,** Wheat plants regenerated from embryo cell cultures, *Jpn. J. Genet.,* 54, 379, 1979.
259. **Yamane, Y.,** Induced differentiation of buckwheat plants from subcultured calluses *in vitro, Jpn. J. Genet.,* 49, 139, 1974.

260. **Vasil, V. and Vasil, I. K.**, Isolation and culture of cereal protoplasts. II. Embryogenesis and plantlet formation from protoplasts of *Pennisetum americanum, Theor. Appl. Genet.*, 56, 97, 1980.
261. **Gressoff, P. M.**, *In vitro* culture of white clover: callus, suspension, protoplast culture and plant regeneration, *Bot. Gaz. (Chicago)*, 141, 157, 1980.
263. **Kao, K. N. and Michayluk, M. R.**, Plant regeneration from mesophyll protoplasts of alfalfa, *Z. Pflanzenphysiol.*, 96, 135, 1980.
264. **Dos Santos, A. V. P., Outka, D. E., Cocking, E. C., and Davey, M. R.**, Organogenesis and somatic embryogenesis in tissues derived from leaf protoplasts and leaf explants of *Medicago sativa, Z. Pflanzenphysiol.*, 99, 261, 1980.

Chapter 6

TREES

R. L. Mott

TABLE OF CONTENTS

I. INTRODUCTION

Forest trees represent a renewable resource of fiber, chemicals, and energy. Forests also hold a firm and necessary place in our conception of what the landscape should be from aesthetic, watershed, and wildlife habitat perspectives. The steadily increasing use of forest products, worldwide, forces the view of forests as a crop to be managed and harvested efficiently. But in most cases, this view of the forest as a crop must also be compatible with other forest uses and functions. Fruit trees give a yearly harvest, but forest crops yield only one relatively low value harvest of bulk vegetative growth after years of growth. These practical and economic considerations severely limit the amount of after-planting cultivations and care which can be given forest crops. The diverse species of gymnosperms, angiosperms, and woody monocots which fall loosely under the forest heading must all grow and develop without the intensive care given the more traditional agricultural drops. Strategies to improve forest production reflect these general considerations.

Forest management practice involves site preparation before planting, controlled burning, and perhaps thinning or the use of mixed species stands to reduce competition for nutrients and space. Management of the understory growth to encourage nitrogen-fixing plants is also a possibility. Direct fertilizer application to forests does not enjoy anywhere near the cost to benefit ratio seen in traditional agricultural crops. However, the recognition of the increased nutrient gathering capacity afforded many forest trees by fungal mycorrhizae associated with the roots offers an indirect opportunity to enhance tree nutrition. Management strategies may be designed to foster mycorrhizal development. Selection and even breeding of superior strains of the fungus itself also offers an effective approach, especially for poorer sites.[1-3] Likewise, a regimine of foliar sprays to control disease and pest infestations is not economically feasible. Here the selection, development, and use of genetically resistant stock are perhaps the only available alternatives. Indeed, for most forest situations, the other coordinate uses and functions of the forest present imperatives against massive spray programs which override even the economic considerations. Overall, the special considerations implicit in the forest crop act to mitigate against the intensive cultivation, fertilization, and spray applications characteristic of traditional agriculture. Genetic improvement emerges as a good and, in some cases, the only obvious approach to improve forest yield.

Forest species present some formidable obstacles to rapid genetic improvement. These center primarily on the long period for growth of trees and on the pronounced juvenile-mature phases of growth typical of forest species. The rapid height growth in the juvenile phase is desired by the forester since seedlings must develop in competitive shading by the parent canopy or the herbs and shrubs of the forest floor. On the other hand, the prolonged juvenile period before flowering and maturity is a handicap to forest tree breeding programs. Trees must reach maturity before superior trees can even be identified with confidence.[4] Once identified, the superior trees can be crossed in the hope of producing even more superior individuals among the seed progeny. Accomplishing the desired crosses is no trivial task, since the trees may be miles apart and cannot be moved, and the flowers usually occur at tree tops far from the unassisted reach of the geneticist. After crosses are made and the seeds collected and planted, the progeny trees themselves must go through the long juvenile period to maturity before superior individuals can be selected and before flowers are available for further crosses. The long time to flowering prevents the sorting through many generations quickly and makes inbreeding and backcrossing schemes impossible in reasonable time periods. The breeder thus seeks methods to induce early flower or cone production to speed up his program. He is also watchful for any device which aids in the early selection of superior traits in the progeny trees.

In contrast to the situation in fruit tree breeding, genetic selection for early flowering is not desirable in forest trees since early flowering trees tend to grow less later on, and the yield is thus reduced.[5] Early flowering in forest tree progeny has been promoted by nongenetic means such as stem girdling, water stress, special nitrogen fertilization, and gibberellin treatments (see the development of this technology for conifer species.)[6-8]

Grafting of young stem cuttings to specific rootstocks which promotes early flowering has been used effectively for fruit trees,[9] but not for forest species. Such devices are only in an early stage of development for forest trees, but they too may ultimately hasten breeding programs.

Some genetic improvement in trees, that must survive and grow unaided in the forest, may be achieved simply by a harvesting practice that leaves some of the best trees standing as a source of seed for reforestation of the site. Substantial and more direct gains have been made through controlled tree breeding programs, especially with the Southern pines of the U.S.[10] Libby[11] published a good review of the comprehensive approaches available to deal with the challenges of tree domestication strategies. Once a mature parent tree is identified as producing a high proportion of desirable seed progeny, the seed orchard is the usual way to get this progeny into reforestation practice. Scions may be taken from the superior parent tree and grafted to young rootstocks to create a large clone of that tree then are placed in a seed orchard for subsequent large-scale production of the desirable seed. The seed, of course, will show the genetic variation typical of meiotic segregation and fertilization as was the case in the parent tree, but the high proportion of superior seed makes it worthwhile. The seed orchard can usually be given extra care and management, but there is still a substantial delay of 10 to 30 years before flowering and seed production from the grafted clone of trees will assume a useful magnitude. Such programs are expensive, but the economics of forest production make yield increases of even 2 to 5% well worth the costs.[12,13] The next round of genetic improvement would involve specific crosses among seed orchard progeny, and here the methods discussed above to hasten flowering become important.

A more direct way to capture the special genetic make up of an elite, mature tree would be through mass vegetative propagation of that particular tree. However, it is characteristic of most forest species that they resist traditional vegetative propagation methods after the mature phase of growth is well underway. While a tree is young, cuttings taken from it will root easily. However, as the tree approaches maturity and is old enough to have proven its worth, it also becomes very difficult or impossible to root the cuttings on a commercial scale. Tissue culture methods of propagation offer a way around this obstacle and offer the promise of larger numbers of propagules in a shorter time than could be obtained with rooted cuttings. Tissue culture methods of micropropagation with apical and axillary buds have been applied successfully to fruit trees in recent years (see Chapter 3), but application to forest species is less well developed. The methods for fruit trees will probably be applicable to some angiosperm forest species, but this chapter will focus primarily on the procedures which have been demonstrated with forest species. Gymnosperms constitute a major forest crop, and special attention will be given to tissue culture methods being developed for this group. Let us first examine some features of traditional vegetative propagation among forest species and then turn our attention to recent advances in tissue culture methodology.

II. TRADITIONAL VEGETATIVE PROPAGATION

Tissue culture propagation is not routinely used for any forest species. The species, the age of the tree, and the particular part of the tree from which cultured explants are taken each can determine success or failure in attempted tissue culture propagation.

Some background in features of tree growth and an evaluation of traditional vegetative propagation experience will be useful when we discuss and evaluate tissue culture methods.

A. Juvenility and Topophysis

During their course of development, plants pass through a series of consecutive stages each of which possesses certain morphologic characteristics. As a tree grows, the entire plant passes through stages recognized as juvenile to mature in growth habit. However, the transitions through these stages are manifest first in the youngest, most recently formed, upper parts of the tree, while the older parts of the tree retain characteristics of earlier stages. Thus, the upper parts of a maturing tree will form flowers while the chronologically older juvenile zone at the base of the tree may retain juvenile characteristics. The upper parts are said to be of more advanced physiological age than the lower and this must also apply to the buds in these zones since they give rise to shoots and branches having the more mature characteristics. Just as the tops of trees are physiologically older than the bases, the lateral branches tend to age physiologically from their bases to the outer tips. Thus while an entire tree may be thought of as having become mature, certain of its buds and branches may yet retain juvenile characteristics according to their location on the tree. Adventitious shoots which arise from cut stumps of mature trees or from burls resulting from injury at the base of standing trees are usually of juvenile form, appropriate to that zone of the tree.

Juvenility in woody plants has complexities beyond the scope of our concern here, and comprehensive and interpretive reviews have been made by Doorenbos,[14] Kozlowski,[15] and Borchert.[16] The juvenile phase in woody plants is often characterized by characteristic leaf shapes, but more generally by vigorous height growth and readiness to form adventitious roots and the inability to form flowers. With increasing physiological age, these characteristics are not lost as a set, but rather one by one. Flowering is usually considered to be the distinctive indicator of maturity, but it can often be experimentally induced on otherwise juvenile trees by applications of gibberellins as mentioned earlier. The giant redwood *Sequoiadendron giganteum* can be made to flower in its first year,[17] but other characteristics of maturity occur much later in the expected 2000-year life span. Likewise, the mature characteristics are not irrevocably stable. Juvenile characters can often be partially regained in new shoots resulting from severe pruning of mature trees. Other influences affect growth and vigor in trees, and these often override the effect of aging. In trees showing discontinuous growth throughout the year, the seasonal fluctuations in vigor of growth and rooting capacity may easily exceed the more gradual changes related to physiological aging.

Physiological aging accompanies tree growth, but the increasing complexity of branch networks also alters and controls the options available to the current year buds. The buds and branches of trees become engaged in increasingly complex networks of correlated growth and inhibition as the tree grows and assumes its characteristic shape. A prime example of correlative inhibition is the phenomenon of apical dominance[18] which dictates that upper, more vertical shoots will grow more vigorously than lateral branches and so on. These correlative inhibitions can affect physiological aging directly as when young vegetative shoots of *Acacia melanoxylon* begin to flower when bent horizontal and as lateral branches revert from adult to juvenile form when bent erect.[16] The correlative inhibitions also can have persistent effects on growth and behavior which go beyond those of juvenility and maturity. Lateral branches of tree tend to grow at a characteristic angle from the vertical — they are said to be plagiotropic as opposed to the vertical, orthotropic growth of the main stem. The plagiotropic nature of branch cuttings often persists when these cuttings are rooted upright in the soil and may persist over many years in the propagated plants even when the shoot is continually restrained in an upright position. Such plagiotropic plants tend to grow laterally along the ground and are bushy rather than tree-like in form. Phenomena such as this, where the new plant seemingly retains the characters of the portion of the plant from which it came, have been called expressions of "topophysis".[16,19] Plagiotropism is not a problem in many tree species, and in others it can follow the classic example shown with juvenility characters in ivy, *Hedra helix,* where adult characters are retained in rooted cuttings, but the cutting reverts to juvenile characters when grafted on juvenile rootstock. It seems then that juvenility and topophysis both have conditional expression and may be persistent under some conditions, but are not stable under all conditions. It seems, however, that procedures for vegetative propagation of forest trees usually must confront one or more of these obstacles.

B. Vegetative Propagation of Hardwoods

Vegetative propagation of hardwood forest species by rooted cuttings and grafting has been reviewed by Farmer[20] and propagation of conifers by Brix and van den Driessche[21] and Girouard.[22] Much information relevant to propagation was presented in the 1975 Symposium on Juvenility in Woody Perennials.[23] The procedures generally conform to standard rooting practice using a mist bench and one or more synthetic auxin treatments perhaps accompanied by a fungicide treatment.[24] In all cases, succulent cuttings from juvenile plants or from basal laterals of older trees are most responsive. The success rate varies greatly with different species and often with genotype and variety.[25,26] The vigor of the parent stock and the season of the year are also of great concern. For most species, only the juvenile cuttings can be rooted with 50% or more efficiency, and mature cuttings usually do not root at all.

Natural vegetative propagation by root suckers occurs in some species of aspen *(Populus),* sweetgum *(Liquidambar),* and some elms *(Ulmus).* These juvenile sucker shoots root easily although there is considerable variation in rooting tendency even among species in these genera. The potential for vegetative propagation can often be deduced from the natural ecology of the tree. Take for example the prolonged juvenile phase typical of climax species like beech *(Fagus)* and fir *(Abies)* which must grow through an existing canopy. Or the ease of rooting even in mature shoots of willow *(Salix)* and poplar *(Populus)* which grow in bottom lands where adventitious rooting has survival value as land rises above the root collar through silting.

Successful rooting of cuttings from juvenile as well as mature trees has been obtained with species of willow, poplar, elm, oak *(Quercus),* yellow poplar *(Liriodendron),* birch *(Betula),* and maple *(Acer).* The use of rooted cuttings of maple is limited since they often lack vigor in their subsequent growth.[27] Yellow poplar, oak, and birch produced easy-rooting, juvenile adventitious coppice shoots from the stumps of cut trees. In species which do not readily produce shoots from stumps, adventitious juvenile shoots may often be obtained by severe pruning of successful grafts of mature scions onto juvenile rootstock. This has been proposed with black walnut *(Juglans)* and cherry *(Prunus).*[20] Elm generates adventitious juvenile shoots from callus at the graft union or from root callus. Aspen, black walnut, cherry, sycamore *(Platanus),* beech *(Fagus),* eucalyptus, chestnut *(Castanea),* and most conifers are among those trees which characteristically root only from juvenile cuttings.

The brief synopsis of vegetative propagation above gives an overview of some important genera of forests. More often than not, success in propagation has depended on the species within the genera and even on the individual variety or genotype selected. Among those species which have been successfully rooted, problems are often encountered with reduced vigor and plagiotropism in the propagated plants especially with material from adult trees. General rules only apply generally. For example, the tropical timber tree *Triplochiton scleroxylon* K. Schum. produces lateral shoots from axillary buds when the juvenile main stem is decapitated. Cuttings from these lateral shoots were easily rooted. Those from the upper portion remained plagiotropic and vegetative while the basal shoots were orthotropic and, unexpectedly, produced precocious flowers on their side branches.[28] Difficulties with plagiotropism are especially acute with many of the conifers, i.e., *Taxus, Araucaria, Sequoia, Pices, Pseudotsuga* and some members of the Cupressaceae. For this reason, as well as the general lack of rooting with cuttings from mature trees, there has been little commercial use of vegetative propagation with most forest conifers.

C. Vegetative Propagation of Conifers

A comprehensive review of the various conifer reforestation programs using clonal rooted cuttings was published in 1977 by Brix and van den Driessche.[21] Clonal propagation by rooted cuttings of *Cryptomeria japonica* has been used for centuries in Japan and provides the only data concerning long term benefits and problems of clonal propagation of forest trees.[29] Major programs for propagation of Norway spruce (*Picea*) with rooted cuttings for operational planting are underway in West Germany[30] and Finland.[31] A major research effort with radiata pine (*Pinus*) is continuing in New Zealand[32] and Australia.[33] Research programs for Douglas-fir (*Pseudotsuga*),[34] loblolly pine (*Pinus*)[35] and western hemlock (*Tsuga*)[36] are being conducted in the U.S.

These programs are relatively new in terms of forest tree harvest cycles, and the majority of the information deals with rooting behavior and with early field performance of the propagules. Long-term field behavior will be forthcoming. Production of rooted cuttings is difficult or impossible for many conifer species, but for those where problems can be overcome, rapid production of select clones (one half million plants from one seedling in 5 years) seems possible. Fully operational cloned planting stock production could be implemented in 10 years given a reduced 4-year field evaluation test of progeny, and the cost of propagules has been estimated at two to three times higher than for seedlings.[21] Time and cost values must be weighed against the impressive tree improvement gains possible through cloning.

The particular problems associated with the production of rooted conifer cuttings vary greatly with species and even individual trees. Generally, cuttings taken from young trees in the juvenile stage root with greater efficiency than cuttings from older trees in the adult stage, e.g., Douglas-fir,[34] radiata pine,[37] and white spruce.[38] For some species this is not so pronounced (e.g., western hemlock[39]), and for others even the juvenile stage shows low rooting efficiency (lodgepole pine[40]) or poor root form (white pine[41]). The rooted cuttings may grow in the field at rates comparable to seedlings (Norway spruce[31]); or they may achieve these rates after a lag period (western hemlock[39]); or they may continue to grow more slowly than seedlings (radiata pine[32]). The rooted cuttings may vary drastically in growth rate and/or assume an irregular, plagiotropic growth form according to the age of parent tree or age and position of the shoot on the parent tree, e.g., Douglas-fir.[34,42] On the other hand, rooted cuttings from adult trees may possess some desirable adult tree characteristics such as resistance to disease, e.g., western red cedar *(Thuja).*[43] Overall, the performance of cuttings may be affected profoundly by tree age, position of the cutting on the tree, tree vigor, and the season when cuttings were taken. Without doubt many of these problems can be overcome for some species by refinements of methodology, but the solution of the general problem areas summarized in Table 1 remain.

The major dilemma with forest improvement through vegetative propagation by rooted cuttings is that cuttings propagate better from juvenile trees, but selection of elite mother trees must come after they have reached the adult stage when propagation

Table 1
GENERALIZED OUTLINE OF PROBLEM AREAS IN THE PROPAGATION OF ELITE CLONES OF CONIFERS BY MEANS OF ROOTED CUTTINGS

Possible improvements	Source of cuttings	Rooting	Propagule behavior in field
Increase selection efficiency for young trees ↕	*Juvenile trees* (poor clone selection at juvenile stage)	+ Good efficiency − Poor root form − Variable with tree − Season dependent	+ Favorable comparison with seedling
Prolong juvenile phase (hedging; rerooting)	↓		
Reversion from adult to juvenile phase ↕	↓		
	↓		
Devise better means of rooting adult tree cuttings	*Adult trees* (better clone selection at adult stage)	− Poor efficiency Age/size of tree Position on tree − Poor root form − Inconsistent rooting − Season dependent	+ Adult wood qualities + Disease resistance − Loss of juvenile growth − Plagiotropism − Abnormal growth

by cuttings becomes a severe problem (Table 1). A practice of repeated shearing of juvenile trees, "hedging", has aided in retaining the juvenile character of cuttings over a longer period.[44] Repeated cycles using plants produced from rooted cuttings as the source of new cuttings, "rerooting", has also helped to maintain juvenility.[45] These practices may be adaptable to a variety of species but, failing that, there remains the possibility of devising more reliable selection procedures for juvenile trees or of using methods which can generate cuttings of juvenile character from the juvenile zone of adult trees.

The emerging application of tissue culture techniques to conifers and other forest species provides an alternative to rooted cuttings for clonal forest propagation. It may become the method of choice for obtaining juvenile clones from mature trees. Unfortunately the present tissue culture capabilities are similar to those of the rooted cutting methods in that rooted propagules are produced most readily from juvenile seedling material. Reliable and routine growth and rooting of the shoots produced in culture, on the scale needed in operational planting, is difficult to achieve. Cost figures per tissue culture propagule are not available, but they are high, and the propagules are just beginning to be field tested and are not ready for operational planting. The tissue culture methods are at a much earlier stage of development than is the case for rooted cuttings. However, there are points of contact between the two technologies where tissue culture might help to resolve some of the problems encountered with cuttings, and where the more extensive information on growth behavior of rooted cuttings may point the way for tissue culture advances.

III. TECHNOLOGY OF TREE TISSUE CULTURE

We have seen that traditional vegetative propagation for trees is beset with general problems associated with the perennial and woody habit of trees and that success is clearly specific to certain species. The same can be said for vegetative propagation by way of tissue culture. Getting adventitious shoot primordia or buds in culture is not so difficult as getting them to elongate and become plants ready for the soil. Correlative inhibitions and bud dormancy are fundamental to the tree habit, and they may be an underlying cause for the general difficulty in getting good development of buds into shoots in culture. Tissue culture treatments may inadvertently trigger a developing shoot to go into dormancy or to slip into the repressed growth situation of apical dominance. The more we learn about these processes in the tree, the better our chances of avoiding related difficulties in culture. On the other hand, the necessity of dealing directly with the complex of factors which control growth and development in vitro may lead to a better understanding of the phenomena in intact plants. Such information might allow better management of tree growth to our benefit. It would seem that this aspect of tissue culture studies is potentially as important as the prospect of mass vegetative propagation.

Where reliable in vitro propagation methods have been developed for a species, they often fail completely on closely related species as is the case, so far, in the eucalyptus group. Generally applicable technology for mass tissue culture propagation, which can apply across forest species, simply is not yet available although forest tree tissues were among the first to be cultured in vitro. The historical flow of progress with woody species has been steady, but it is also disjunct in that workers have apparently abandoned woody species in favor of the more tractable herbaceous species. For this reason, much of the older literature is still useful in that many observations and accomplishments are yet to be broadly tested and woven into the fabric of reliable knowledge.

Let us attempt to gain an understanding of current technology for tree species looking back a little at the historical context and, where possible, relating this to the general features of tree growth previously discussed. We will not cover all of the literature, for our purpose is to assess current technology. One aspect of gaining access to current technology is to discover who is working with trees. A list of world tree tissue culturists was compiled in the October 1978 IAPTC Newsletter,[46] and periodic updates of specific crops in culture, including trees, and the respective workers are compiled by the Tissue Culture Working Group of the American Society of Horticulture Science. The review articles cited in the following discussion can serve as starting places for those seeking more detailed contact with the literature of certain areas. Likewise this overview will not be burdened with specific detail of growth regulator (phytohormone) concentrations used in each case. Trees, like herbaceous plants, show a bewildering array of growth regulator requirements specific to the species, the nature of the cultured explant material, and preferences of the investigator. The general rules of thumb expressed in Murashige's 1974 review[47] apply adequately to trees. We will focus our examination particularly on forest species, since Chapter 3 deals with fruit crops in particular. Fruit tree technology can be expected to have direct application to many dicotyledonous forest trees. Since there is no commercially viable use as yet, our discussion will center largely on gaining perspective on what has been accomplished with forest species and what might reasonably be expected in the near future. Gymonosperms are a major forest group, and the emerging technology for this group will be discussed in more detail in a following section.

A. Embryo Culture

One of the simplest tissue culture tasks was also the first to be put to applied use. Developing or mature embryos may be excised from the ovule or seed and placed aseptically onto a nutrient medium to continue development into viable plants. Plants may thus be recovered from hybrid genetic crosses where the embryo would otherwise abort due to incompatibilities with the mother plant or to incompatibilities with the special nutritive tissue of the endosperm in angiosperms or the female gametophyte in gymnosperms. By this device, genetic crosses between otherwise incompatible varieties or species may often be recovered.

The large embryos of pine and other gymnosperms were among the first to be cultured as early as 1924[48] and 1936.[49] The methods and applied uses for embryos and ovule culture were reviewed in 1965 by Maheshwari and Rangaswamy.[50] These methods have found much use in fruit tree breeding and in breeding of herbaceous crops as well. The media required to support continued development of mature embryos are usually quite simple, containing mineral salts, sucrose, and minor additions of growth regulators. As the embryo is excised at earlier and less well-developed stages, the media required to replace the natural nutrition must be changed to include more complex mixtures of endosperm fluids, yeast and malt extracts, amino acid mixtures, and critical concentrations of growth regulators. The complex media requirements can occasionally be circumvented by culture of the earlier embryo stages within the excised ovule as was done for embryo recovery in *Pinus* hybrids.[51] In this case only agar-solidified water medium was required, other nutrients being contained within the ovule itself. The ovule tissues may be replaced by cultured nurse tissue derived from hypocotyls. *Pinus* embryos excised from the ovule even at an early stage of cleavage retained their suspensor apparatus and developed normally when kept in intimate contact with the nurse culture.[52] The technology for culture of excised, zygotic embryos is generally applicable across species and has become a routine tool for geneticists in many breeding programs. However, culture of the less mature stages becomes increasingly difficult the earlier the stage, and, for many tree species, achieving this goal is a research project in itself.

Excised embryo culture has also been used to gain knowledge of the physiological and morphological processes in plant embryo development.[53] The media and other environmental conditions may be altered and the effects on embryo development monitored. This information is useful in attempting to understand the complexities of normal embryo development *in situ*, deep within the ovule tissues. It is also useful in gaining a better understanding of processes relevant to somatic embryogenesis for propagation. Work on embryo culture has been reviewed and evaluated in recent books dealing with *in situ* embryo development in angiosperms, 1976[54] and gymnosperms, 1978.[55]

B. Organ Culture

Just as entire embryos can develop and grow in vitro, so may excised plant organs be maintained and grown in culture. Organ culture of tree roots was attained[56] as early as 1947. Simple media of salts, vitamins, and auxin were usually sufficient to permit continued growth of excised, aseptic roots. Aseptic root culture led to studies[57] on conifer root physiology and metabolism of specific amino acids by 1959. It was also recognized that excess auxin brought on branching of secondary roots which resembled the characteristic fleshy, bifurcate branch roots of mychorrizal fungal associations.[58,59] By 1966 the establishment of typical mychorrizal infection of aseptic pine roots in culture was achieved.[60] Shortly thereafter, successful nodulation with nitrogen fixing *Rhizobium* bacteria in cultured roots of locust, *Robinia pseudoucacia*, was demonstrated.[61] Sterile plant root culture was reviewed, and the general use of root cultures for the study of nitrogen fixation by beneficial symbiont microorganisms was brought up-to-date recently.[62] The interactions between roots and disease organisms such as fungi and nematodes were mentioned as well. Root cultures of herbaceous crop plants have figured prominently in the study of symbiont and disease interaction with host roots, but the technology is directly applicable to tree root studies as well. Forest soils are usually nitrogen poor, and fertilization is usually impractical, so nitrogen fixation has special importance to the forest crop. Host tissues apparently produce metabolites which favor expression of nitrogenase activity in the microbial symbiont.[63] Some cells within root and callus tissues of the woody *Comptonia* possess barriers to invasion by the mycrobial symbiont or pathogen.[62] It would seem that root culture and perhaps even callus cultures could play an increasingly significant part in gaining useful information about the interactions between other organisms and their tree hosts. This could include associations of tree roots with mycorrhizal fungi or with the bacterial symbionts responsible for nitrogen fixation or even with the disease pathogens. The genotype uniformity of clonal root cultures, and indeed even clonal plants produced in culture, brings a simplifying aspect to such studies, and may even lead to efficient in vitro selection of superior host genotypes.

Gains in the nitrogen-fixing capacity within the forest ecological complex could, by way of increased nitrogen availability to the trees, bring about crop yield increases to rival those expected from general tree improvement through breeding programs or clonal vegetative propagation. Direct nitrogen fixation by the trees themselves would be even more efficient. Emerging tissue culture techniques offer the promise of an indirect, but broadly applicable approach to this objective. Fungus mycorrhizal associations are common in forest tree roots. If the fungus involved were to possess nitrogen-fixing capacity, the organic nitrogen would be available directly to the tree via the mycorrhizal contact. If the fungus had many potential species of host trees, the nitrogen-fixing capacity of this hypothetical fungus would be available to each of those species. Steps to create a synthetic symbiosis between a mycorrhizal fungus, *Rhizopogon,* and a free-living, nitrogen-fixing bacterium, *Azotobacter,* have already been taken.[64,65] The bacterium was introduced into spheroplasts of the fungus produced by enzymatic removal of the fungal cell wall. Five strains of the fungus were subsequently isolated which grew on nitrogen-deficient medium and demonstrated nitrogen-fixing capacity. The bacteria were apparently incorporated and multiplied in the fungal cytoplasm as modified L-forms of the *Azotobacter.* The final step of establishing the mycorrhizal association with the natural pine host has not yet been reported nor has the competitive vigor of this synthetic fungus been assessed relative to other mycorrhizal fungi in the soil. Caution must be exercised in the creation and deployment of such synthetic organisms, but the potentials seem clear enough.

Turning our attention back to tree organ culture, it would seem that there are several, as yet unrealized, indirect applications of this technology as well. Mention has already been made of the difficulties with achieving reliable shoot growth and rooting of trees in culture. Clever use of isolated organs in culture should be instructive in gaining the information to surmount these problems. Studies with surgically altered embryos relative to rooting and further shoot growth suggest the potential, e.g., with *Fraxinus,*[66] *Pinus,*[67] and *Tsuga.*[68] Likewise, the induction in culture of reproductive organs may be instructive for understanding the process as part of in vivo tree response.[69] The potential of organ culture is just beginning to show for study of tree disease resistance.[70]

C. Callus Culture

Excised embryo culture has been usefully applied by geneticists to recover individual plants in otherwise difficult crosses. Organ culture is a useful investigative tool with much unused potential. However, propagation requires more than just subsistence and growth of an organ in culture. Mass vegetative propagation requires proliferation of the explant in culture so that many units capable of regenerating plants can be obtained from one individual. The explant itself may produce multiple shoots directly from its surfaces in a continuing process. Alternatively, cell proliferation may occur as cells undergo division in a more random way from one or more tissues of the explant to product a growing callus. The callus may then be cut into smaller pieces and subcultured for continued proliferation either in liquid or agar-solidified media. Regeneration of organized shoots or embryos from this more or less disorganized callus then has the potential for producing large numbers of plants. First let us examine the requirements and behavior of tree callus in culture and then turn to the specifics of organogenesis and embryogenesis.

1. Early Work

Callus growth from cultured cambial explants of conifers and willow in the 1930s were among the first from any species.[71,72] Isolated and randomly oriented tracheid formation was noted even in these first cultures. The cultures declined in a short time and did not subculture. Adventitious shoot production from callus as well as tracheid formation within callus was noted even in the earliest cambial callus cultures of elm, *Ulmus*.[73] The shoots failed to grow out to become plants. Some measures of control of adventitious shoot production from *Ulmus* callus had been ascribed to adenine and inositol by 1951.[74] Shoot regeneration from callus was not restricted to angiosperms. The first cultures of conifers (*Sequoia*) which were capable of continuous subculture were established in 1950 from shoot-producing burls at the base of the tree.[75] Bud production and tracheid differentiation was obvious during the first few subcultures. The early work with tree species was covered in the general tissue culture review by Gautheret,[76] and the early consideration of practical forestry use was addressed by Geissbühler and Skoog.[77] By 1955, some ten species of angiosperms including *Salix, Populus, Ulmus, Castenea, Quercus, Syringa, Betula, Fagus, Tilia,* and *Theobroma,* as well as the gymnosperms *Pinus, Abies,* and *Sequoia,* were reported in callus culture. Both root and bud-like shoot primordia production had been noted repeatedly from callus, and the difficulty in achieving full development of these buds into viable shoots which could be rooted was recognized even in this early literature. The tendency of cambial callus to form isolated tracheids and the production of tannins and phenolic substances inhibitory to further growth were also recognized. Most cambial explants required auxin for callus development although some, *Betula* and *Castanea,* initiated callus which could be subcultured without auxin. The habituation of certain callus lines to the point where they lost the need for exogenous auxin was also demonstrated in several species. In short, by the mid-fifties the potential for vegetative propagation of trees had been shown, but the problems which remain today were also recognized, e.g., the tendency toward phenolic production and the associated decline of subcultures, the difficulty with continued growth of the adventitious buds, and an extreme species specificity in response. No viable tissue culture trees were produced from callus cultures until the late sixties as we shall see.

The improvements in methods, media, and a better appreciation of growth regulator requirements for plant tissue culture in general have been shared by tree tissue culture in particular.[78] It is no longer noteworthy to report the mere establishment of callus cultures from a new tree species, but rather attention now must focus on plantlet regeneration, some controlled morphogenesis, or differentiation from cultured tissues. The body of literature leading to our current understanding for trees can be seen in the general tissue culture review in 1963 by White[79] and the excellent 1968 review of adventitious bud formation in plants by Broertjes et al.,[80] and in the annotated literature lists for woody species by Winton in 1972 to 1974,[81] by Durzan and Campbell in 1974,[82] and by Brown and Sommer in 1975.[83] These literature lists show well over 200 woody species at least in callus culture representing over 40 genera and 20 families. This work led to reviews along the way speculating as to the uses in forestry for the emerging capabilities in tissue culture of woody angiosperms by Jaquiot in 1966,[84] and of conifers in 1965 by Haissig,[85] and again in 1968 by Konar and Guha.[86]

2. Cell Differentiation

Callus formation usually occurs from cells near the vascular traces in explants from tree stems, roots, and needles. Its initiation and continued culture requires only exogenous auxin stimulation. Occasionally gibberellic acid was also helpful as for *Thuja orientalis*.[87] The gibberellic acid probably was involved in stimulating initial cambial activity in a dormant cambium prior to callus formation.[88]

The internal morphology of growing callus cultures of woody plants has been described in some detail for *Pinus banksiana*,[89] for *Picea glauca*,[90] and in several angiosperm genera.[84] White[91] gave an excellent description of the sequence of cell differentiation to be seen in *Picea* callus and related this to sequences of cell differentiation in the plant. One is left with the view that cells in cambium-derived callus have not progressed far along the path of dedifferentiation. Meristematic islands of cells are randomly dispersed among larger cells and the callus is usually quite friable. Nests of randomly oriented tracheids are common in cambial callus as if it were a distorted carryover of normal cambial activity. Tracheid formation was easily induced by buds grafted onto the callus, and this led to the classical experiments with *Syringa* callus showing that auxin and sucrose act to induce tracheid formation.[92] Physical pressure in *Populus* callus[93] and dehydration in *Fraxinus*[94] have also been implicated in vascular differentiation for trees. A full discussion of this line of investigation is beyond the scope of our purposes in this review, but the lessons should not be overlooked when confronted with the common failure of proper vascular hook up between the shoots and the roots formed on callus masses, e.g., as was observed with *Betula*.[95]

3. Polyploidy

Callus cultures of forest trees, similar to callus cultures of herbaceous plants, display genetic instabilities: polyploidy, anuploidy, etc. (see the Reviews by Bonga[96] and D'Amato[97]). On the other hand, polyploid individual trees are not common among forest species, and they are especially rare in conifers.[98] Traditional genetic studies indicate that in some angiosperms the occasional polyploid individuals are more vigorous, e.g., *Populus*, but for most forest genera, they are weak.[12] Even the homozygosity increases by self-pollination often tend to reduce vigor.[99] Thus, while it is evident that polyploidy and other chromosomal instabilities do occur in tree callus cultures, it seems reasonable to expect that the apparently strong tendency away from polyploidy and homozygosity displayed in forest trees should ensure that regenerated plants will be normal diploid. Among variable ploidy in callus cells of *Prunus*, only diploid plants were produced.[100] By the same token, the failure to regenerate buds or the difficulty in growing these buds into plants observed in several genera may have its basic cause in ploidy problems of the cultured cells. Little concrete information exists, but it would seem prudent for investigators to monitor chromosome numbers especially where regeneration is not forthcoming.

Two phenomena seem worthy of mention with respect to polyploidy. *Picea glauca* cultures showed that the callus growth rate increased with increasing ploidy level of the callus cells.[101] Differences in degree of polyploidy can also be extreme with little obvious effect on callus or suspension growth.[102] Thus, the common practice of selecting good growing subcultures for regeneration work could be hazardous, at least in some species. Some gymnosperm cultures seem remarkably stable. Konar and Nagmani[103] reported stable diploid cultures of *Pinus* during an initial 10 weeks on a variety of growth regulator concentrations. It may be that some conifer species will reflect in culture the tendency toward retaining diploidy shown by the trees. Venketeswaren and Huhtenen[104] observed stable diploid cultures of *Pseudotsuga* in liquid suspension or on agar-solidified media during 8 months. More information is needed with regard to freshly initiated vs. the subcultured callus in tree species. As it looks now, it would seem wise to expect ploidy and other chromosome problems in callus cultures but not in the derived plants.

4. Retention of Organ Traits

Calli initiated from different tree parts of linden or basswood (*Tilia*) did not differ in their growth or media requirements,[105] and the same was so for almond (*Prunus*).[100] On the other hand, initial tissue selection is usually the most critical step for success of procedures leading to plantlet regeneration in culture.[47] Whether callus retains features related to juvenility and maturity of the parent plant which will carryover to plants regenerated from the callus is not clear. Ivy (*Hedra helix*) provides a classic example of easily monitored juvenile and mature traits. The trailing juvenile vine with palmate-lobed leaves becomes erect with entire, ovate leaves and produces flowers when mature. The callus derived from mature vs. juvenile parent stock showed persistent differences in vigor through several subcultures.[106] The juvenile was the more vigorous. The juvenility or maturity of plants derived from the respective calli have not been tested since only sporadic shoot formation has been observed in ivy callus and then only for callus from juvenile stems.[107] Shoots and plants have recently been formed at high frequency directly from the surface of excised hypocotyl and cotyledons of ivy exposed to 0.5 mg/ℓ each of cytokinin, 6-benzylaminopurine (BAP) and auxin α-napthaleneacetic acid (NAA).[108] The plants produced showed clear juvenile traits as would be expected. Callus derived from embryos and seedlings produced shoots and roots at low frequency (1 in 50), and these too were juvenile as expected. No direct shoot regeneration from callus of mature plants has yet been reported so the question remains unanswered. However, embryogenesis has been observed from callus which was derived from mature stems, and the embryos grew into plants having juvenile traits.[109] Thus, maturity in ivy is not an irreversible property of its individual cells, and embryogenesis effectively established a return to juvenile characters. Callus may retain some aspects of tree behavior, however. This was shown for almond trees susceptible to high-temperature bud death. Callus produced from several plant parts was also temperature-sensitive over approximately the same range.[110] Reports such as this hold the possibility that some useful tree traits may be screened and selected in callus culture prior to vegetative propagation.

D. Organogenesis

Organogenesis and plantlet regeneration in culture were common enough with herbaceous plant species by 1974 that generalizations in methods and media requirements could be made.[47] Although regeneration for some tree species could be mentioned at that time, they were few and not routine. Shoot and root production from callus was not uncommon even among the earliest attempts at callus tree culture,[73,75] but, with the exception of aspen, substantial progress in organogenesis and plantlet production for woody species in culture was not evident until the mid-seventies. The early progress can be followed in the annotated literature lists up to 1974 by Brown and Sommer[84] and Durzan and Campbell.[82] Various aspects received special attention in later reviews in 1975 by Pierik,[111] in 1976 by Winton and Huhtinen,[78] in 1977 by Abbot[112] and Bonga,[96] and in 1978 by Mott[113] and Winton.[114] Woody species have received increasing attention during this period giving a rapidly increasing list of trees from which plants have been regenerated. The methods range from micropropagation from node and bud cultures to shoot organogenesis from callus and suspension cultures as shown for forest related species in Table 2. Examples of these methods will be discussed to give a feel for the state of the emerging art.

Table 2
FOREST TREES PROPAGATED BY ORGANOGENSIS AND ORGAN CULTURE

Species[a]	Tissue source	Mode of propagation[b]	Plants obtained	Ref.
Abies balsamaea L.	Adult tree, resting bud	X	No	115
Acacia koa Gray	Root sucker stem tip	R	Yes	116
Araucaria cunninghamii Ait.	Juvenile nodes	G	Yes	117
Betula pendula Roth.	Juvenile stem	R	Yes	118
	Anther, microspores	R	Yes	120
Broussonetia kazinoki Sieb	Juvenile stem/hypocotyl	R/X	Yes	119
Cryptomeria japonica D. Don	Hypocotyl segments	X	No	121
Eucalyptus sp.	Lignotubers	R X	Yes	122
	Juvenile & adult nodes	G M R	Yes	123, 124
	Cotyledon, hypocotyl	X	Yes	125
Gleditsia triacanthos L.	Stem	R	Yes	126
Paulownia tiawaniana	Stem	R	Yes	127
Picea sp.	Hypocotyl segments	X	Yes	128, 129
	Resting buds, needles juvenile and adult trees	X	Yes	130, 131, 132
Pinus sp.	Cotyledon, hypocotyl, juvenile buds and needles	X G	Yes	129, 133, 134, 135, 136
	Juvenile needle fascicles	X G	Yes	137
Populus sp.	Root sucker stem tip	R	Yes	138, 139, 140
	Female catkin	G	No	141
	Adult, dormant, or leafy buds	M R	Yes	142
Pseudotsuga menziesii (Mirb.)	Cotyledon, hypocotyl	X R	Yes	130, 143, 144, 145
	Cotyledon callus	X R	Yes	146
	Juvenile buds	G	Yes	147
Prunus amygdalus Batsch	Juvenile stem	R	Yes	100
Santalum album L.	Hypocotyl	X	Yes	148
Ulmus americana L.	Hypocotyl suspension culture	R	Yes	149
Thuja plicata Donn.	Cotyledon, shoot tip, 4—10 year trees	X	Yes	150
Tsuga heterophylla Sarg.	Cotyledon	X	Yes	151
Sequoia sempervirens Endl.	Juvenile basal sprouts of adult tree	X R	Yes	75, 152

[a] Where several species within the same genera were used, this is indicated by sp.

[b] Key to mode of propagation: R = regenerative callus; X = multiple shoots from primary explant; G = growth of existing buds apical or axillary; and M = micropropagation via recycled nodes.

In 1964, Mathes[153] obtained leafy shoots on cambial callus derived from juvenile tiploid quaking aspen (*Populus tremuloides* Michx.) maintained in the dark on a nutrient medium plus 10% coconut milk, 1 mg/ℓ of NAA, and 0.8 mg/ℓ of kinetin. The shoots made substantial development, but failed to elongate and form plants. In 1968, Wolter[154] reported similar results with diploid aspen callus which had been cultured with inositol, kinetin, and 2,4-dichlorophenoxyacetic acid (2,4-D) for 7 years. When firm white callus was selected from among the tan friable callus and exposed to NAA (0.5 mg/ℓ), nongeotropic roots formed within 8 weeks. On the other hand, shoots were initiated from the underside of the callus at the agar-solidified medium interface within 2 to 3 months when kinetin was provided at a critical concentration (0.14 to 0.18 mg/ℓ) in the absence of NAA. These shoots were excised and rooted on media lacking growth regulators, but they failed to grow afterwards. In 1970, Winton[138] succeeded in obtaining four trees growing in soil from seven rooted shoots which had been recovered from many trials using similar media and growth regulator treatments on callus which had been subcultured for 2 years. The long culture times and the extremely low efficiency of these methods prevent commercial application for propagation. The methods have been improved somewhat and applied to other *Populus* species,[139,140] but the efficiency still remains low.

A few plants were produced from liquid suspension cultures of *Ulmus*.[149] Callus suspension cultures were derived from seedling hypocotyl segments of *Ulmus* in the presence of NAA, kinetin, and inositol and were subcultured 9 months on the same medium. Red and green nodules were produced under dim fluorescent light after 7 weeks on an agar-solidified medium with the NAA, kinetin, inositol, and thiamine removed. After an additional 1-year period of subculture, multiple leafy shoots were produced from the nodules. These shoots were excised and rooted during an additional 2 months.[149] Again we see the very long culture times. This author is not aware of any subsequent work which significantly reduces the time or increases the dependability of this procedure for elm.

A few plantlets were obtained in 1969 from a teratomaceous, partially organized, callus from a lignotuber of *Eucalyptus* within weeks in culture by manipulation of growth regulator concentrations.[122] Callus from juvenile stem sections of birch (*Betula*) produced shoots within 2 months.[118] These adventitious shoots were excised and rooted during an additional 2 weeks after which they were transferred to soil with 70% survival. The time required in culture in these two cases was less than might be expected from the two preceding examples, but the sequence of events was quite different as well. An already partially organized callus-like mass of roots and shoots was obtained initially in *Eucalyptus*. Stem sections of *Betula* placed on media with indoleacetic acid (IAA) and kinetin in the dark produced a white, firm callus which quickly produced nongeotropic roots while attached to the parent tissue on the same medium. The roots in time became geotropic. When cultures at this stage were transferred to light, without media change, only those growths which had also produced roots, produced shoots. The shoots were unconnected to the roots so plantlets were not formed. The existing roots soon died after shoots were produced. When elongated shoots were excised, they were rooted easily in response to 2,4-D in the absence of kinetin. These sequences are clearly different than was seen above for aspen and elm, and they apparently depend upon existing organs which are part of the callus-like growths.

The above examples all came from juvenile material of easily propagated species. In the cases where the callus did not show immediate organogenesis, the callus intermediate steps seemed to require long periods of culture as callus or suspension cultures, as if a difficult dedifferentiation of the callus must first occur. The shorter times occurred with *Betula* and *Eucalyptus* which root easily in the juvenile phase, and the shoots came from callus predisposed to root immediately in culture. The long dedifferentiation period seen for aspen and elm seems to have been avoided for *Betula* and *Eucalyptus* by taking advantage of the natural tendencies of plant organs in culture. An extension of this approach using buds has been successfully applied as a better alternative than the callus route for clone propagation in *Populus*.[142] Axillary buds from leafy or dormant stems were excised, surface sterilized and caused to grow out on a nutrient medium with the cytokinin, BAP at 0.2 mg/ℓ. Node sections of these shoots were excised and again cultured with BAP (0.1 mg/ℓ) and NAA (0.02 mg/ℓ) until axillary and adventitious shoots proliferated and grew out to a stage where they could be excised, rooted, and transferred to soil. The number of shoots obtained in the experiments suggests that with this micropropagation method, a million plants could be produced from one bud in 1 year. The plants initially showed juvenile leaf form, but reverted to mature leaf shape after some growth in soil. Similar possibilities were suggested by culture of female catkin primordia of black cottonwood (*Populus*).[141] Attempts to use buds of conifers in a similar way were tried as early as 1959[155] and later,[156] but they have been notably unsuccessful (see Reviews by Bonga[115,157]).

Considerable work has been done towards development of axillary bud micropropagation procedures which could apply in general to *Eucalyptus*. Cresswell and Nitsch[158] established cloned plants of *E. grandis* by aseptic culture of excised juvenile nodes. On a nutrient medium containing 1 mg/ℓ of indolebutyric acid (IBA), the axillary buds at the nodes were activated and subsequently grew out and rooted to form plants capable of transfer to soil. By traditional methods, *E. grandis* cuttings will root only from nodes below number 15, e.g., young plants less than 30-cm high. The node culture methods extended this to at least node 80 and up to 5-years old, but several difficulties exist with the method. First, surface sterilization of the explants from field grown trees was a severe problem, as is the case when similar approaches are attempted on most forest species. Smooth, young stem internodes with undamaged epidermis can be cleaned for culture, but the crevices and irregularities of leaf axils tend to trap and protect microorganisms from the sterilant solutions. The problem was especially severe when coppice shoots arising near the soil were used.[123] The problem was avoided somewhat by steps to keep new growth clean in the field. Growing shoot tips were treated with insecticide and fungicide sprays followed by allowing new shoot extension to occur within a plastic bag.[124] A second problem with node culture concerns the initial secretion of growth-retarding, brown phenolic substances from cut ends of the explant.[123] This was reduced to acceptable levels when explants were soaked up to 3 hr in distilled water before planting on culture media and then were kept in the dark for an initial 7 days and then only in weak light (approximately 1500 lx) thereafter, until cultures were established. Phenolic secretion from established cultures and subcultures was much less severe.

The capacity of *E. grandis* node explants to root was dependent on seasonal effects in the parent plant. Short days and cold nights were beneficial, as was any abrupt change in light or temperature just prior to node collection.[123] The method gave little rooting success on five other *Eucalyptus* species. de Fossard et al.[124] tried similar methods with four species utilizing nodes from in vitro grown seedlings, and in addition, nodes from the crown of adult trees and nodes from the juvenile coppice shoots formed at the base of adult trees. Several media were used containing IBA (0.1 μM) ± kinetin or BAP (0.1 μM) with various organic supplements. Seedling nodal cultures of all species formed roots, but no nodal cultures from adult trees formed roots. Coppice nodes of *E. grandis* rooted to form good plants, and this method is suggested for other species as well. Seedling node cultures often initiated a regenerative callus which continued to produce buds when subcultured in the presence of cytokinin. Some adult nodes of all four species also produced multiple buds but with reduced vigor, and reliable rooting has not yet been obtained.

In view of the frequent establishment of regenerative callus from node cultures both from juvenile and mature shoots, one might expect shoot formation from callus on a routine basis. Callus is easily obtained from *Eucalyptus* stem explants, however, no shoot formation from subcultured callus has been observed. Lee and de Fossard[159] tested nearly 200 cytokinin-auxin combinations on *Eucalyptus* callus without substantial morphogenetic effect. However, callus from axillary buds and recently from embryonic materials can generate shoots in response to appropriate exogenous growth regulators. Lakshmi Sita[125] obtained profuse rooting from excised *Eucalpytus* cotyledons and hypocotyls with auxin alone, but with zeatin (1 mg/ℓ) plus IAA (0.2 mg/ℓ) a slow-growing compact red callus developed. Dark green protuberances appeared after 4 weeks, and by 5 weeks shoots formed. Excised shoots rooted in the presence of IBA (1 mg/ℓ) to form viable plants. Similar treatments on leaf and shoot explants from 1-year-old plants produced only callus which subcultured easily, but did not regenerate shoots or roots.

Node culture methods similar to those developed for *Ecualyptus* have been successfully applied to a conifer timber tree of Australia, *Araucaria cunninghamii* Ait.[117] This tree has strongly plagiotropic side branches which remain plagiotropic when used for rooted cuttings. Thus, for this species, only the orthotropic main stem and the bud traces in leaf exils of the main stem can be used for vegetative propagation by rooted cuttings or grafting. Main stem nodes from juvenile, 18-month-old *Araucaria* trees formed axillary buds which grew out in culture. Rooting was difficult and only a few plants were produced. The problems were similar to those observed with *Eucalyptus.*

Micropropagation methods using shoot apices and needle fascicles have been applied with limited success to other conifers, e.g., *Picea,*[130,132] *Pinus,*[133,136] and *Pseudotsuga.*[147] All other reports of plantlet regeneration in conifers, with the exception of one report from cotyledon-derived *Pseudotsuga* callus,[146] involve adventitious bud formation directly on cultured embryo, seedling organs, or on primary callus derived from them (Table 1). The developing technology for conifer propagation will be treated in more detail in a subsequent section of this chapter. All conifer tissue culture regeneration has been with juvenile materials, with the exception of micropropagation from terminal buds of *Picea*[132] and basal juvenile sprouts from adult *Sequoia* trees.[152]

Monocot forest species have not received much attention. However, techniques are being developed to produce regenerative callus from embryos and inflorescences of the oil palm, *Elaeis quineensis* Jacq.,[160] and the date palm, *Phoenix dactylifera* L.[161,162]

As noted earlier in this chapter, root and shoot formation has been a common place occurrence on callus from trees. Adventitious shoot and root production has been reported from callus and explants of a number of tree species.[82,96] These cases are not listed in Table 1 because viable plants have yet to be produced from the bud-like shoots formed in culture. The problem has consistently been one of achieving subsequent growth of the shoots and establishing functional roots on them.[82,96]

E. Embryogenesis

Somatic cells in culture may undergo organogenesis to produce shoots and roots, or they may engage in somatic embryogenesis to produce embryos which then grow similarly to natural zygotic embryos and produce the whole plant. These somatic embryos are often called embryoids. Unfortunately, this term has also been applied to any structures which resemble proembryos or embryos, but whose development into plants may have not been confirmed. Special attention is required when assessing literature reports of embryogenesis, for in some cases the putative embryos, embryoids, show no further development, and are therefore as yet of no use for propagation. Asexual embryogenesis occurs naturally *in situ* in many plants (some 329 species) and has been attained in tissue culture from many species (132) including some (14) not known to exhibit the phenomenon naturally. The literature has been reviewed often; most recently in a comprehensive review of natural and tissue culture asexual embryogenesis by Tisserat et al.[163] and by Kohlenbach.[164] Generalizations in methods and requirements for in vitro somatic embryogenesis are compiled in these reviews, but it is also apparent that the phenomenon is especially sensitive to the species and to the initial explant selection. Zygotic embryos, nucellus, and ovule tissues are especially propitious starting materials for most species.

Reports of somatic embryogenesis in tree species are sparse in comparison to herbaceous plants. Embryo-like structures have been reported from tree cultures, and some have actually been grown out to form plants, but with the exception of *Citrus* and *Coffea,* there has been simply too little work to make this a reliable avenue for vegetative propagation, (Table 3). Practical use in propagation demands that sufficient information be available about the external prerequisites for somatic embryogenesis so that useful genotypes can be induced and released at will, with synchrony and in sufficient amounts.

The usual procedure involves establishing an actively dividing callus from a suitable explant, usually some part of a zygotic embryo and usually in response to exogenous 2,4-D and kinetin. The callus or suspension culture continues to grow well with these growth regulators and may tend to produce roots in later subcultures with reduced cytokinin and auxin concentrations. When auxin is further reduced, or omitted, shoot buds and embryos may form in some cultures. The nongreen callus with high embryogenetic tendency can then be recognized. It may then be isolated and subcultured on slightly higher auxin media containing reduced nitrogen where the callus proliferates as nodular structures in a friable cell matrix. These cultures repeatedly give rise to many embryos when the growth regulators are removed. This general scheme can be quite varied among species, but most cases involve the steps with high auxin concentrations for induction and auxin removal for embryo development[164] and reduced nitrogen is apparently mandatory.[180] Somatic embryogenesis may take place directly on the surface of cultured embryo explants, as in arbor vitae (*Biota orientalis* = *Thuja orientalis*),[165] or after a series of subcultures of callus from the diploid embryo or haploid female gametophyte, as in *Zamia*.[167] Difficulty with further growth of the embryos was a problem in both of these cases. *Coffea* stem segments cultured in the dark on kinetin and 2,4-D each at 0.1 mg/ℓ produced a compact callus among a generally more white, friable callus. When pieces of the compact callus fell isolated onto the media, they grew into yellowish globules which then developed into embryos on the same media.[174] Callus from *Coffea* leaf explants, on the other hand, required "conditioning" by exposure to high cytokinin vs. auxin ratios (kinetin 18.4 μM/2,4-D 4.5 μM) for later maximum capacity to produce embryos, followed by an "induction period" on reduced concentrations (kinetin 2.3 μM/NAA 0.27 μM), followed by a medium free of growth regulators for embryo development.[175] Histological studies[181] and scanning electron microscope studies[182] throughout this sequence suggest that a population of spherical cells were formed among more elongate cells during the conditioning period. These cells were the source of embryos, and the frequency of embryogenesis was related to selection of cells which avoided an apparent mitotic blockage in the induction media. Just how much the difference in the procedures for somatic embryogenesis from stem-derived callus vs. leaf mesophyll-derived callus may reflect requirements for induction from different callus sources is not clear. Likewise, the basis for the high cytokinin requirement in *Coffea* is not known. But, high cytokinin is usually consid-

Table 3
SOMATIC EMBRYOGENESIS IN TREES AND WOODY SPECIES

Species	Tissue source	Ref.
Biota orientalis Endl.	Embryo cotyledon ± callus	165
Zamia integrifolia Ait.	Female gametophyte, embryo	166, 167
Hevea braziliensis Muell.	Somatic tissue of anther	168
Ilex aquifolium L. (sp.)	Embryo cotyledon and hypocotyl, directly without callus	169
Loranthaceae		
Dendrophthoe falcata (Lf) Ettings	Embryo	170
Arceutholoium pusillum Bieb.	Embryo	171
Elaeis quineensis Jacq.	Embryo	172
Prunus sp.	Anther, microspores	173
Coffea so.	Stems, anthers, leaves	174, 175
Citrus sp.	Nucellus, embryo	163
	Callus regenerated from protoplast	176
Santalum album L.	Embryo	177
Ribes rubrum L.	Ovule	178
Vitis vinifera L.	Ovule	179

ered to inhibit somatic embryogenesis.[163] Woody plants may prove to have different growth regulator requirements in this regard. Gibberellic acid was necessary for induction of embryogenesis in *Santalum*,[177] and *Vitis* required both gibberellic acid and cytokinin for further development of embryos after formation.[179]

With the increasing attention to woody and forest species, we should see more reports of somatic embryogenesis. This route may even permit vegetative propagation in species still recalcitrant to organogenesis in culture.

F. Haploid Propagation

In addition to its use for vegetative propagation, tissue culture offers the possibility of generating haploid callus and plants for special use in protoplast fusion genetics and traditional breeding programs. Pollen is an easily accessible source of haploid cells in both gymnosperms and angiosperms, but each pollen grain is potentially of a different genotype. The unfertilized egg of angiosperms is not accessible, and the endosperm is triploid. The comparable nutritive tissue of gymnosperm seeds, the female gametophyte, is an ideal material since it is composed of haploid cells all of the same genotype. We will briefly examine the state of the art in utilizing these sources of haploid cells. This approach for trees has been reviewed by Bonga[183] and Winton and Stettler.[184]

Although haploid plants have been produced from some 13 families and 92 species,[163] there are few reports of success with trees. Sondahl and Sharp[175] report dihaploid embryos from anthers of *Coffea*. Callus and plantlet formation from anthers have been reported for *Populus*[185] and for *Prunus*,[186] but polyploidy was a problem, and no certain haploid plants were confirmed. Callus cultures have been derived from pollen of several gymnosperm genera: *Ephedra*,[187] *Pinus*,[157] *Ginkgo*,[188] *Taxus*,[189] *Juniperus*,[190] and *Picea*.[192] These callus cultures did not undergo significant organogenesis, and the ploidy was variable from 1*N* to in excess of 8*N* within one culture.

By contrast, even the earliest culture of female gametophyte of *Zamia* in 1948 regenerated a few shoots and roots on a simple medium, but they failed to develop further.[166] Later, haploid roots, shoots, and embryos were obtained from *Zamia*.[167] The female gametophyte of *Ginkgo*[188] taken just before fertilization produced a callus which under the influence of kinetin and auxin produced many roots and a few abortive shoots. Bud-like growths, which grew to have eight cotyledon-like leaves, were produced from callus of the female gametophytes of *Picea*.[192] The callus was derived in low frequency from gametophytes harvested just after fertilization. The haploid sprouts did not grow further or produce terminal buds. These last cases represent about the closest steps toward production of haploid gymnosperm plants. Haploid individuals are rare in gymnosperms, but they do exist, and haploid conifer trees have been isolated in natural crosses.[193,194] This at least proves that the haploid state in this group need not be incompatible with normal development.

IV. TECHNOLOGY OF CONIFER TISSUE CULTURE

The gradual commercial application of tissue culture to herbaceous plants brought serious attention to tissue culture as a possible means of propagating conifers. The possibility became more immediate with the successful production of multiple adventitious shoots on cultured longleaf pine (*Pinus palustris* mill.) embryos by Sommer et al.[195] Growth regulators, primarily cytokinin, were required to induce the development of shoot primordia directly from the cotyledon surface of cultured embryos. The shoot primordia failed to continue growth to form elongated shoots when left on the mother tissue. Upon excision and transfer to other media lacking growth regulators, some shoots grew and rooted to form complete plantlets suitable for growth in soil. Coincident with this work, Isikawa[121] reported spontaneous adventitious shoot formation on cultured hypocotylis of *Cryptomeria japonica* D. Don., but these were not rooted to form plantlets. *C. japonica* is an exception among conifer trees in having long been vegetatively propagated by rooted cuttings.[29] These early papers dealing with tissue culture propagation of conifers displayed many of the problems of conifer tissue culture which still plague workers today in hardwood as well as conifer propagation schemes.

Substantial progress has been made with conifer cultures since 1975. The increasing list of species which have produced adventitious shoots, but not necessarily entire plants, in culture from embryonic or young seedling tissues includes about ten pine species, Sequoia, Douglas-fir, Western hemlock, two spruces, Western red cedar, Cupressus species, Araucaria, and perhaps a few others.[114] In addition, continued investigations have brought improvements in the methods and procedures which bring about greater reliability and which extend the range of suitable plant parts that can serve as the initial starting materials. The diagram in Figure 1 outlines the several options now demonstrated among the conifers, and this diagram will serve as a guide in discussing the current methodology and possibilities.

A. Clones from Seedlings

The pathway A→E of Figure 1 constitutes the most generally applicable route for in vitro plantlet production among the conifers. In this discussion the bud-like outgrowths or adventitious shoot primordia which are initiated in culture on the surface of the parent tissue will be referred to loosely as buds, recognizing that they may or may not have a complement of bud scales. These shoot primordia or buds subsequently may elongate to form shoots with a proper stem, leaves, etc. The pathway A→E is basically the sequence first used with longleaf pine.[195] Further research has brought better definition of the requirements for certain species (Douglas-fir, loblolly pine, slash pine), but for conifers, in general, the routine application of these methods still poses some real problems.

One has little choice in starting materials since embryonic or at least young seedling tissues are required for this procedure. Bud production varies greatly among seedlings used as starting material, and much difficulty is often encountered in getting the buds to grow into shoots which can then be rooted to form complete plants. Also, the bud

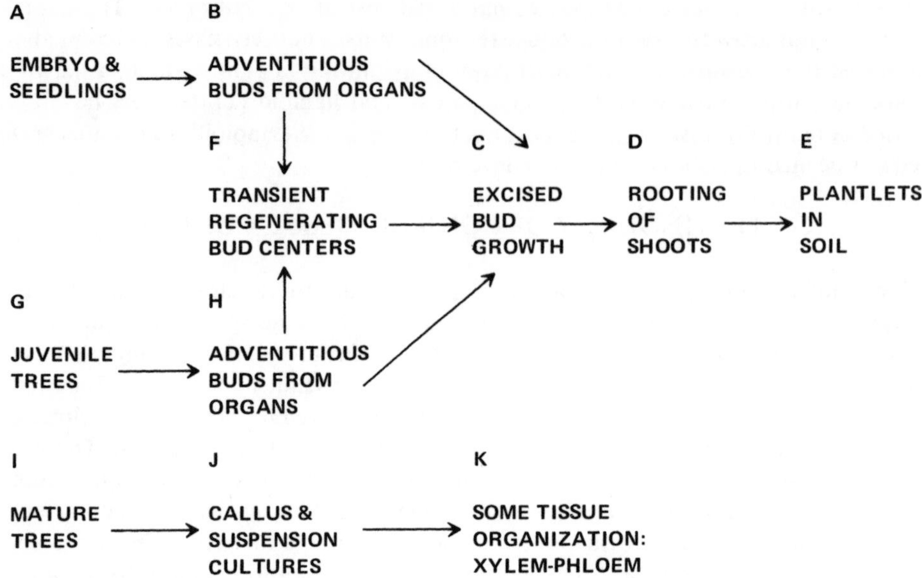

FIGURE 1. Possible plantlet regeneration schemes for conifers using tissue culture methods.

production in culture by this sequence seems to be a one-shot process in that cultures cannot be sustained with continuous bud production capability. The pathway has little prospect for commercial application since the many distinct steps require much handling of materials making it an expensive way to produce clones. Since organs can produce only a limited number of buds, and given the frequent shoot growth and rooting problems, one can expect to produce clones consisting of only tens of members. However, this method does generate the modest numbers of buds and rooted plants necessary to develop more efficient methods for the latter steps of shoot growth, rooting, and transfer to soil. This is an important use since there is every reason to expect methods for maintaining bud growth, inducing rooting of the elongated shoots, and transfer of the plants to soil will be similar whether the buds originate from embryos and seedlings (Steps A→B→C), from older but still juvenile trees (Steps G→-H→C), or in much larger numbers from mature trees by way of callus and suspension cultures (Steps I→J→C).

Let us examine more closely the steps in the most common sequence A→B→C→-D→E of Figure 1 and, following that, we will consider the progress and potential represented by the other pathways which are also shown.

Bud production from explants of embryos and seedlings (Step A→B) has been successful for many species as mentioned earlier. Successful growth and rooting of these buds to produce significant numbers of complete plantlets capable of growth in soil (Steps C→D→E) have been more difficult. As the technology is emerging for conifers, the reports of successful plant production with various species range from cases where only one or two plants were produced, merely showing that it could be done, to cases where the methods have been sufficiently refined to yield large numbers of plants. Small numbers of rooted plantlets have been reported for longleaf pine,[195] white spruce, *Picea glauca* (Muench) Voss,[128] radiate pine, *Pinus radiata* D. Don,[135] and *Pinus pinaster* Sol.[136] Better production giving sufficient numbers of elongated shoots to allow calculation of a best rooting percentage between 20 to 80% have been obtained for Douglas-fir, *Pseudo tsuga menziesii* (Mirb.) Franco,[143] loblolly pine, *Pinus taeda*

L.,[133,196] Western red cedar, *Thuja plicata* Donn.,[150] Norway spruce, *Picea abies* L. Karst.,[130,131] and for the Western hemlock, *Tsuga heterophylla* L.,[151] which rooted easily upon direct rransfer to soil. Sufficient production has been obtained with loblolly pine to permit calculation of average percentages of bud formation, elongation, and rooting with consideration of seed family and among clone variation.[133,196-198] Data on bud production by clone and overall rooting percentage of 3 to 11% were reported for Douglas-fir.[145] The loblolly pine study has recently been extended to include calculation of the percent plant establishment in soil according to clone and under different conditions.[199-201] Similar data are generally lacking for other species. The following discussions will rely heavily on loblolly pine for examples since I am most familiar with this species. Other species will be mentioned where they add information or offer possibilities not yet demonstrated for pine.

Bud initiation (Step B) is induced on excised cotyledons or hypocotyls by a relatively high exogenous cytokinin concentration, near 1 mg/ℓ of BAP, and the process shows little, if any, requirement for auxin. Buds commonly form directly on the surface of cotyledons or hypocotyls within 6 to 10 weeks. If the germinated embryo or seedling is excised from its place in the female gametophyte of the seed and cultured intact, higher concentrations of cytokinins are necessary (10 mg/ℓ), and the process can be strongly enhanced by environment and other conditions which suppress normal embryo and seedling growth.[196] Much of this interaction between seedling growth and bud formation was avoided, and higher numbers of buds were produced when cotyledons were excised and cultured separately.[133,196] Buds have been shown to initiate from the peripheral cell layers of the organ including the epidermis in loblolly pine,[198] Douglas-fir,[202] Norway spruce,[131] and white spruce.[203] From gross comparison with reports for other species, it seems safe to presume that this is also the case for all conifers responding in Steps A→B.

Once formed, some of the buds grow out within a few weeks to form shoots 5- to 10-mm high while still attached to the parent cotyledon in some species and fewer do so in others. In general, buds fail to develop into elongated shoots or are overtaken by callus growth when left on the parent tissue on the initiation medium. Most workers found that shoot growth was fostered when the buds were excised and transferred to a medium of reduced-strength mineral salts without growth regulators.

In the case of Douglas-fir, a portion of the buds formed on the bud initiation medium grew out to 2-cm high in 2 to 4 weeks if the parent cotyledon supporting the buds was transferred to media lacking growth regulators.[143,146] Some bud growth was also observed for white spruce when cotyledons were transferred to media lacking growth regulators.[203] In all other cases, and especially for loblolly pine,[133] radiata pine,[135] and Western red cedar,[150] bud growth required removal of the buds from the parent tissues and subculture of the upright excised buds to media lacking growth regulators. Even when buds are excised and placed on fresh media, lack of bud growth remains a serious problem and is the main reason that so few of the species, for which adventitious buds have been obtained, have actually produced rooted clones of plantlets in the soil. The need for more information about factors influencing bud growth is clear.

It seems evident that the transition from initiation of adventitious buds to their subsequent growth and elongation to produce a stem is not a trivial step. Nor does the difficulty rest solely in the tissue culture procedures leading to adventitious bud formation since similar difficulties have been reported in obtaining growth in culture of terminal buds excised after normal formation on the tree.[115,204] Growth regulator maladjustments would seem to be a prime candidate for the cause, since application of cytokinins to the terminal buds of trees growing in soil effectively stopped growth of slash pine.[205] Information on the internal concentrations of auxins in young and old buds,[206] on the metabolism of auxin and cytokins in various tissues,[207,208] and on the subtleties of auxin transport[209] is beginning to fill in the gaps. Concern for the special nutrient requirements of growing shoots in soil[210] might well find useful application to the problems in culture. We need to bring the information of bud growth in woody perennial species into focus, especially those species that normally set short branch buds which in nature do not grow out.[211] Of course, even long branch buds which do grow out, normally do not do so immediately after formation, but rather await some period of dormancy. Tissue culture procedures may inadvertently trigger some dormancy routine in many of the buds produced. Investigation of the immediate controls of this aspect in terms readily applicable to cultured shoot primordia is lacking in the literature.

With few exceptions, rooting of the shoots grown in vitro from adventitious buds (Step D) is also a troublesome step and part of the difficulty may reflect inadequate shoot elongation at Step C. Complexities in growth regulator metabolism and distribution within the developing shoot[212] may also play a part. Vitamin D analogues have recently been shown to aid rooting in *Populus*[213] and may aid the process in conifers as well. Rooting is generally not successful until the shoots reach 5 mm or more in height. Rooted shoots have been produced only in a few conifer species as noted above, which bespeaks the problem, but it also means that our information base is small. Cheng and Voqui[143] reported 80% rooting of a select, small group of Douglas-fir buds which had attained 1 to 2 cm in height. Low sucrose concentration (0.5%) and the auxin NAA 0.25 μM were required with a temperature of 19°C. Rooting was substantially reduced and much callus occurred for shoots on the same medium, but with a temperature of 24°C. Root primordia appeared in 4 weeks after which the NAA was withdrawn to foster further root growth. A 70% rooting of shoots more than 5 mm in height has been reported for loblolly pine.[133] In this species rooting was accomplished at 21°C, and both BAP (0.44 μM) and NAA (0.54 μM) were required. The growth regulator combination was critical, and no strong temperature sensitivity was noted.

Rooting has been a severe problem with radiata pine[135] and occurred only sporadically in response to auxin in Norway spruce.[130] Up to 50% rooting within 4 to 8 weeks was reported for shoots 5- to 30-mm high in response to 50 μM indolebutyric acid in the case of Western red cedar.[150] Tissue culture shoots of Western hemlock have rooted easily when placed directly in soil mix supplied with 15 μM NAA.[151] This ease of rooting seems to be unique to those species like Western hemlock which root easily by traditional methods from stem cuttings.

In summary, one can say that most workers have been concerned that the shoots should be of sufficient size to permit rooting, and beyond that, success seems to bear a strong relation to species.

B. Plantlet Transfer to Soil

The literature is generally lacking in reports of transfer of tissue culture-produced, rooted propagules of conifers to soil (Step D→E). Generally successful transfer to soil and subsequent growth in the greenhouse was reported for Douglas-fir.[143] Good survival was noted for Western red cedar which was initially established in a mist chamber and then grown under greenhouse conditions.[150] In both cases the plants were said to appear normal, but specific data were not given. The only systematic study of the transfer to autotrophic growth in soil which has been reported is that for loblolly pine.[199-201] Plantlets rooted in vitro on agar-solidified agar in deep petri plates or specimen jars under 9000 lx fluorescent-incandescent light were transplanted directly to soil mixtures after a wash to remove adhering agar. It was necessary to maintain the freshly transferred plants at high humidity initially, either by conventional mist bench procedure or in pots covered with clear plastic tents. Survival rates of 60 to 100% were recorded for average quality planting stock. The mortality was related to shoot and root vigor classification of plantlets at time of transfer, and plantlet classifications were defined which gave nearly 100% survival when placed directly in the greenhouse with minimal attention to humidity.[199] Using conventional mist bench methods, average tissue-culture plantlets showed 38% greater mortality than comparable soil-germinated seedlings and 50% less subsequent growth in the soil, although the specific rates of growth had become equivalent after 24 weeks in soil. Some irregularities in growth form were observed in the tissue-culture plants including precocious fascicled-needle development and some curved stems.[201] Root growth and ectomycorrhizal development were monitored and found to be normal. The uniformity in growth that one might expect among plantlets of a clone could not be demonstrated statistically. There was considerable variation in the shoot height growth which occurred during the initial 28-week period of observation. Of course, the plantlets lacked the pronounced and quite uniform hypocotyl elongation which was an important part of seedling height growth. Further work will be required to reduce the variation in growth, but one can be encouraged that the tissue culture-produced clones can attain growth rates and growth form comparable to seedlings. Many of these plants are now growing in the field where observations can be made throughout the life cycle.

C. Clones from Juvenile Trees

The work with sequence A→E of Figure 1 has indicated that A→B is successful for most conifer species. Steps C and D are troublesome as yet and show strong species dependence. Transfer to soil and subsequent growth in soil does not seem to present severe problems, based on the limited data available. Sufficient plantlets have not been obtained from enough other species to make a general statement. The sequence serves well to provide materials for further development of methods and identification of problems in the latter steps, but it suffers from the low numbers of individuals produced in a clone and the need to use very young seedlings or embryos as initial material. Each embryo or seedling carries an unproven genotype. At the young stages which must be used in this propagation scheme, the seedlings are too young for judgments to be made of whether or not they carry tree traits worthy of cloning. Clone production from mature trees is, of course, the ideal. Sound judgments of desirable traits can be made on mature trees. A change from the juvenile to the mature phase of growth usually occurs when conifers become sexually mature and produce their first seeds. With the onset of the mature phase of growth in most conifers, the ability to root from cuttings decreases dramatically and problems of plagiotropic growth and loss of vigor becomes pronounced in plants produced by conventional vegetative means with cuttings or needle fascicles.[21] Some progress has been made in methods for tissue culture propagation from established trees up to 5-years old, but still in the juvenile stage. These methods apply to trees still in the juvenile phase, but it is a step in the right direction.

The pathway G→H→C→D→E of Figure 1 simply inserts older, but still juvenile-starting materials in place of embryos in the sequence already discussed for seedlings and embryos. The procedures are much the same, and the key appears to be the selection of tissues from capable species, which are conductive to organogenesis. Bud formation from needles and succulent stem sections of Douglas-fir 2- to 4-years old has been reported but without data.[214] Data showing bud formation from very young needles of Norway spruce have been reported,[215] and buds derived from callus-like swellings at cultured needle bases have been reported for *Pinus radiata*[216] and Western red cedar.[150] This technology is not broadly based since several workers have mentioned in passing that they were unable to obtain adventitious bud formation from even young needles of a number of species. Excised young needle fascicles have been shown to resume growth in culture at their terminal bud in loblolly pine,[137] radiata pine,[216] and *Pinus pinaster* Sol.[136] Multiple adventitous buds were also generated from needle fascicles cultured on media with relatively high cytokinin concentrations in the case of loblolly pine.[137] Shoot apices from juvenile trees have also been made to grow out directly to produce rooted plants in culture for Norway spruce[130] and to grow and support a micropropagation scheme utilizing successive axillary bud subculture for Douglas-fir.[147] Arnold and Eriksson have shown adventitious bud production from resting terminal and lateral buds of 5-year- to even 50-year-Norway spruce.[132] This is the only report of consistent propagation by a scheme applicable to mature conifer trees. The latter work gives promise that methods developed for juvenile trees can be extended to include mature trees, at least for some species.

There has also been some attention given to producing the sexual structures directly from excised buds placed in culture. Such a procedure might aid in making crosses with trees not yet mature enough to flower. Vegetative shoot tips from mature trees of Western red cedar have been induced to form male strobili in the presence of gibberellic acid during 8 weeks culture in vitro, but meiosis failed to occur.[69] Neither exogenous auxin nor gibberellic acid supplied at several concentrations caused significant modification in the male strobili, nor could female strobili be induced. Flowering is usually taken as the prime indicator of maturity in plants; however, the reproductive organs do generate the gametophyte pollen and egg which generate the zygotic embryo, the epitome of juvenility. The strobili of conifers and the pollen and female gametophyte tissue they produce are likely sources of meristematic tissues to be used in regeneration of plants by tissue culture. The pollen of *Pinus resonosa* and female gametophyte tissue of *Pinus nigra* and *Pinus mugo* have been used in attempts to generate haploid plants. Trachial differentiation and occasional "embroy-like" growths were obtained from the callus, but they failed to develop further. Viable haploid plant generation in tissue culture has not been reported among the conifers.

Thus, the option of using select meristematic portions of juvenile trees in the propagation scheme G→E is developing and can be expected to become generally applicable to conifer species as more work is done. This scheme still suffers from the low number of plants that can be obtained per initial explant. Although a young tree can provide more needles, fasicles, buds, etc., than can be obtained from embroys or seedlings, decontamination of these parts from field grown trees presents a serious problem. It would be much more efficient if more buds were produced continuously from a single initial explant by repeated subculturing.

D. Regenerative Subcultures

The prospect of adjusting the media and culture conditions so that explants, once induced into a bud-initiating mode, could be proliferated in that mode and subsequently produce many more buds has often been realized for herbaceous plants.[47] This prospect is noted as Step F in Figure 1 where it could apply either to seedling- or juvenile-tree starting material. Continued proliferation of bud-producing cultures has not been achieved in conifers from either starting material. In fact, the weight of reports suggests that bud initiation occurs directly from surfaces of the primary explant and is rapidly lost as subculture continues. One aspect of the conifer propagation schemes brings some confusion into the literature. This relates to the commonly observed sequence where early stages of bud formation occur in response to high exogenous cytokinin concentrations, but subsequent growth of buds occurs only when cytokinin is removed. Should a mass of initiating shoot primordia form close together, they can be mistaken for callus proliferation in appearance, and upon excision and subculture to media without cytokinin, the growth of already initiated buds can easily be mistaken for new bud production. The ultimate distinction between this process and continual bud initiation from a truely regenerative callus lies in the failure of the former process to continue producing buds beyond one or two subcultures. Short-term bud production in subculture, perhaps of the former kind, has been observed in loblolly pine,[133,198] in Norway spruce,[131] and radiata pine.[135] The continued bud initiation obtained in Douglas-fir which may extend to three or four successive subcultures may have been of the second kind.[144-146] The bud production observed in proliferated tissue obtained from juvenile basal trunk burls of mature *Sequoia* was probably of the former kind, since the capacity for bud production decayed rapidly in subculture.[75,152]

There seem to be enough reports to suggest that the regenerative subculture approach is feasible for conifers, and surely the success with herbaceous plants can support an optimistic view. The distinction between development of already initiated buds in subculture and regenerative callus, which can be maintained for long periods in subculture, may dissolve into a semantic argument, but for the present, caution in interpretation of results seems warranted. The real test is whether a culture can be made to continue to generate viable plants of uniform genotype and growth.

E. Clones from Callus

Maintenance of subcultures which continue to initiate buds would greatly increase the numbers of cloned plantlets that could be produced from an initial explant. That would overcome one hurdle to mass propagation of cloned planting stock. The methods may still depend on proper selection of initial meristematic organs, and they may even require that juvenile-starting materials be used which would still preclude clonal propagation from time-tested and mature, superior trees. The latter problem would be resolved if methods for juvenile trees could be extended to include select organs from mature trees as has been done for Norway spruce.[132] An alternative scheme (Steps I→J→C→D→E of Figure 1) would solve both problems. Instead of seeking bud production directly from some selected plant part, such as a cotyledon, needle, or bud, one would seek to initiate an unorganized callus growth which could be rapidly multiplied in subculture and at some later time be induced to form multiple adventitious shoots or embryos. Regenerative callus of this kind was produced in the case of aspen and elm as mentioned earlier. The attractions of this scheme are the rapid multiplication of the cell population from which buds could be developed and perhaps the lesser need to be so selective in the choice of initial tissue. The problems of surface sterilization of mature-tree explants from the field would be more tractable and the expense of propagation would be decreased as callus cultures could be handled in bulk. An additional attraction is that bud regeneration from callus would open the door to the emerging protoplast technology since the callus produced from recovered protoplasts could then be expected to generate buds and plants. Protoplasts obtained from Douglas-fir have been recovered to produce some limited callus growth.[217]

Callus can be derived from almost all parts of any age tree.[218,219] Callus has been obtained from nearly all conifer species. It originates similarly for all conifers from tissues near the vascular traces, cambium, and vascular parenchyma, in response to auxin with little or no effect by exogenous cytokinin. The callus obtained from conifers, with the exception of that from burls of sequoia,[75] does not show the tendency toward organogenesis of shoots and roots which is common for angiosperm trees. The callus of all conifers thus seems similar in its origins, in its appearance and lack of strong organogenesis tendency. Since the callus appears to be similar, one might reasonably expect that methods developed to foster organogenesis and plantlet production from the common callus of one species would be readily extended to the similar callus of many other species.

Sporadic bud production from callus obtained from embryonic or seedling explants has been reported to extend over one to six subcultures by Douglas-fir.[146] Similar response has not been obtained for other species, even when essentially the same methods were used. There is no report in the conifers of viable shoots or plants being derived from callus which was not initially engaged in bud production at its first primary culture. We have seen that this is not the case for angiosperms. A stem callus of aspen showed no evidence of organogenesis until the passage of several subcultures.[138,153] The methods used for aspen have been available since 1970, but there has been no report of any similar response for conifer callus cultures. Even considering the reported instances of somewhat extended retention of bud production capacity through a few subcultures, nothing even approaching mass propagation of buds from callus has been reported for conifers. That is not to say that considerable cell differentiation has not been observed in callus cultures. Likewise, early stages of differentiation and perhaps even organogenesis may be reflected in biochemical differentiation which was detected in explants undergoing organogenesis as compared to nonorganogenic callus of Douglas-fir.[220] In many tree species, differentiation of individual cells and even rudementary tissue organization (Step K of Figure 1) often occurs without the overall organization leading to organ development. The vascular elements associated with protuberances of early bud formation observed in Norway spruce cotyledons[131] have also been recognized in callus protuberances of radiata pine.[221] These protuberances subsequently were involved in bud production with Norway spruce, but the protuberances from radiata pine callus did not give rise to buds. Even so, xylem and phloem were recognized in the vascularized central area of the protuberances of radiata pine.[222] Differentiation of vascular tissues within callus has been recognized frequently in both anglosperm and conifer trees. Such differentiation is not specifically a product of tissue culture since it has been easily recognized in the natural, *in situ* callus growths on trees as well, e.g., *Pinus echinata* Mill.[223] The occurrence of vascular differentiation in nodules on otherwise disorganized callus is encouraging even though the callus was of embryonic origin and already in the process of shoot production. The occasional reference to embryo-like structures in both primary and callus or suspension cultures is likewise suggestive that it will not be long before mass propagation from callus cultures will be available for conifers and other forest trees.

V. PROSPECTIVE APPLICATIONS IN FORESTRY

A. Overview of Propagation

From our examination of the various possibilities in the developing field of tree tissue culture, it is evident that much has been accomplished since the early establishment of cambial callus cultures prior to 1954. Substantially increased attention has been given to tree species since plantlets were first regenerated in the mid-seventies from subcultured stem callus for angiosperm trees and directly from primary explants of seedlings for conifer trees. The list of tree species having produced cloned plants through tissue culture continues to increase. As more examples have been reported, generalities have begun to emerge among tree culture methods. Devices such as node and bud culture have been used to circumvent the distressingly persistent inability to achieve stable regenerative callus from trees. The steps of shoot growth and rooting still appear to be very species-dependent, but progress is being made in these areas as well. It appears that methods applicable to juvenile trees will ultimately develop to include mature trees on a more general basis. Overall, no great surprises or unique methods have emerged from the work with trees which offer significant contrast to methods for herbaceous plants. An exception may be the problematical bud growth phase and some hints of requirements for more complex growth regulator sequences. The outlook seems to be that trees may require a little extra care and precision in culture methods, but we can expect with confidence that additional experimentation will make available the entire array of possibilities already accepted for herbaceous plants.

B. Clone Uniformity

One can look with optimism toward a capability for mass clonal propagation of forest trees, but other opportunities presented by the developing tissue culture methods should not be overlooked. Forest trees mature slowly over several years, and commonly must grow on a variety of sites, each presenting a different degree of hazard. The use of even the small numbers of clonal propagules, which can be produced with current methods from seedlings, could greatly aid genotype evaluation in that the same genotype could be planted and monitored on several sites. Thus, site hazard-genotype interactions would be more clearly delineated when growth evaluation is made after the prescribed number of years. Of course, the members of a clone must be shown to have uniformity of phenotype under a given environment situation. Few clones of tissue-culture produced trees have yet been tested in this respect. Clones of black poplar, obtained by tissue culture means, have shown considerable within-clone variation in height growth and branching vigor in the first growing season.[224] A part if not all of this variation may be associated with a variable degree of virus elimination from the plants within a clone[225] although the variable ploidy which was observed in root tip cells may indicate a more serious problem. The loblolly pine plants from tissue culture showed some within-clone variation in the vigor of early growth,[201] but this may settle down in subsequent years growth as is commonly observed to be the case for seedlings. Optimism is warranted since fruit trees propagated in vitro by meristem culture, studied over 4-years-growth and more, have shown no obvious mutations or abnormalities.[226] The clonal propagations of many trees by rooted cuttings testify to marked stability of clones.[26] Some caution seems justified until tree clones have had more study, since induced variation has been observed with other plants and is sensitive to the different tissue culture methods which might be used.[227,228]

C. Germplasm Storage

Tissue culture techniques of long-term, low-temperature storage of cultures may possibly be of use in programs to retain the broad genetic base of the wild population of trees as genetic management becomes more intensive. Recent reviews by Dougall,[229] Withers,[230] and Bajaj[231] summarize the general progress in this area. Very little has been done with tree species since mass propagation is not yet readily available. The prospects for long-term seed storage at liquid nitrogen temperatures[232] should not be overlooked for trees in this respect, perhaps coupled with clonal multiplication by tissue culture after recovery.

D. Trait Selection In Vitro

Disease and insect invasion cause serious reduction in forest yields. But the very nonuniformity of site and microclimate which helps to prevent destruction of entire stands of trees acts, along with the sheer size of most trees, to make careful study of the disease interaction most difficult. Clonal tissues and plantlets growing in otherwise sterile bottles, in controlled environments, on defined media present opportunities for careful study of the interaction between disease organism and host. Further, any detection and selection methods for disease resistance in tree tissues at this level would fit efficiently with mass clonal propagation schemes. The application of tissue culture techniques for disease resistance selection has been the object of considerable research, especially with agricultural crop species, and in a few cases, it has been used successfully (see Reviews by Ingram[233] and Bretell and Ingram[234]). The advantages of this approach for forest trees, where resistant stock is often the only way to combat disease problems, may be sufficient to merit an intensive research effort. Some efficiency in genotype selection can be attained even by simple screening of seedlings in vitro as was done for tolerance to phosphate deficiency in birch.[235] The immediate extension to cloned plantlets is obvious.

Coculture of tree organs or callus tissue with beneficial mychorrhizal and nitrogen-fixing symbionts was mentioned earlier. Pioneering work with coculture of disease organisms has also been used as a way to culture obligate fungal pathogens of trees[236-238] and to study disease interaction.[239,240] Reliable methods have been reported for obtaining axenic pine-infective basidiospores of *Cronartium fusiforme,* a rust pathogen of loblolly pine.[70] From these spores, axenic fungus cultures were obtained, and infection of loblolly pine organs was achieved in culture. Thus, the basic aspects of infection and disease development become open to in vitro study for this forest tree disease. Clonal propagation of trees for correlation of field tests with in vitro tests of host resistance now adds another dimension to such studies. A similar approach may be applied for host resistance to insect pests. Methods have been reported for rearing larvae of Southern pine beetle (*Dendroctonus frontallis* Zimmerman) from egg to adult solely on callus cultures of its loblolly pine host, in the complete absence of the natural associate fungi of the beetle.[241] The way is open to investigate in vitro the complex relationship of host cells and associate fungi in supplying nutrients and metabolic precursors for beetle life cycle completion.

These and other tissue culture possibilities hold special promise for trees, which are so refractive to study in the forest situation and whose size and long reproductive cycles place severe limitations on more traditional approaches.

REFERENCES

1. **Marx, D. H.,** Mycorrhizae and establishment of trees on strip-mined land, *Ohio J. Sci.,* 75, 288, 1975.
2. **Marx, D. H., Morris, W. G., and Mexal, J. G.,** Growth and ectomycorrhizal development of loblolly pine seedlings in fumigated and nonfumigated nursery soil infested with different fungal symbionts, *For. Sci.,* 24, 193, 1978.
3. **Anderson, R. L. and Barry, P. J., Eds.,** Forest Insect and Disease Conditions in the South 1978, Forestry Rep. SA-FR4, Forest Service-Southeastern Area, State and Private Forestry, U.S. Department of Agriculture, Atlanta, Ga., 1979, 25.
4. **Cannell, M. G. R. and Last, F. T.,** *Tree Physiology and Yield Improvement,* Academic Press, New York, 1976, 1.
5. **Heybroek, H. M. and Visser, T.,** Juvenility in fruit growing and forestry, *Acta Hortic.,* 56, 71, 1976.
6. **Ross, S. D. and Pharis, R. P.,** Promotion of flowering in the *Pinaceae* by gibberellins. I. Sexually mature, non-flowering grafts of Douglas-fir, *Physiol. Plant.,* 36, 182, 1976.
7. **Pharis, R. P. and Kuo, C. C.,** Physiology of gibberellins in conifers, *Can. J. For. Res.,* 7, 299, 1977.
8. **Ross, S. D. and Greenwood, M. S.,** Promotion of flowering in *Pinaceae* by gibberellins. II. Grafts of mature and immature *Pinus taeda, Physiol. Plant.,* 45, 207, 1979.

9. **Visser, T.**, The effect of rootstocks on growth and flowering of apple seedlings, *J. Am. Soc. Hortic. Sci.*, 98, 26, 1973.
10. **Zobel, B. J.**, Increasing southern pine timber production through tree improvement, *So. J. Appl. For.*, 1, 3, 1977.
11. **Libby, W. J.**, Domestication strategies for forest trees, *Can. J. For. Res.*, 3, 265, 1973.
12. **Libby, W. J., Strittler, R. F., and Seitz, F. W.**, Forest genetics and forest-tree breeding, *Annu. Rev. Genet.*, 3, 469, 1969.
13. **Carlisle, A., and Teich, A. H.**, The costs and benefits of tree improvement programs, *Can. For. Serv. Publ.*, 1302, 14, 1971.
14. **Doorenbos, J.**, Juvenile and adult phases in woody plants, in *Encyclopedia Plant Physiology*, Vol. 15 (Part 1), Ruhland, W., Ed., Springer-Verlag, Berlin, 1965, 1222.
15. **Kozlowski, T. T.**, *Growth and Development of Trees*, Vol. 1, Academic Press, New York, 1971, 94 and 117.
16. **Borchert, R.**, The concept of juvenility in woody plants, *Acta Hortic.*, 56, 21, 1976.
17. **Pharis, R. P. and Morf, W.**, Precocious flowering of coastal and giant redwood with gibberellins A_3, $A_{4/7}$ and A_{13}, *BioScience*, 19, 719, 1969.
18. **Phillips, I. D. J.**, Apical dominance, *Annu. Rev. Plant Physiol.*, 26, 341, 1975.
19. **Robbins, W. J.**, Topophysis, a problem in somatic inheritance, *Proc. Am. Philos. Soc.*, 108, 395, 1964.
20. **Farmer, R. E., Jr.**, Vegatitive propagation and the genetic improvement of North American hardwoods, *N. Z. J. For. Sci.*, 4, 211, 1974.
21. **Brix, H. and van den Driessche, R.**, Use of Rooted Cuttings in Reforestation. A Review of Opportunities, Problems and Activities, Rep. 6, British Columbia Forestry Service/Canadian Forestry Services Joint Rep., Victoria, 1977, 1.
22. **Girouard, R. M.**, Vegetative Propagation of Pines by Means of Needle Fascicles — A Literature Review, Inf. Rep. Q-X-23, Canadian Forestry Service, Centre de Recherche Forestiere des Laurentides Quebec Region, Canada, 1971, 16.
23. **Zimmerman, R. H.**, Symposium on juvenility in woody perennials, *Acta Hortic.*, 56, 1, 1976.
24. **Hartmann, H. T. and Kester, D. E.**, *Plant Propagation — Principles and Practices*, 3rd ed., Prentice-Hall, Englewood Cliffs, N.J., 1975, 271.
25. **Hinds, H. V. and Krugman, S. L.**, Vegetative propagation, *N. Z. J. For. Sci.*, 4, 119, 1974.
26. **Longman, K. A.**, Tree biology research and plant propagation, *Proc. Int. Plant Propagator's Soc.*, 25, 219, 1975.
27. **Donnelly, J. R. and Yawney, A. W.**, Some factors associated with vegetatively propagating sugar maple stem cuttings, *Proc. Int. Plant Propagator's Soc.*, 22, 413, 1972.
28. **Longman, K. A.**, Some experimental approaches to the problem of phase change in forest trees, *Acta Hortic.*, 56, 81, 1976.
29. **Toda, R.**, Vegetative propagation in relation to Japanese forest tree improvement, *N. Z. J. For. Res.*, 4, 410, 1974.
30. **Kleinschmit, J.**, A programme for large-scale cutting propagation of Norway spruce, *N. Z. J. For. Sci.*, 4, 359, 1974.
31. **Lepistö, M.**, Successful propagation by cuttings of *Picea abies* in Finland, *N. Z. J. For. Sci.*, 4, 367, 1974.
32. **Sweet, G. B. and Wells, L. G.**, Comparison of the growth of vegetative propagules and seedlings of *Pinus radiata*, *N. Z. J. For. Sci.*, 4, 399, 1974.
33. **Shelbourne, C. J. A. and Thulin, I. J.**, Early results from a clonal selection and testing programme with radiata pine, *N. Z. J. For. Sci.*, 4, 387, 1974.
34. **Ross, S. D.**, Production, propagation, and shoot elongation of cuttings from sheared 1-year-old Douglas-fir seedlings, *For. Sci.*, 21, 298, 1975.
35. **van Buijtenen, J. T., Toliver, J., Bower, R., and Wendel, M.**, Mass production of loblolly and slash pine cuttings, *Tree Planters' Notes*, 26, 4, 1975.
36. **Boyd, C. C.**, Rooting 20 Million Western Hemlock: Production Alternatives and Research Needs, presented at Western Hemlock Conference, Seattle, Washington, May 1976.
37. **Libby, W. J., Brown, A. G., and Fielding, J. M.**, Effects of hedging radiata pine on production, rooting, and early growth of cuttings, *N. Z. J. For. Sci.*, 2, 263, 1972.
38. **Rauter, R. M.**, A short-term tree improvement programme through vegetative propagation, *N. Z. J. For. Sci.*, 4, 373, 1974.
39. **Brix, H. and Barker, H.**, Rooting Studies of Western Hemlock Cuttings, Inf. Rep. BC-X-87, Canadian Forestry Service, Pacific Forest Research Center, Victoria, 1975, 13.
40. **Longman, K. A., Coutts, M. P., and Bowen, M. R.**, Tree physiology, Rep. on For. Res., Forestry Commission, London, 1972, 83.
41. **Thomas, J. E. and Riker, A. J.**, Progress on rooting cuttings of white pine, *J. For.*, 48, 474, 1950.

42. **Copes, D. L.,** Comparative leader growth of Douglas-fir grafts, cuttings, and seedlings, *Tree Planters' Notes*, 27, 13, 1976.
43. **Segaard, B.,** Leaf blight resistance in *Thuja*. Experiments on resistance to attack by *Didymascella thujima* (Dur.) Maire (*Keithia thujima*) on *Thuja plicata* Lamb., *Yearbook*, Royal Veterinary and Agriculture College, Copenhagen, 1956, 30.
44. **Libby, W. J. and Hood, J. V.,** Juvenility in hedged radiata pine, *Acta Hortic.*, 56, 91, 1976.
45. **Fielding, J. M.,** Factors affecting the rooting and growth of *Pinus radiata* cuttings in the open nursery, *Bull. Aust. For. Timber Bur.*, 45, 1969.
46. **Newsletter, Yamada, Y., Ed.,** International Association for Plant Tissue Culture, Kyoto University, Japan, 1978, 22.
47. **Murashige, T.,** Plant propagation through tissue cultures, *Annu. Rev. Plant Physiol.*, 25, 135, 1974.
48. **Schmidt, A.,** Über die Chlorophyll-bildung in Kamferenembryo, *Bot. Arch.*, 5, 260, 1924.
49. **LaRue, C. D.,** The growth of plant embryos in culture, *Bull. Torrey Bot. Club*, 63, 365, 1936.
50. **Maheshwari, P. and Rangaswamy, N. S.,** Embryology in relation to physiology and genetics, *Adv. Bot. Res.*, 2, 219, 1965.
51. **Stone E. and Duffield, J.,** Hybrids of sugar pine by embryo culture, *J. For.*, 48, 200, 1950.
52. **Thomas, J. J.,** Etude compare des developments in situ et in vitro des embryons de pins: influence des culturenourrices sur les embryons isoles durant le premieres phases de leur developments, *Soc. Bot. France, Mem. Coll. Morphologie*, 147, 1973.
53. **Monnier, M.,** Culture of zygotic embryos, in *Frontiers of Plant Tissue Culture 1978*, Thorpe, T. A., Ed., The International Association of Plant Tissue Culture, Calgary, Canada, 1978, 277.
54. **Raghavan, V.,** *Experimental Embryogenesis in Vascular Plants*, Academic Press, New York, 1976, 603.
55. **Singh, H.,** *Embryology of Gymnosperms*, Gebrüder Borntraeyer, Berlin, 1978, 302.
56. **Slankis, V.,** Influence of the sugar concentration on the growth of isolated pine roots, *Nature (London)*, 160, 645, 1947.
57. **Barnes, R. L. and Naylor, A. W.,** *In vitro* culture of pine roots and the use of *Pinus serotina* roots in metabolic studies, *For. Sci.*, 5, 158, 1959.
58. **Slankis, V.,** Effect of α-napthalene acetic acid on dichotomous branching of isolated roots of *Pinus silvestris, Physiol. Plant.*, 3, 40, 1950.
59. **Slankis, V.,** The role of auxin and other exudates in mycorrhizal symbiosis of forest trees, in *The Physiology of Forest Trees*, Thimann, K. V., Ed., Roland Press, New York, 1958, 427.
60. **Fortin, J. A.,** Synthesis of mycorrhizae on explants of the root hypocotyl of *Pinus silvestris* L., *Can. J. Bot.*, 44, 1087, 1966.
61. **Maheshwari, R.,** Applications of plant tissue culture in the study of physiology of parasitism, *Proc. Indian Acad. Sci., Sect. B*, 59, 152, 1969.
62. **Torrey, J. G.,** *In vitro* methods in the study of symbiosis, in *Frontiers of Plant Tissue Culture 1978*, Thorpe, T. A., Ed., The International Association of Plant Tissue Culture, Calgary, Canada, 1978, 373.
63. **Reporter, M., Raveed, D., and Norris, G.,** Binding of *Rhizobium japonicum* to cultured soybean root cells; morphological evidence, *Plant Sci. Lett.*, 5, 73, 1975.
64. **Giles, K. L. and Whitehead, H. C. M.,** Uptake and continued metabolic activity of *Azotobacter* within fungal protoplasts, *Science*, 193, 1125, 1976.
65. **Giles, K. L. and Whitehead, H. C. M.,** The localization of introduced *Azotobacter* cells within the mycelium of a modified mycorrhiza (Rhizopogon) capable of nitrogen fixation, *Plant Sci. Lett.*, 10, 367, 1977.
66. **Bulard, C. and Monin, J.,** Etude du comportement d'embryons de *Fraxinus excelsior* L. preleves dans des graines dormantes et cultives *in vitro, Phyton (Buenos Aires)*, 20, 115, 1963.
67. **Greenwood, M. S. and Berlyn, G. P.,** Sucrose-indole-3-acetic acid interactions on root regeneration by *Pinus lambertiana* embryo cuttings, *Am. J. Bot.*, 60, 42, 1973.
68. **Smith, D. R. and Thorpe, T. A.,** Rooting in cuttings of *pinus radiata* seedlings: fluctuations in relation to selection day, *Bot. Gaz. (Chicago)*, 137, 128, 1977.
69. **Coleman, W. K. and Thorpe, T. A.,** *In vitro* culture of western red cedar (*Thuja plicata*). II. Induction of male strobili from vegetative shoot tips, *Can. J. Bot.*, 56, 557, 1978.
70. **Amerson, H. V. and Mott, R. L.,** Technique for axenic production and application of *Cronartium fusiforme* basidiospores, *Phytopathology*, 68, 673, 1978.
71. **Gautheret, R. J.,** Culture du tissu cambial, *C. R. Acad. Sci.*, 198, 2195, 1934.
72. **Gautheret, R. J.,** Nouvelles recherches sur la culture du tissu cambial, *C. R. Acad. Sci.*, 205, 572, 1937.
73. **Gautheret, R. J.,** Nouvelles recherches sur la bourgeonment du tissu cambial d'*Ulmus compestris* cultive *in vitro, C. R. Acad. Sci.*, 210, 744, 1940.
74. **Jacquiot, C.,** Action du méso-inositol et de l'adenine sur le formation de bourgeous par le tissu cambial d'*Ulmus compestris* cutivi *in vitro, C. R. Acad. Sci.*, 233, 815, 1951.

75. **Ball, E. A.**, Differentiation in a callus culture of *Sequoia sempervirens, Growth,* 14, 295, 1950.
76. **Gautheret, R.**, The nutrition of plant tissue cultures, *Annu. Rev. Plant Physiol.,* 6, 433, 1955.
77. **Geissbühler, H. and Skoog, F.**, Comments on the application of plant tissue cultivation to propagation of forest trees, *Tappi,* 40, 257, 1957.
78. **Winton, L. and Huhtinen, O.**, Tissue culture of trees, in *Modern Methods in Forest Genetics,* Miksche, J. P., Ed., Springer-Verlag, Berlin, 1976, 243.
79. **White, P. R.**, *The Cultivation of Animal and Plant Cells,* Ronald Press, New York, 1963, 1.
80. **Broertjes, C., Haccius, B., and Weidlich, S.**, Adventitious bud formation on isolated leaves and its significance for mutation breeding, *Euphytica,* 17, 321, 1968.
81. **Winton, L. L.**, Bibliography of somatic callus cultures from deciduous trees, *Genet. Physiol. Notes,* No. 17, 1972; Bibliographic addendum of tree callus cultures, *Genet. Physiol. Notes,* No. 18, 1973; Second addendum of tree callus cultures, *Genet. Physiol. Notes,* No. 19, 1974.
82. **Durzan, D. and Campbell, R.**, Prospects for the mass production of improved stock of forest trees by cell and tissue culture, *Can. J. For. Res.,* 4, 151, 1974.
83. **Brown, C. L. and Sommer, H. E.**, An Atlas of Gymnosperms Cultured *In Vitro,* 1924-1974, Georgia Forest Research Council, Macon, Ga., 1975.
84. **Jacquiot, C.**, Plant tissues and excised organ cultures and their significance in forest research, *J. Inst. Wood Sci.,* 16, 22, 1966.
85. **Haissig, B.**, Organ formation *in vitro* as applicable to forest tree propagation, *Bot. Rev.,* 31, 607, 1965.
86. **Konar, R. N. and Guha, C.**, Culture of gymnosperm tissue *in vitro, Acta Bot. Neerl.,* 17, 258, 1968.
87. **Khatoon, K. and Mahmood, A.**, The effect of applied growth hormones on cell proliferation in the cambial zone of *Thuja orientalis* L., *Pak. J. Bot.,* 4, 117, 1972.
88. **Wereing, P., Hanney, C., and Digby, J.**, The role of endogenous hormones in cambial activity and xylem differentiation, in *The Formation of Wood in Forest Trees,* Zimmerman, M. H., Ed., Academic Press, New York, 1964, 323.
89. **Unger, J. W.**, A Study of the Jack Pine Seedlings and the Origin and Morphology of Callus Growth *In Vitro* from Seedling Segments, Ph.D. dissertation, University of Wisconsin, Madison, 1953.
90. **White, P. R.**, Sites of callus production in primary explants of spruce tissue, *Am. J. Bot.,* 54, 1055, 1967.
91. **White, P. R.**, Some aspects of differentiation in cells of *Picea glauca* cultivated *in vitro, Am. J. Bot.,* 54, 334, 1967.
92. **Wetmore, R. H. and Rier, J. P.**, Experimental induction of vascular tissues in callus of angiosperms, *Am. J. Bot.,* 29, 195, 1963.
93. **Brown, C. L.**, The influence of external pressure on the differentiation of cells and tissues cultured *in vitro,* in *The Formation of Wood in Forest Trees,* Zimmermann, M. H., Ed., Academic Press, New York, 1964, 389.
94. **Doley, D. and Leyton, L.**, Effects of growth regulating substances and water potential on the development of wound callus in *Fraxinus, New Phytol.,* 69, 87, 1970.
95. **Huhtinen, O. and Yahyaoglu, Z.**, Das früle Blühen von aus Kalluskulturen her augogenen Pflanzchen bei der Birke (*Betula pendula* Roth.), *Silvae. Genet.,* 23, 1, 1974.
96. **Bonga, J. M.**, Applications of tissue culture in forestry, in *Plant Cell, Tissue and Organ Culture,* Reinert, J. and Bajaj, Y. P. S., Eds., Springer-Verlag, New York, 1977, 93.
97. **D'Amato, F.**, Chromosome number variation in cultured cells and regenerated plants, in *Frontiers of Plant Tissue Culture 1978,* Thorpe, T. A., Ed., The International Association of Plant Tissue Culture, Calgary, Canada, 1978, 287.
98. **Stanley, R. G.**, Biochemical approaches to forest genetics, in *International Review of Forestry Research,* Vol. 3, Romberger, J. A. and Mikola, P., Eds., Academic Press, New York, 1970, 253.
99. **Sorensen, F. C. and Miles, R. C.**, Self pollination effects on Douglas-fir and ponderosa pine seeds and seedlings, *Silvae Genet.,* 23, 135, 1974.
100. **Mehra, A. and Mehra, P. N.**, Organogenesis and plantlet formation *in vitro,* in almond, *Bot. Gaz. (Chicago),* 135, 61, 1974.
101. **de Torok, D.**, The cytological and growth characteristics of tumor and normal clones of *Picea glauca, Cancer Res.,* 28, 608, 1968.
102. **Wright, K. and Northcote, D. H.**, Differences in ploidy and degree of intercellular contact in differentiating and non-differentiating sycamore calluses, *J. Cell Sci.,* 12, 37, 1973.
103. **Konar, R. N. and Nagmani, R.**, Chromosome numbers in callus cultures of *Pinus gerardiana* Wall., *Curr. Sci.,* 41, 714, 1972.
104. **Venketeswaren, S. and Huhtenen, O.**, Cytology of callus of Douglas-fir, *Pseudotsuga mensiesii* (Mirb.), in *Abstr., 4th Int. Congr. of Plant Tissue and Cell Culture,* Poster No. 1105, Thorpe, T. A., Ed., University of Calgary, Canada, 1978, 102.
105. **Barker, W. G.**, Behavior *in vitro* of plant cells from various sources within the same organism, *Can. J. Bot.,* 47, 1334, 1969.

106. **Stoutemeyer, V. T. and Britt, O. K.,** The behavior of tissue cultures from English and Algerian ivey in different growth phases, *Am. J. Bot.,* 52, 805, 1965.
107. **Magrum, L. and Steponkus, P. L.,** Differentiation of callus cultures of *Hedra helix* L. var. Thorndale, *Hortic. Sci.,* 7, 207, 1972.
108. **Banks, M. S., Christensen, M. R., and Hackett, W. P.,** Callus and shoot formation in organ and tissue cultures of *Hedra helix* L., English ivey, *Planta,* 145, 205, 1979.
109. **Banks, M. S.,** Phase change in *Hedra helix,* in *Abstr., 4th Int. Congr. of Plant Tissue and Cell Culture,* Poster No. 140, Thorpe, T. A., Ed., University of Calgary, Canada, 1978, 45.
110. **Kester, D. E., Tabachnik, L., and Nequeroles, J.,** Use of micropropagation and tissue culture to investigate genetic disorders in almond cultivars, *Acta Hortic.,* 78, 95, 1978.
111. **Pierik, R. L. M.,** Vegetative propagation of horticultural crops *in vitro* with special attention to shrubs and trees, *Acta Hortic.,* 54, 71, 1975.
112. **Abbot, A. J.,** Propagating temperature woody species in tissue culture, *Sci. Hortic.,* 28, 155, 1977.
113. **Mott, R. L.,** Tissue culture propagation of conifers, in Propagation of Higher Plants through Tissue Culture, Hughes, K. W., Henke, R., Constantin, M., Eds., Technical Information Center, U.S. Department of Energy, Springfield, Va., 1978.
114. **Winton, L. L.,** Morphogenesis in clonal propagation of woody plants in *Frontiers of Plant Tissue Culture 1978,* Thorpe, T. A., Ed., The International Association of Plant Tissue Culture, Calgary, Canada, 1978, 419.
115. **Bonga, J. M.,** Organogenesis in *in vitro* cultures of embryonic shoots of *Abies balsamea* (Balsam Fir), *In Vitro,* 13, 41, 1977.
116. **Skolmen, R. G. and Mapes, M. O.,** *Acacia koa* gray plantlets from somatic callus tissue, *J. Hered.,* 67, 114, 1976.
117. **Haines, R. J. and de Fossard, R. A.,** Propagation of hoop pine (*Araucaria cunninghamii* Ait.) by organ culture, *Acta Hortic.,* 78, 297, 1977.
118. **Huhtinen, O.,** Early flowering of birch and its maintenance in plants regenerated through tissue culture, *Acta Hortic.,* 56, 243, 1976.
119. **Ohyama, K. and Oka, S.,** Bud and root formation in hypocotyl segments of *Broussonetia kazinoki* Sieb *in vitro,* in *Abstr., 4th Int. Congr. of Plant Tissue and Cell Culture,* Poster No. 116, Thorpe, T. A., Ed., University of Calgary, Canada, 1978, 33.
120. **Huhtinen, O.,** Callus and plantlet regeneration from anther cultures of *Betula pandula* Roth., in *Abstr., 4th Int. Congr. of Plant Tissue and Cell Culture,* Poster No. 1740, Thorpe, T. A., Ed., University of Calgary, Canada, 1978, 169.
121. **Isikawa, H.,** *In vitro* formation of adventitious buds and roots on the hypocotyl of *Cryptomeria japonica, Bot. Mag.,* 87, 73, 1974.
122. **Aneja, S. and Atal, C. K.,** Plantlet formation in tissue cultures from lignotubers of *Eucalyptus citiodora* Hook., *Curr. Sci.,* 38, 69, 1969.
123. **Durand, C. R., Cresswell, R., and Nitsch, C.,** Factors influencing the regeneration of *Educalyptus grandis* by organ culture, *Acta Hortic.,* 78, 149, 1978.
124. **de Fossard, R. A., Barker, P. A., and Bourne, R. A.,** The organ culture of nodes of four species of eucalyptus, *Acta Hortic.,* 78, 157, 1978.
125. **Lakshmi Sita, G.,** Morphogenesis and plant regeneration from cotyledonary cultures of Eucalyptus, *Plant Sci. Lett.,* 14, 63, 1979.
126. **Rogozinska, J. H.,** The influence of growth substances on the organogenesis of honey locust shoots (Polish), *Acta Soc. Bot. Pol.,* 37, 485, 1968.
127. **Fu, M. L.,** Plantlets from paulownia tissue culture, in *Abstr., 4th Int. Congr. of Plant Tissue and Cell Culture,* Poster No. 1736, Thorpe, T. A., Ed., University of Calgary, Canada, 1978, 167.
128. **Campbell, R. A. and Durzan, D. J.,** Vegetative propagation of *Picea glauca* by tissue culture, *Can. J. For. Res.,* 6, 240, 1976.
129. **Webb, K. J. and Street, H. E.,** Morphogenesis *in vitro* of *Pinus* and *Picea, Acta Hortic.,* 78, 259, 1978.
130. **Chalupa, V.,** Organogenesis in Norway spruce and Douglas fir tissue cultures, *Commun. Inst. For. Cech.,* 10, 79, 1977.
131. **von Arnold, S. and Eriksson, T.,** Induction of adventitious buds on embryos of Norway spruce grown *in vitro, Physiol. Plant.,* 44, 283, 1978.
132. **von Arnold, S. and Eriksson, T.,** Induction of buds on buds of Norway spruce (*Picea abies*) grown *in vitro, Physiol. Plant.,* 45, 29, 1979.
133. **Mehra-Palta, A., Smeltzer, R. H., and Mott, R. L.,** Hormonal control of induced organogenesis: experiments with excised plant parts of loblolly pine, *Tappi,* 61, 37, 1978.
134. **Brown, C. L. and Sommer, H. E.,** Bud and root differentiation in conifer cultures, *Tappi,* 60, 72, 1977.
135. **Reilly, K. and Washer, J.,** Vegetative propagation of radiata pine by tissue culture: plantlet formation from embryonic tissue, *N. Z. J. For. Sci.,* 7, 199, 1977.

136. **David, A. and David, H.,** Manifestations de diverses potentialities organogenes d'organes ore de fragments d'organes de pin maritime (*Pinus pinaster* Sol.) en culture *in vitro, C. R. Acad. Sci.,* 284, 627, 1977.
137. **Mehra-Palta, A., Smeltzer, R. H., and Mott, R. L.,** Hormonal Control of Induced Organogenesis from Excised Plant Parts of Loblolly Pine (*Pinus taeda* L.), Tappi Conf. Papers, Forest Biology/Wood Chemistry Conference, Madison, Wis., 1977, 15.
138. **Winton, L. L.,** Shoot and tree production from aspen tissue cultures, *Am. J. Bot.,* 57, 904, 1970.
139. **Berbee, F. M., Berbee, J. G., and Hildebrandt, A. C.,** Induction of callus and trees from stem tip cultures of a hybrid poplar, *In Vitro,* 7, 269, 1972.
140. **Venverloo, C. J.,** The formation of adventitious organs. I. Cytokinin-induced formation of leaves and shoots in callus cultures of *Populus nigra* L. "Italica", *Acta Bot. Neerl.,* 22, 390, 1973.
141. **Bawa, K. S. and Stettler, R. F.,** Organ culture with black cottonwood; morphogenetic response of female catkin primordia, *Can. J. Bot.,* 50, 1627, 1972.
142. **Whitehead, H. C. M. and Giles, K. L.,** Rapid propagation of poplars by tissue culture methods, *N. Z. J. For. Sci.,* 7, 40, 1977.
143. **Cheng, T. Y. and Voqui, T. H.,** Regeneration of Douglas fir plantlets through tissue culture, *Science,* 198, 306, 1977.
144. **Cheng, T. Y.,** Adventitious bud formation in culture of Douglas fir (*Pseudotsuga menziesii* (Mirb.) Franco.), *Plant Sci. Lett.,* 5, 97, 1975.
145. **Wochok, Z. S. and Abo El-Nil, M.,** Conifer tissue culture, *Proc. Int. Plant Propogator's Soc.,* 27, 131, 1977.
146. **Winton, L. L. and Verhagen, S. A.,** Shoots from Douglas fir cultures, *Can. J. Bot.,* 55, 1245, 1977.
147. **Boulay, M. and Franclet, A.,** Recherches sur la propagation vegetative du Douglas: *Pseudotsuga menziesii* (Mirb.) Franco. Possibilities d'obtantion de plants viables a partir de la culture *in vitro* de pieds-mieres juveniles, *C. R. Acad. Sci.,* 284, 1405, 1977.
148. **Rao, P. S. and Bapat, V. A.,** Vegetative propagation of sandalwood plants through tissue culture, *Can. J. Bot.,* 56 1153, 1978.
149. **Durzan, D. J. and Lopushanski, S. M.,** Propagation of American elm via cell suspension cultures, *Can. J. For. Res.,* 5, 273, 1975.
150. **Coleman, W. K. and Thorpe, T. A.,** *In vitro* culture of western red cedar (*Thuja plicata* Donn.). I. Plant formation, *Bot. Gaz. (Chicago),* 138, 298, 1977.
151. **Cheng, T. Y.,** Vegetative propagation of western hemlock (*Tsuga heterophylla*) through tissue culture, *Plant Cell Physiol.,* 17, 1347, 1976.
152. **Ball, E. A.,** Cloning *in vitro* of Sequoia sempervirens, in *Abstr. 4th Int. Congr. of Plant Tissue and Cell Culture,* Poster No. 1726, Thorpe, T. A., University of Calgary, Canada, 1978, 163.
153. **Mathes, M. C.,** The *in vitro* formation of plantlets from isolated aspen tissue, *Phyton,* 21, 137, 1964.
154. **Wolter, K. E.,** Root and shoot initiation in aspen callus cultures, *Nature (London),* 219, 509, 1968.
155. **Al-Talib, K. and Torrey, J.,** The aseptic culture of buds of *Pseudotsuga taxifolia, Plant Physiol.,* 34, 630, 1959.
156. **Chalupa, V. and Durzan, D.,** Growth and development of dormant buds of conifers *in vitro, Can. J. For. Res.,* 3, 196, 1973.
157. **Bonga, J. M.,** *In vitro* culture of microsporophylls and megagametophyte tissue of *Pinus, In Vitro,* 9, 270, 1974.
158. **Cresswell, R. J. and Nitsch, C.,** Organ culture of *Eucalyptus grandis* L., *Planta,* 125, 87, 1975.
159. **Lee, E. C. M. and de Fossard, R. A.,** The effects of various auxins and cytokinins on the *in vitro* culture of stem and lignotuber tissues of *Eucalyptus bancroftii* Maiden, *New Phytol.,* 73, 709, 1974.
160. **Jones, L. H.,** Plant cell culture and biochemistry: studies for improved vegetable oil production, in *Industrial Aspects of Biochemistry,* Spender, B., Ed., Federation of European Biochemical Society, Elsevier, Amsterdam, 1974, 813.
161. **Ammar, S. and Benbadis, A.,** Multiplication vegetative du Palmier-datter (*Phoenix dactylifera* L.) par la culture de tissus de jeunes plantes issues de semis, *C. R. Acad. Sci.,* 284, 1789, 1977.
162. **Eeuwens, C. J. and Blake, J.,** Culture of tissues from coconut, oil and date palms with a view to vegetative propagation of selected palms, in *Abstr., 4th Int. Congr. of Plant Tissue and Cell Culture,* Poster No. 1738, Thorpe, T. A., Ed., University of Calgary, Canada, 1978, 168.
163. **Tisserat, B., Esan, E. B., and Murashige, T.,** Somatic Embryogenesis in angiosperms, *Hortic. Rev.,* 1, 1, 1979.
164. **Kohlenbach, H. W.,** Comparative somatic embryogenesis, in *Frontiers of Plant Tissue Culture 1978,* Thorpe, T. A., Ed., The International Association of Plant Tissue Culture, Calgary, 1978, 59.
165. **Konar, R. N. and Oberoi, Y. P.,** *In vitro* development of embryoids on the cotyledons of *Biota orientalis, Phytomorphology,* 15, 137, 1965.
166. **LaRue, C. D.,** Studies on growth and regeneration in gametophytes and sporophytes of gymnosperms, *Brookhaven Symp. Biol.,* 6, 187, 1954.

167. **Norstog, K. and Rhamstine, E.**, Isolation and culture of haploid and diploid cycad tissues, *Phytomorphology*, 17, 374, 1967.
168. **Paranjothy, K. and Othman, R.**, Embryoid and plantlet development from cell cultures of *Hevea*, in *Abstr., 4th Int. Congr. of Plant Tissue and Cell Culture*, Poster No. 134, Thorpe, T. A., Ed., University of Calgary, Canada, 1978, 42.
169. **Hu, C. Y. and Sussex, I. M.**, *In vitro* development of embryoids on cotyledons of *Ilex aquifolium*, *Phytomorphology*, 21, 103, 1971.
170. **Johri, B. M. and Singh Bajaj, Y. P.**, Growth responses of globular proembryos of *Dendrophthoe falcata* (L.f.) Ettings. in culture, *Phytomorphology*, 15, 292, 1965.
171. **Bonga, J. M.**, Formation of holdfasts, callus, embryoids and haustorial cells in *in vitro* cultures of dwarf mistletoe *Arceuthobium pusillum*, *Phytomorphology*, 21, 140, 1971.
172. **Rabechault, H., Ahee, J., and Guenin, M. G.**, Colonies cellulaires et formes embryoides obtenues *in vitro* À partir de cultures d'embryons de Palmier À l'huile (*Elaeis quineensis* Jacq. var. dura Becc.), *C. R. Acad. Sci.*, 270, 3067, 1970.
173. **Zenkteler, M., Misiura, E., and Ponitka, A.**, Induction of androgenic embryoids in the *in vitro* cultured anthers of several species, *Experientia*, 31, 289, 1975.
174. **Staritsky, G.**, Embryoid formation in callus tissues of coffee, *Acta Bot. Neerl.*, 19, 509, 1970.
175. **Sondahl, M. R. and Sharp, W. R.**, High frequency induction of somatic embryos in cultured leaf explants of *Coffea arabica* L., *Z. Pflanzenphysiol.*, 81, 395, 1977.
176. **Vardi, A., Spiegel-Roy, P., and Galun, E.**, Citrus cell culture: isolation of protoplasts, planting densities, effect of mutagens and regeneration of embryos, *Plant Sci. Lett.*, 4, 231, 1975.
177. **Lakshmi Sita, G., Raghava Ran, N. V., and Vaidyanathan, C. S.**, Differentiation of embryoids and plantlets from shoot callus of sandalwood, *Plant Sci. Lett.*, 15, 265, 1979.
178. **Zatykó, J. M.**, Complete red currant (*Ribes rubrum* L.) plants from adventive embryos induced *in vitro*, *Curr. Sci.*, 48, 456, 1979.
179. **Mullins, M. G. and Srinivasan, C.**, Somatic embryos and plantlets from an ancient clone of the grapevine (cv. Cabernet-Sauvignon) by apomixis *in vitro*, *J. Exp. Bot.*, 27, 1022, 1976.
180. **Wetherell, D. F. and Dougall, D. K.**, Sources of nitrogen supporting growth and embryogenesis in cultured wild carrot tissue, *Physiol. Plant.*, 37, 97, 1976.
181. **Sondahl, M. R., Spahlinger, D. A., and Sharp, W. R.**, A histological study of high frequency and low frequency induction of somatic embryoids in cultured leaf explants of *Coffea arabica* L., *Z. Pflanzenphysiol.*, 94, 101, 1979.
182. **Sondahl, M. R., Salisbury, J. L., and Sharp, W. R.**, SEM characterization of embryogenic tissue and globular embryos during high frequency somatic embryogenesis in *Coffea* callus cells, *Z. Pflanzenphysiol.*, 94, 185, 1979.
183. **Bonga, J. M.**, Vegetative propagation: tissue and organ culture as an alternative to rooting cuttings, *N. Z. J. For. Sci.*, 4, 253, 1974.
184. **Winton, L. L. and Stettler, R. F.**, Utilization of haploidy in tree breeding, in *Haploids in Higher Plants: Advances and Potential*, Kasha, K. J., Ed., University of Guelph, Ontario, 1974, 259.
185. **Sato, T.**, Callus induction and organ differentiation in anther culture of poplars, *J. Jpn. For. Soc.*, 56, 55, 1974.
186. **Michellon, R., Hugard, J., and Jonard, R.**, Sur l'isolement de colonies tissulaires de Pecher (*Prunus persica* Batschy cultivars *Dixired* et *Nectared IV*) et d'Amandier (*Prunus amygdalus* Stokes, cultivar At) À partir D'antheres cultivées *in vitro*, *C. R. Acad. Sci.*, 278, 1719, 1974.
187. **Konar, R. N.**, A haploid tissue from the pollen of *Ephedra foliata* Boiss., *Phytomorphology*, 13, 170, 1963.
188. **Tulecke, W.**, The pollen of *Ginkgo biloba: in vitro* culture and tissue formation, *Am. J. Bot.*, 44, 602, 1957.
189. **Zenkteler, M. A. and Guzowska, J.**, Cytological studies on the regenerating mature female gametophyte of *Taxus baccata* L. and mature endosperm of *Tilia platyphyllos* Scop. in *in vitro* culture, *Acta Soc. Bot. Pol.*, 39, 161, 1970.
190. **Duhoux, E. and Norreel, B.**, Sur l'isolement de colonies tissulaires d'origine pollinique A′ partir de cones males du *Juniperus chinensis* L. du *Juniperus communis* L. et du *Cupressus anzonica* G., cultivés *in vitro*, *C. R. Acad. Sci.*, 279, 651, 1974.
191. **Huhtinen, O.**, Production and Use of Haploids in Breeding Conifers, IUFRO Genetics-SA-BRAO Joint Symp., Forest Experiment Station of Japan, International Union of Forestry Research Organization, Tokyo, 1972, 1.
192. **Steinhauer, A. and Huhtinen, O.**, Haploid tissue culture in forest trees: callus and shoot regeneration from megagametophyte cultures of *Picea abies*, in *Abstr., 4th Int. Congr. for Plant Tissue and Cell Culture*, Poster Nos. 1733 and 1738, Thorpe, T. A., Ed., University of Calgary, Canada, 1978, 166.
193. **Illies, Z. M.**, Auftreten Haploids keimlinge bei *Picea abies*, *Naturwissenshaften*, 51, 442, 1964.
194. **Polheim, F.**, *Thuja gigantea gracilis* Beissn. ein Haplont unter den Gymnospermen, *Biol. Rundsch.*, 6, 84, 1968.

195. **Sommer, H. E., Brown, C. L., and Kormanik, P. P.**, Differentiation of plantlets in long-leaf pine (*Pinus palustris* Mill.) tissue cultures *in vitro, Bot. Gaz. (Chicago)*, 136, 196, 1975.
196. **Mott, R. L., Smeltzer, R. H., Mehra-Palta, A., and Zobel, B. J.**, Production of forest trees by tissue culture, *Tappi*, 60, 62, 1977.
197. **Smeltzer, R. H., Mehra-Palta, A., and Mott, R. L.**, Influence of Parental Tree Genotype on the Potential for *In Vitro* Clonal Propagation from Loblolly Pine Embryos, Tappi Conf. Papers, Forest Biology/Wood Chemistry Conference, Madison, Wis., 1977, 5.
198. **Mott, R. L., Smeltzer, R. H., and Mehra-Palta, A.**, An Antomical and Cytological Perspective on Pine Organogenesis *In Vitro*, Tappi Conf. Papers, Forest Biology/Wood Chemistry Conference, Madison, Wis., 1977, 9.
199. **Kelly, C. W., III.**, Intra-Clonal Variation in Tissue Cultured Loblolly Pine (*Pinus taeda* L.), M. S. thesis, North Carolina State University, Raleigh, 1978.
200. **Leach, G. N.**, Early growth of tissue culture produced loblolly pine (*Pinus taeda* L.) plantlets in soil, M. S. thesis, North Carolina State University, Raleigh, 1978.
201. **Leach, G. N.**, Growth in soil of plantlets produced by tissue culture, loblolly pine, *Tappi*, 62, 59, 1979.
202. **Cheah, K. T. and Cheng, T. Y.**, Histological analysis of adventitious bud formation in cultured Douglas fir cotyledon, *Am. J. Bot.*, 65, 845, 1978.
203. **Campbell, R. A. and Durzan, D. J.**, Induction of multiple buds and needles in tissue cultures of *Picea glauca, Can. J. Bot.*, 53, 1652, 1975.
204. **Romberger, J. A. and Tabor, C. A.**, The *Picea abies* shoot apical meristem in culture. I. Agar and autoclaving effects, *Am. J. Bot.*, 58, 131, 1971.
205. **Varnell, R. J. and Vasil, I. K.**, Experimental studies on the shoot apical meristem of seed plants. II. Morphological and cytochemical effects of kinetin applied to the exposed meristem of *Pinus elliottii, Am. J. Bot.*, 65, 47, 1978.
206. **Caruso, J. L., Smith, R. G., Smith, L. M., Cheng, T. Y., and Daves, G. D., Jr.**, Determination of Indol-3-acetic acid in Douglas fir using a deuterated analog and selected ion monitoring: comparison of micro-quantities in seedling and adult tree, *Plant Physiol.*, 62, 841, 1978.
207. **Riov, J., Cooper, R., and Gottlieb, H. E.**, Metabolism of auxin in pine tissues: napthaleneacetic acid conjugation, *Physiol. Plant.*, 46, 133, 1979.
208. **Van Staaden, J.**, A comparison of the endogenous cytokinins in the leaves of four gymnosperms, *Bot. Gaz. (Chicago)*, 139, 32, 1978.
209. **Zajaczkowski, S. and Wodzicki, T. J.**, On the question of stem popularity with respect to auxin transport, *Physiol. Plant.*, 44, 122, 1978.
210. **Ingstaad, T.**, Mineral nutrient requirements of *Pinus silvestris* and *Picea abies* seedlings, *Physiol. Plant.*, 45, 373, 1979.
211. **Owens, J. N. and Molder, M.**, Bud development in *Larix occidentalis*. I. Growth and development of vegetative long shoot and vegetative short shoot buds, *Can. J. Bot.*, 57, 687, 1979.
212. **Greenwood, M. S., Harlow, A. C., and Hodgson, H. D.**, The role of auxin metabolism in root meristem regeneration by *Pinus lambertiana* embryo cuttings, *Physiol. Plant.*, 32, 198, 1974.
213. **Buchala, A. J. and Schmid, A.**, Vitamin D and its analogues as a new class of plant growth substances affecting rhizogenesis, *Nature (London)*, 280, 230, 1979.
214. **Cheng, T. Y.**, Propagating woody plants through tissue culture, *Am. Nurseryman*, May 14, 1978, 4.
215. **von Arnold, S. and Eriksson, T.**, Bud induction on isolated needles of Norway spruce (*Picea abies* (L.) Karst.) grown *in vitro, Plant Sci. Lett.* 15,363, 1979.
216. **Reilly, K. and Brown, C. L.**, *In Vitro* Studies of Bud and Shoot Formation in *Pinus radiata* and *Pseudotsuga menziesii*, Georgia Forest Research Paper No. 86, Georgia Forest Research Council, Macon, 1976, 1.
217. **Kirby, E. G. and Cheng, T. Y.**, Colony formation from protoplasts derived from Douglas fir cotyledons, *Plant Sci. Lett.*, 14, 145, 1979.
218. **Brown, C. L. and Lawrence, R. H.**, Culture of pine callus on a defined medium, *Forest Sci.*, 14, 62, 1968.
219. **Harvey, A. E. and Grasham, J. L.**, Procedures and media for obtaining tissue cultures of 12 conifer species, *Can. J. Bot.*, 47, 547, 1969.
220. **Hasegawa, P. M., Yasuda, T., and Cheng, T. Y.**, Effect of auxin and cytokinin on newly synthesized proteins of Douglas-fir cotyledons, *Physiol. Plant.*, 46, 211, 1979.
221. **Washer, J., Reilly, K. J., and Barnett, J. R.**, Differentiation in *Pinus radiata* callus culture: the effect of nutrients, *N. Z. J. For. Sci.*, 7, 321, 1977.
222. **Barnett, J. R.**, Fine structure of parenchymatous and differentiated *Pinus radiata* callus, *Ann. Bot. (London)*, 42, 367, 1978.
223. **Wodzicki, T. J. and Humphreys, W. J.**, Cytodifferentiation of maturing pine tracheids: the final stage, *Tissue Cell*, 4, 525, 1972.

224. **Lester, D. T. and Berbee, J. G.,** Within-clone variation among black poplar trees derived from callus culture, *Forest Sci.,* 23, 122, 1977.
225. **Berbee, J. G., Martin, R. R., and Castello, J. D.,** Detection and elimination of viruses in poplars, in Intensive Plantation Culture, Gen. Tech. Rep. NC-21, Forestry Service, U.S. Department of Agriculture, St. Paul, Minn., 1976, 85.
226. **Boxus, P. and Quoirin, M.,** Comportment eu pepiniere d'arbres frutiers issus de culture "*in vitro*", *Acta Hortic.,* 78, 373, 1978.
227. **Green, C. E.,** Prospects for crop improvement in the field of cell culture, *Hortic. Sci.,* 12, 131, 1977.
228. **Skirvin, R. M.,** Natural and induced variation in tissue culture, *Euphytica,* 27, 241, 1978.
229. **Dougall, D. K.,** Preservation of germplasm, in Propagation of Higher Plants through Tissue Culture, Hughes, K. W., Henke, R., Constantin, M., Eds., Technical Information Center, U.S. Department of Energy, Springfield, Va., 1978, 213.
230. **Withers, L. A.,** Freeze-preservation of cultured cells and tissues, in *Frontiers of Plant Tissue Culture 1978,* Thorpe, T. A., Ed., The International Association of Plant Tissue Culture, Calgary, 1978, 297.
231. **Bajaj, Y. P. S.,** Technology and prospects for cryopreservation of germplasm, *Euphytica,* 28, 267, 1979.
232. **Starrwood, P. C. and Roos, E. E.,** Feed storage of several horticultural species in liquid nitrogen (−196°C), *Hortic. Sci.,* 14, 628, 1979.
233. **Ingram, D. S.,** Growth of plant parasites in tissue culture, in *Plant Tissue and Cell Culture,* Street, H. E., Ed., University of California Press, Berkeley, 1973, 392.
234. **Brettell, R. I. S. and Ingram, D. S.,** Tissue culture in the production of novel disease-resistant crop plants, *Biol. Rev.,* 54, 329, 1979.
235. **Pelham, J. and Mason, P. A.,** Aseptic cultivation of sapling trees for studies of nutrient responses with particular reference to phosphate, *Ann. Appl. Biol.,* 88, 415, 1978.
236. **Cutter, V. M.,** Studies on the isolation and growth of plant rusts in host tissue cultures and upon synthetic media. I. *Gymnosporangium, Mycologia,* 51, 248, 1959.
237. **Harvey, A. E. and Grasham, J. L.,** Growth of the rust fungus *Cronartium ribicola* in tissue cultures of *Pinus monticola, Can. J. Bot.,* 47, 663, 1969.
238. **Harvey, A. E. and Grasham, J. L.,** Production of the nutritional requirements for growth of *Cronartium ribicola* by a non-host species, *Can. J. Bot.,* 49, 1517, 1971.
239. **Schneider, A. and Reverdy, F.,** Comportement de *Taphrina deformans* var. *Persicae* dans des tissus de Pêcher parasites cultivés *in vitro, C. R. Acad. Sci.,* 277, 2169, 1973.
240. **Chaumont, D., Taris, B., and Harada, H.,** Essais de cultures dissociées entre des parasites obligatoires et des tissus ou des cellules dissociees (de leur hôte) cultivées *in vitro.* Essais de cultures associees de *Podosphaera leucotricha* (Ell. et Ev.) Salm., parasite obligatoire du Pommier, et decellules dissociees du parenchyme foliaire de *Malus pumila* Mill., *C. R. Acad. Sci.,* 278, 465, 1974.
241. **Mott, R. L., Thomas, H. A., and Namkoong, G.,** *In vitro* rearing of southern pine beetle larvae on tissue-cultured loblolly pine callus, *Ann. Entomol. Soc. Am.,* 71, 564, 1978.

Indexes

SUBJECT INDEX

A

B

C

D

H

I

J

K

L

M

N

O

P

T

V

W

X

Y

Z

TAXONOMIC INDEX

A

B

C

D

E

M

N

O

P